NMR 27

Basic Principles and Progress

Editors: P. Diehl E. Fluck H. Günther
R. Kosfeld J. Seelig

In-Vivo Magnetic Resonance Spectroscopy II: Localization and Spectral Editing

Guest-Editor: M. Rudin

With contributions by
J.J.H. Ackermann, N. Askenasy,
B.A. Berkowitz, C.S. Bosch, P.A. Bottomley,
D. Bourgeois, M. Decorps, D. Freeman,
J.A. Helpern, H.P. Hetherington,
J. den Hollander, R. Hurd, O. Kaplan, T. Kushnir,
P.R. Luyten, J.H. Marien, G. Navon,
R.J. Ordidge, M. Rudin, A. Sauter, P. Styles

With 115 Figures

Springer-Verlag
Berlin Heidelberg NewYork
London Paris Tokyo
Hong Kong Barcelona
Budapest

ISBN-13: 978-3-642-77210-8 e-ISBN-13: 978-3-642-77208-5
DOI: 10.1007/978-3-642-77208-5

Softcover reprint of the hardcover 1st edition 1992

Typesetting: Thomson Press (India) Ltd, New Delhi
Offsetprinting: Colordruck Dorfi GmbH, Berlin; Bookbinding: Lüderitz & Bauer, Berlin
51/3020 5 4 3 2 1 0 Printed on acid-free paper

Foreword

Since the first report on NMR in living cells in 1974, in vivo magnetic resonance spectroscopy (MRS) has evolved into a potent technique for studying metabolic processes in an undisturbed biological system. In principle, MRS should provide well-resolved spectra of a selected region of interest allowing the identification and quantitation of as many metabolites as possible. However, a number of methodological problems have to be solved in order to generate useful MRS data. For example, the spectra should contain only little or no contribution from signals from outside the region of interest. Furthermore, proper quantification of metabolite concentrations is often difficult since the large number of compounds contributing to the spectra produce overlapping resonances. Similarly, the resolution of the spectra is often impaired by variations in tissue magnetic susceptibilities, leading to large linewidths. Many metabolites occur at low concentrations and a poor signal-to-noise ratio is another major obstacle in proper quantification. The use of radiofrequency (RF) probes with an inhomogeneous RF field profile, such as the surface coil, causes problems when defined RF excitation has to be used, e.g. for proper signal inversion.

During the last 10 to 15 years, a number of solutions to the above mentioned problems have been developed and a proper review of the methodological achievements in in vivo MRS has long been overdue. Rather than writing a textbook, we decided to invite the acclaimed experts to review their respective speciality. We did not intend to discuss in vivo MRS from its basics, but rather refer to textbooks and review articles which have covered these aspects. Basic principles e.g. in NMR probe design or data processing are considered as 'prior knowledge', with which the reader should be familiar to some extent.

The book is divided into three volumes. Volume I contains sections on MR probe construction, design of RF pulses, and spectrum analysis. Volume II is devoted to localization techniques, i.e. methods for selecting a region of interest, and techniques for spectral editing. Volume III finally describes potential applications of in vivo MRS. Due to the rapid growth of the field during the last few years it would have been an almost 'infinite' task to provide a comprehensive coverage of all ongoing activities. We thus have asked the respective authors to focus on methodological and conceptional aspects in order to illustrate the potential of MRS more than the breadth of application.

We would like to thank all contributors for their efficient and pleasant collaboration and for keeping within the short deadlines imposed on them by the editors. We are also indebted to Springer-Verlag, in particular Dr. R. Stumpe and Ms. P. Frank, for their assistance in general and the editorial supervision.

Basel, September 1991

M. Rudin
J. Seelig

Guest-Editor

Dr. M. Rudin

Preclinical Research 386/206
Sandoz Pharma Ltd.
Ch-4002 Basel

Table of Contents

Localization Techniques

Spectral Editing

Table of Contents of Volume 26
In Vivo Magnetic Resonance Spectroscopy I:
Probeheads and Radiofrequency Pulses, Spectrum Analysis

Table of Contents of Volume 28

In Vivo Magnetic Resonance Spectroscopy III:
In-Vivo MR Spectroscopy: Potential and Limitations

Localization Techniques

A crucial step in in vivo MRS is the proper definition of the region-of-interest (ROI). The goal is to obtain a spectrum, which originates exclusively from this region, with the best achievable sensitivity. Localization techniques are therefore extensively discussed in this volume.

The first localization method introduced was 'topical MR' involving magnetic field profiling [1]. A distinct step forward was made by the introduction of techniques which took advantage of B_1 gradients associated with the inhomogeneous RF field of surface coils. The conventional surface coil spectroscopy is discussed by Bosch and Ackerman. This simple technique is the method of choice for many applications, since it combines experimental simplicity with good sensitivity. There are, however, severe limitations as far as positioning of the ROI and contamination by unwanted signal contributions are concerned. Conventional surface coil spectroscopy is limited to the study of superficial structures. Moreover, the boundary of the ROI is poorly defined. Depth resolution may be obtained with rotating frame techniques, which explicitely exploit the B_1 field variation to achieve spatial encoding of the MR signal. The potential and the limitations of rotating frame spectroscopy are discussed by Styles. Alternatively, depth resolution is provided by the Depth REsolved Surface coil Spectroscopy (DRESS), a technique that uses both B_1 and B_0 gradient. This topic is reviewed by Bottomley.

Techniques to select a single cubic-shaped volume element (voxel) have also been introduced, the first being Volume Selective Excitation spectroscopy (VSE) [2]. The rely on the combination of frequency selective RF pulses and B_0 gradients. The most popular single voxel technique for phosphorus MRS is Image Selected In vivo Spectroscopy (ISIS) reviewed by Ordidge and Helpern. Localization methods suitable for proton MRS are discussed by Decorps and Bourgeois. The single voxel techniques may be easily extended to allow the simultaneous measurement of multiple voxels by applying multiple frequency excitation pulses as described by Mueller in the first volume of this series.

Phosphorus spectroscopic imaging (SI) was described some years ago [3, 4], but the inherently long measurement times were prohibitive for many applications. Only recently these techniques have gained some broader acceptance. Proton SI offers some advantages over phosphorus SI due to the significantly longer T_2 values for the metabolites of interest and the inherently better sensitivity. This allows the design of experiments providing images with 8 × 8 or 16 × 16 pixel resolution, showing the distribution of metabolites such as *N*-actylaspartate, creatine, choline, lactate in human brain. Similar spectroscopic images with a resolution of 12 × 12 pixels (voxel volume 11μl) were recently obtained for rat brain [5]. The potential of proton SI is discussed by den Hollander et al..

1. Gordon RE, Hanley PE, Shaw D, Gadian DG, Radda GK, Styles P, Bore PJ, Chan L (1980) Nature 287: 367
2. Aue WP, Mueller S, Cross TA, Seelig J (1984) J Magn Reson 56: 350
3. Bendel P, Lai CM, Lauterbur PC (1980) J Magn Reson 38: 343
4. Maudsley AA, Hilal SK, Simon HE, Wittekoek S (1984) Radiology 153: 745
5. Van Waals JJ, Bergman AH, van den Boogert HJ, Heerschap A, van der Kogel AJ, Ruifrok ACC, Bernsen HJJA (1991), NMR Biomed 4: 125

Surface Coil Spectroscopy

Coleen S. Bosch and Joseph J. H. Ackerman

Department of Chemistry, Washington University, One Brookings Drive, St. Louis, MO 63130, USA

Table of Contents

The surface coil is examined in its role as an NMR transmitter coil and as a receiver coil. Magnetic-field and signal-amplitude distributions are presented, illustrating the signal localization achieved by surface-coil excitation and/or detection. The effects of experimental pulsing conditions on the interrogated volume of the surface coil are considered. Techniques developed to improve the signal localization capabilities or to compensate for the inhomogeneous transverse field of the surface coil are discussed. Various coil designs developed for local signal detection are described. The high sensitivity and localized volume of signal detection inherent to the surface coil offer distinct advantages for spectroscopic and imaging investigations of biological tissues. Representative surface-coil spectra obtained in vivo from various tissues and for several nuclides illustrate the broad range of applications that can be addressed by in vivo surface-coil NMR spectroscopy.

NMR Basic Principles and Progress, Vol. 27

1 Introduction

Since the introduction of the surface coil in 1980 [1] a large number of applications of surface-coil nuclear magnetic resonance (NMR) spectroscopy to problems in biological and medical research have been reported. The surface coil, in combination with the use of low-energy radio frequencies (RF) inherent to NMR, provides a noninvasive tool to study local molecular environments and cellular dynamics in intact tissue systems. The use of isotopically-labeled metabolic substrates has further increased the usefulness of surface-coil NMR to biological applications.

In its simplest form, the surface coil consists of a small, flat, circular loop of wire that is brought to electrical resonance at a particular nuclide's frequency and positioned adjacent to a sample for NMR signal excitation and/or detection (Fig. 1). In the plane of the coil, the magnetic field lines produced by current in the coil are parallel to the coil axis. Beyond the coil plane, the field lines diverge, wrap around the coil, and eventually reconverge, as shown in Fig. 2(a). The coil is generally positioned on the sample such that the coil axis is perpendicular to the main static magnetic field $\mathbf{B}_0$ (Fig. 2(b)). In this orientation, the magnetic field produced by the coil is nearly perpendicular to $\mathbf{B}_0$ for regions near the coil, where the flux density is greatest. Although the interrogated volume[1] of the

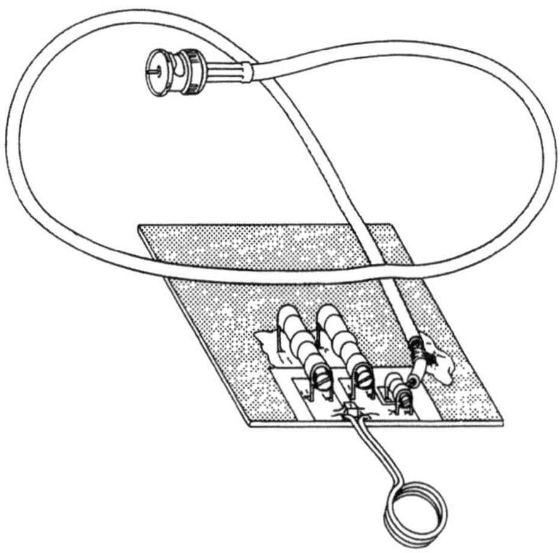

Fig. 1. Multi-turn surface coil with associated fixed and variable tuning elements and 50 Ω coaxial transmission line

[1] Since the excited sample volume and the detected sample volume are not equivalent in surface-coil NMR, the phrase "interrogated volume" will be used to refer to the volume over which signal is detected (i.e. combined excitation and reception profiles); the phrase "excitation volume" will only refer to the volume over which spins are excited

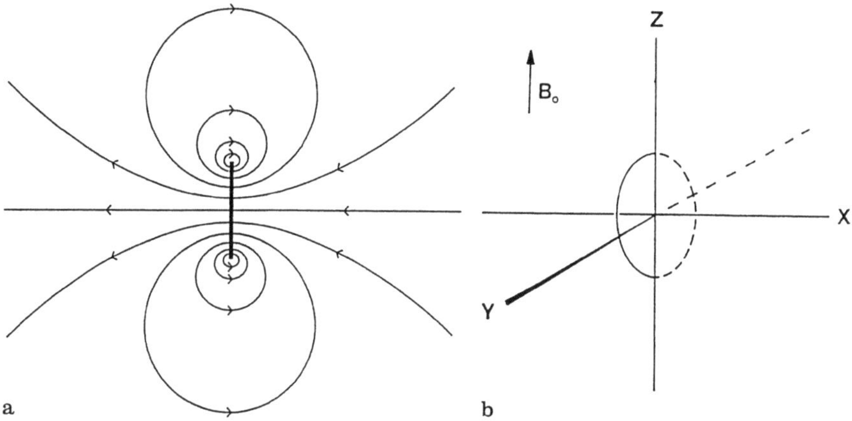

Fig. 2a, b. a Magnetic field lines produced by current in the surface coil. **b** Orientation of a single-turn, circular, surface coil relative to the laboratory frame of reference. The surface coil lies in the *yz* plane with the static magnetic field $\mathbf{B}_0$ in the *z* direction. The center of the coil is at $x = y = z = 0$. The coil loop crosses the *y* and *z* axes at $\pm r$, the coil radius. The *x* axis (coil axis) represents depth into the sample

surface-coil experiment is ultimately a function of coil and sample geometry, as well as experimental pulsing conditions, it is convenient to regard it as roughly hemispherical with a penetration depth approximately equal to the radius of the coil.

The utility of the surface coil arises from its simple design, ease of implementation, and inherent spatial selectivity allowing localized observation of a specific tissue or organ in vivo. Additionally, and significantly, by nature of its design the surface coil achieves a high signal-to-noise ratio relative to other NMR coils over the same volume of detection. The primary disadvantage of the surface coil is the extremely inhomogeneous transverse magnetic field produced by current in the coil. While this "disadvantage" is intimately coupled to the "advantage" of localized signal detection, it produces a wide distribution of flip angles within the sample upon surface-coil excitation, thus, resulting in ill-defined volumes of excitation and detection.

Considerable effort has gone into the design of specialized pulsing schemes to enhance the advantages of the surface coil (i.e. improve surface-coil signal localization) and to mitigate its disadvantages (i.e. compensate for the inhomogeneous RF magnetic field). Inclusion of linear $\mathbf{B}_0$ field gradients to further focus and define the NMR signal has found wide application in localized surface-coil spectroscopy.

The surface coil has been used broadly in both human and animal investigations including studies of phosphorus metabolite energetics, carbohydrate metabolism, blood flow, tumor metabolism, and tissue-water dynamics. The simplest use of the surface coil in these studies has been as a single-frequency-tuned, single-coil-mode transmitter and receiver. Other applications have included its use in concert with other coils, or with more complex resonance schemes to allow multi-frequency excitation and to enhance signal detection.

The advent of the surface coil has not only had a significant impact on NMR spectroscopy, but has also made major contributions to medical imaging. The high sensitivity and limited detection volume of the surface coil allow high-resolution images of excellent quality to be obtained in reasonable time. Surface-coil imaging of "superficial" tissues such as spinal cord, liver, kidney, heart, orbit, joints, and tumors has proven to be important in clinical applications.

2 Surface Coil Theory

2.1 *Surface Coil Receiver*

An oscillating magnetic dipole, positioned at point q in space relative to a loop of wire, will induce a time-varying voltage in the loop whose intensity is directly related to the strength of the magnetic field $\mathbf{B}_1$ that would be produced at point q by unit current (DC) in the loop [2, 3]. This relationship between the induced signal voltage and the (hypothetical) magnetic field produced by unit current, known as the principle of reciprocity [2], describes the oscillating signal voltage $S(t)_q$ induced in an NMR receiver coil by a precessing magnetic moment $\mathbf{M}$ at q:

$$S(t)_{\mathrm{q}} = -\frac{\partial}{\delta t}(\mathbf{B}_1(q)\cdot\mathbf{M}(q,t)), \tag{1}$$

where $\mathbf{B}_1$ is the magnetic field produced by unit current flowing through the receiver coil.

For the NMR experiment with the static magnetic field $\mathbf{B}_0$ in the z direction, the components of $\mathbf{B}_1$ and $\mathbf{M}$ which contribute to an NMR signal are those in the transverse, or xy, plane. Evaluating Eq. (1) for transverse magnetization, the signal induced in the NMR coil from a moment at point q can be written as

$$S(t)_{\mathrm{q}} = (B_1)_{\mathrm{xy}}(q)M_{\mathrm{xy}}(q)\omega_0 \sin(\omega_0 t + \varphi(q)), \tag{2}$$

where $(B_1)_{\mathrm{xy}}(q)$ and $M_{\mathrm{xy}}(q)$ are the magnitudes of the xy components of the field and magnetization at point q, ω_0 is the precession frequency of the magnetic moment about $\mathbf{B}_0$, and the phase offset $\varphi(q)$ is the phase angle (at $t = 0$) between the xy component of $\mathbf{B}_1$ and the xy component of $\mathbf{M}$. (The phase angle φ is fundamental to the signal excitation and detection processes and is independent of nuisance phase offsets introduced by the spectrometer.) The total signal induced in the NMR receiver coil (i.e. the integral of Eq. (2) over all space) and the volume over which the majority of the signal is obtained (the "interrogated volume") is, thus, dependent on how $(B_1)_{\mathrm{xy}}$ and M_{xy} vary with spatial coordinates.

The spatial distribution of the magnetic field $(B_1)_{\mathrm{xy}}$ describes completely the receptivity pattern of the NMR receiver coil. The spatial distribution of $(B_1)_{\mathrm{xy}}$ can be determined for any coil geometry through application of the Biot–Savart law, which describes the magnetic field contribution at point q produced by a current element a distance d away. $(B_1)_{xy}$ has been evaluated for the surface coil by numerous investigators (see, for example [1, 4–9]). The magnitude of the calculated $(B_1)_{xy}$ distribution produced by unit current in a surface coil of radius r, oriented in the yz plane as shown in Fig. 2(b), is illustrated in Fig. 3. Field-magnitude surfaces and the corresponding iso-field contours are shown for the xy plane at $z = 0$ (Fig. 3(a, b)), and for the xz plane at $y = 0$ (Fig. 3(c, d)). The magnitude of the field has been normalized relative to a value of 100 at the coil center ($x = y = z = 0$). The axes are given in units of coil radii. Regions of extremely intense $(B_1)_{xy}$ exist near the coil wires, indicating regions of very high reception sensitivity. The marked inhomogeneity of the transverse magnetic field of the surface coil is evidenced by the rapid fall-off of $(B_1)_{xy}$ with increasing distance from the coil. The region of significant $(B_1)_{xy}$ produced by the surface coil extends further laterally (in the y direction) and axially (in the x direction) than it does longitudinally (in the z direction).

The shape of the $(B_1)_{xy}$ distribution for the xy plane is distinctly different from the shape for the xz plane. This results from the fact that the components of $\mathbf{B}_1$ transverse to $\mathbf{B}_0$ in the xy plane are $(B_1)_x$ and $(B_1)_y$, i.e., $(B_1)_{xy} = [(B_1)_x^2 + (B_1)_y^2]^{1/2}$, whereas the component of $\mathbf{B}_1$ transverse to $\mathbf{B}_0$ in the xz plane is only $(B_1)_x$. Furthermore, as the $\mathbf{B}_1$ field lines wrap around the coil in the xz plane (Fig. 2(a)), at some point they become parallel to the z axis. At this point there is no transverse field component, and, thus, no signal will be detected. These points of zero-$(B_1)_x$ form nodal curves in the xz plane which radiate outward from the coil wires. These nodal curves, evident in Fig. 3(c), are indicated by the dashed lines in the corresponding contour map (Fig. 3(d)). Beyond these nodal curves the transverse component of the field is reversed in direction.

2.2 *Surface Coil Transmitter*

The magnetic field used for signal excitation is produced by RF current in the transmitter coil. The magnetic field describing the receptivity pattern of the receiver coil, in contrast, is produced by hypothetical unit DC current in the receiver coil. The reception $(B_1)_{xy}$ of a coil described in the laboratory frame of reference is equivalent to the excitation $(B_1)_{xy}$ of the same coil at $t = 0$ described in a reference frame rotating at the frequency of the driving current, with the possible differences of a constant amplitude factor and/or a constant phase offset between the laboratory and rotating frames of reference at all points q [4][2]. Thus, the $(B_1)_{xy}(q)$ describing the reception pattern of the surface coil also determines its excitation pattern.

[2] This relationship holds for samples in which RF attenuation due to sample conductivity can be ignored and for coils whose length is substantially smaller than the RF wavelength so that the phase change of the current around the coil is negligible

When a sample is excited on or near resonance by the application of a transverse RF magnetic field $\mathbf{B}_1$, the magnetization produced in the xy plane, M_{xy}, is related to $(B_1)_{xy}$ through the flip angle α:

$$M_{xy} = M_z \sin\alpha, \tag{3}$$

$$\alpha = \gamma t_p I (B_1)_{xy} \tag{4}$$

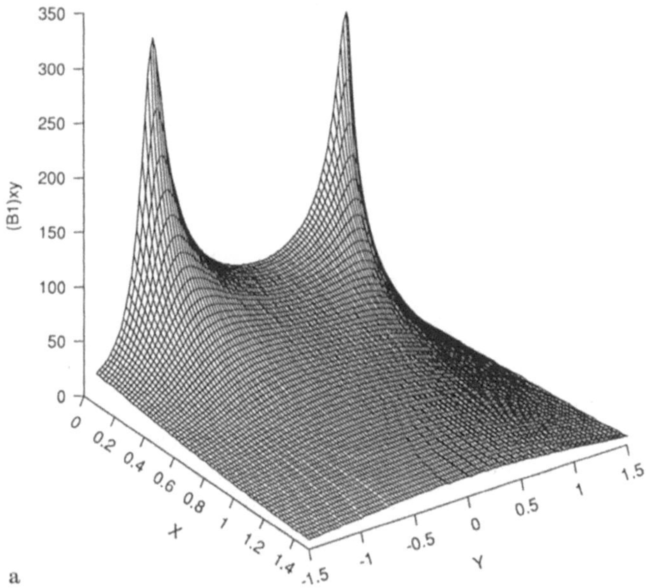

a

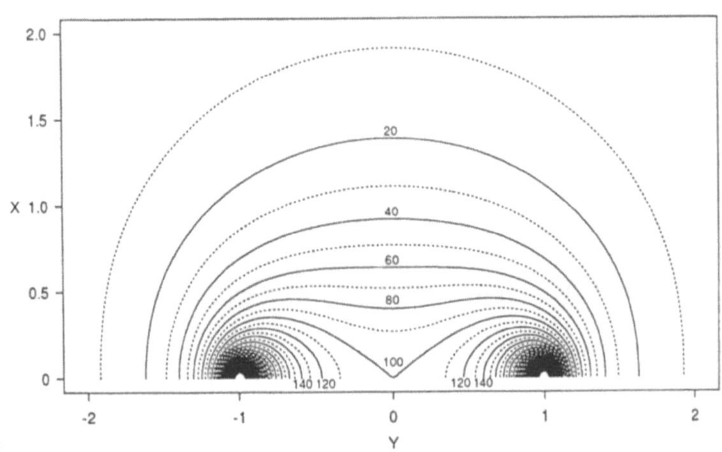

b

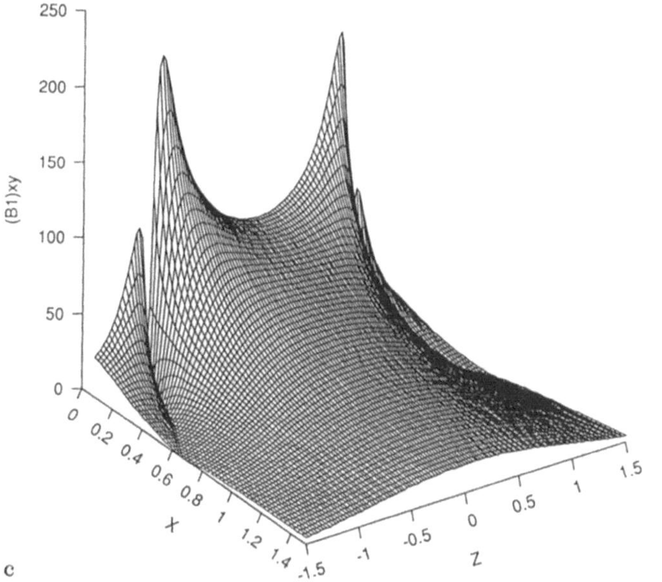

c

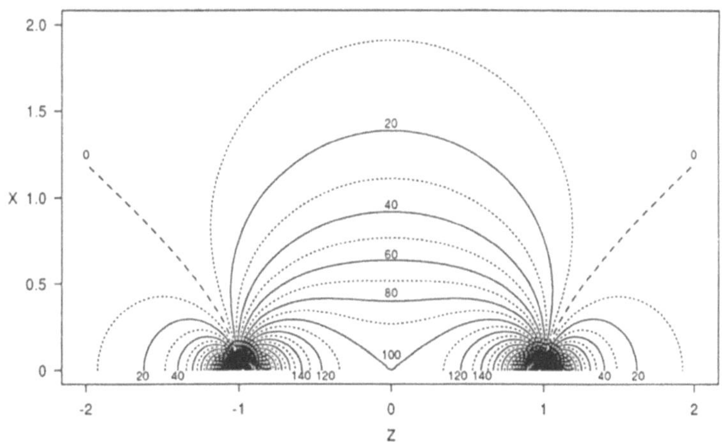

d

Fig. 3 a–d. Magnitude of calculated $(B_1)_{xy}$ as a function of spatial coordinates for a 1-turn, circular surface coil lying in the *yz* plane. The *x* axis represents depth into the sample. The axes are given in units of coil radii. The $(B_1)_{xy}$ intensity is normalized relative to a value of 100 at the coil center. $(B_1)_{xy}$ surfaces (plotted for $0.1r \leqq x \leqq 1.5r$) and corresponding iso-field contours are plotted for (**a, b**) the *xy* plane at $z = 0$, and (**c, d**) the *xz* plane at $y = 0$

where M_z is the amplitude of the (longitudinal) magnetization just prior to excitation, γ is the magnetogyric ratio, t_p is the transmitter pulse width, and I is the current through the coil. When a surface coil is used as the transmitter, the highly inhomogeneous $(B_1)_{xy}$ of the surface coil results in a wide distribution of flip angles within the sample:

$$\alpha(q) = \gamma t_p I (B_1)_{xy}(q). \tag{5}$$

This distribution of spin flip angles produces a dependence of the transverse magnetization amplitude on spatial coordinates:

$$M_{xy}(q) \propto \sin \alpha(q). \tag{6}$$

Although surface-coil excitation does not produce a single flip angle over the entire sample, the spins in given restricted regions of the sample, corresponding to regions of iso-$(B_1)_{xy}$, can be given specific flip angles through the appropriate selection of t_p. For example, if t_p is selected such that the spins in the center of the coil ($x = y = z = 0$) receive a 180° flip angle, then spins approximately 0.77 radii deep along the axis of the coil ($x = 0.77\,r$, $y = z = 0$), which experience the same t_p but half the field strength, receive a 90° flip angle. However, since the iso-$(B_1)_{xy}$ contours of the surface coil do not lie in parallel bands within the sample, but, rather, curve back to the sample surface, the sample region receiving a 90° flip angle will also curve back to the sample surface in all directions, forming a 90°-flip-angle "shell". Additionally, any sample positioned close to the coil (i.e. in regions of high $\mathbf{B}_1$) experiences net flip angles corresponding to 270°, 450°, etc., resulting in strong signal contributions of rapidly-varying amplitudes. These effects are illustrated in Fig. 4, in which the signal-amplitude distribution (i.e. the contribution to total signal amplitude as a function of spatial coordinates) obtained following a single pulse producing a 180° flip angle at the center of the coil is plotted for the xy plane. The intense spikes near the coil and curved maximum-signal-amplitude "ridge" in Fig. 4 illustrate that selection of t_p alone is inadequate to optimize signal detection at a specific depth using the surface coil.

2.3 Surface Coil Transmitter and Receiver

The NMR signal produced in the coil by a magnetic dipole at point q (Eq. (2)) is a function of both the receiver $(B_1)_{xy}(q)$ and the transmitter $(B_1)_{xy}(q)$ (via $M_{xy}(q)$, Eqs. (5) and (6)). When the same coil is used for both transmission and reception (the so-called single-coil-mode experiment), the receiver $(B_1)_{xy}(q)$ is equivalent to the transmitter $(B_1)_{xy}(q)$ and a dual dependency of the signal on $(B_1)_{xy}(q)$ is realized. This dual dependency on the inhomogeneous $(B_1)_{xy}(q)$ is the primary factor determining the interrogated volume of the single-coil-mode surface-coil experiment, and results in an interrogated (detected) volume that is different from the excitation volume.

The signal-amplitude distribution achieved as a function of spatial coordinates as a result of surface-coil excitation and reception with a homogeneous

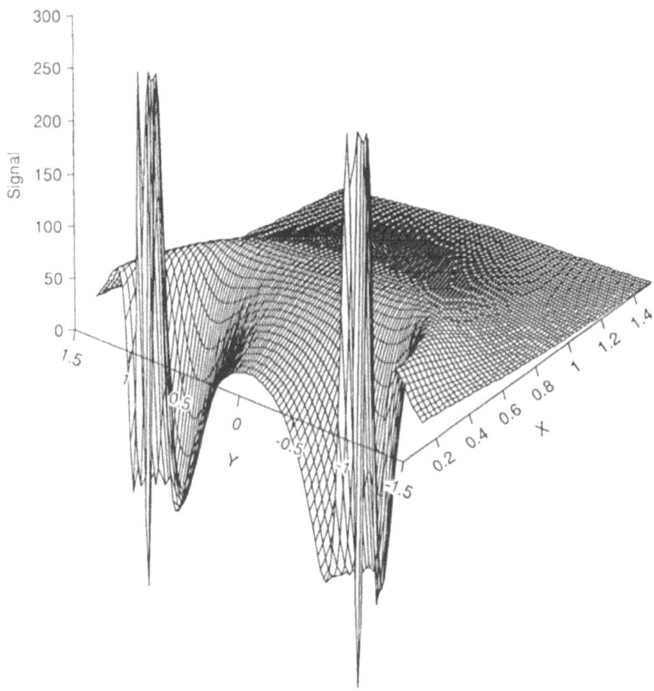

Fig. 4. Calculated signal-amplitude distribution map for a single pulse from a 1-turn, circular surface coil lying in the *yz* plane (used as both transmitter and receiver coil) in which a 180° flip angle has been produced at the coil center. The axes are given in units of coil radii. Signal amplitudes are normalized relative to an amplitude of 100 that would result from a 90° flip angle at the coil center

sample is described by

$$S(q) \propto (B_1)_{xy}(q) \sin \alpha(q), \quad (7)$$

$$\alpha(q) = \alpha_0 \frac{(B_1)_{xy}(q)}{(B_1)^0_{xy}}, \quad (8)$$

where $\alpha(q)$ is expressed relative to the flip angle experienced at the center of the coil, α_0, with $(B_1)^0_{xy}$ representing the field at the center of the coil.

Signal-amplitude distribution maps calculated for a single pulse with surface-coil excitation and reception of a homogeneous sample at equilibrium conditions are shown in Fig. 5, and represent the interrogated volume of the single-coil-mode surface-coil experiment. The signal amplitudes have been scaled relative to an amplitude of 100 for a 90° flip angle at the coil center. The coil orientation is the same as that for Fig. 3. Signal-amplitude distribution surfaces and corresponding iso-amplitude contours are shown for the xy plane at $z = 0$ (Fig. 5(a, b)) and the xz plane at $y = 0$ (Fig. 5(c, d)). The differences in shape between the xy and *xz* planes and the presence of the nodal curves in the *xz* plane result from the same causes as the corresponding features of the $(B_1)_{xy}$ distribution of the surface coil (Fig. 3). The large negative spikes near the coil wires in the amplitude maps correspond to regions of intense $\mathbf{B}_1$, which produce spin flip angles between 180° and 360° when a flip angle of 90° is produced at the coil center. Although the regions of significant signal intensity extend approximately

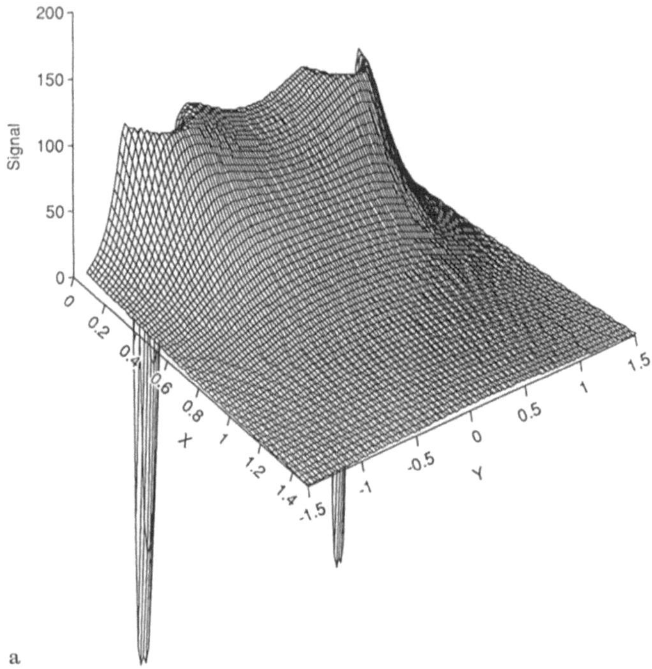

a

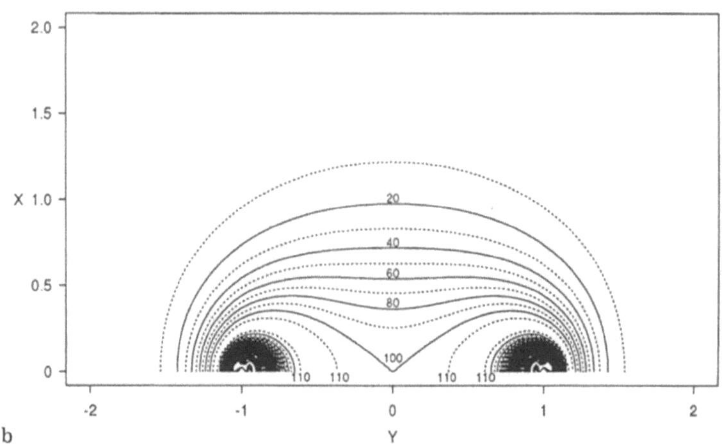

b

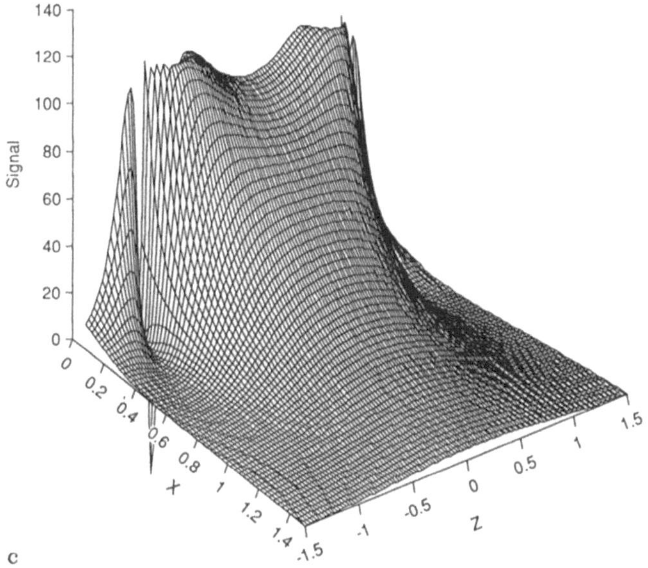

c

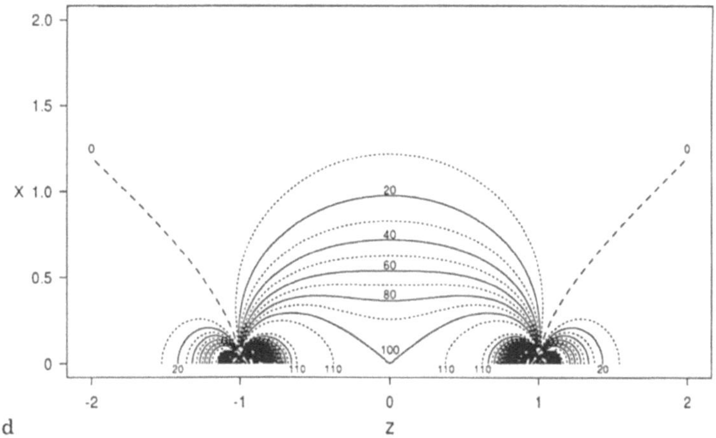

d

Fig. 5 a–d. Calculated signal-amplitude distribution maps for a single pulse from a 1-turn, circular surface coil lying in the *yz* plane used as both transmitter and receiver. The axes are given in units of coil radii. All signal amplitudes are normalized relative to an amplitude of 100 for a 90° flip angle at the coil center. Signal-amplitude surfaces and corresponding iso-amplitude contours are plotted for (**a, b**) the *xy* plane at $z = 0$, and (**c, d**) the *xz* plane at $y = 0$

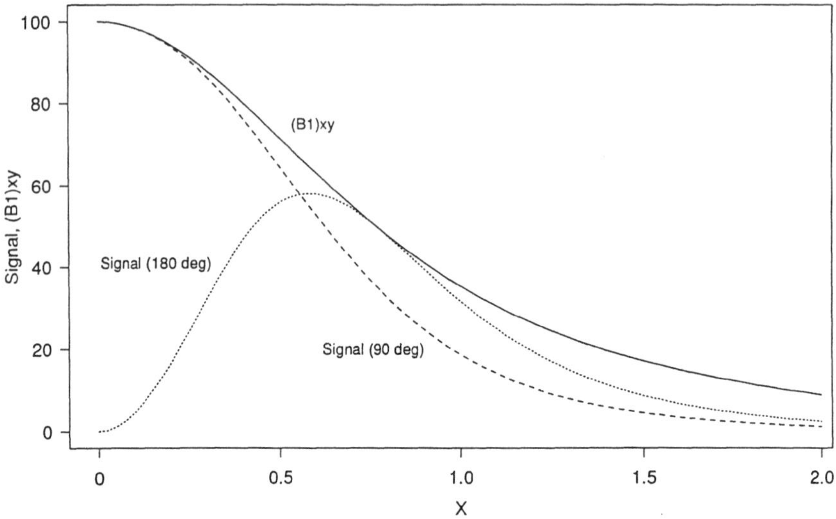

Fig. 6. Comparison of signal amplitude and $(B_1)_{xy}$ for a one-dimensional x-axis sample as a function of distance along the x axis for a single pulse from a 1-term, circular surface coil used as both transmitter and receiver. The x axis is given in units of coil radii. The *solid line* is the relative axial $(B_1)_{xy}$ magnitude. The *dashed line* corresponds to the signal produced for a 90° flip angle achieved at the coil center, normalized to 100; the *dotted line* corresponds to the signal produced for a 180° flip angle achieved at the coil center

one coil radius from the center of the coil, describing the interrogated volume of the surface coil as "hemispherical" in shape is only a very rough approximation.

The effect of the dual dependency of the signal amplitude on $(B_1)_{xy}$ is seen in the more rapid fall-off of the signal-intensity profiles with distance than that of the $(B_1)_{xy}$ profiles. This is apparent in Fig. 6, in which the magnitude of the field and the amplitude of the signal produced by a 90° flip angle at the coil center have been plotted as functions of the distance along the x axis (i.e. for a "one-dimensional x-axis sample"). The axial signal amplitude produced by a 180° flip angle at the coil center is also plotted, for comparison.

Although $(B_1)_{xy}$ for the surface coil is highly inhomogeneous in both magnitude and direction, when the surface coil is used as both transmitter and receiver on or near resonance, the phase angle between the excitation field and the reception field is the same at every point q [1, 3, 4]. Since the phase of M_{xy} is determined by the phase of the excitation field $(B_1)_{xy}$, the phase angle between the receiver $(B_1)_{xy}$ and M_{xy} will be constant at every point over the sample volume, and $\varphi(q) = \varphi$ in Eq. (2) [4]. Thus, although some parts of the sample may experience negative signal amplitudes due to flip angles greater than 180°, the NMR signals from all parts of the sample will be in phase and will add coherently over the excitation volume of the surface coil.

2.4 Surface Coil Receiver, Homogeneous Transmitter

The inhomogeneous-$(B_1)_{xy}$ excitation profile and the resultant wide distribution of flip angles limit the usefulness of the surface coil in more complex applications that depend upon well defined flip angles over the volume of detection. One approach to circumvent this problem, while maintaining the high signal sensitivity and limited detection volume of the surface-coil receiver, is surface-coil reception in the presence of homogeneous excitation [5]. This can be accomplished through use of a homogeneous-$\mathbf{B}_1$ transmitter coil, such as a saddle coil, and a surface-coil receiver positioned orthogonally to each other to minimize magnetic coupling between the two coils.

For homogeneous-$\mathbf{B}_1$ excitation, the magnetization produced in the xy plane will be constant in amplitude (assuming a homogeneous sample) and direction over the sample. However, the direction of the reception (surface-coil) $\mathbf{B}_1$ is not constant over the volume of detection. Thus, in contrast to the single-coil-mode excitation and detection experiment, the phase angle between the excitation field and the reception field is not the same at every point q in the homogeneous-excitation, surface-coil-reception experiment. Therefore, the phase relationship between the xy components of the reception $\mathbf{B}_1$ and $\mathbf{M}$ will not be constant over space, and $\varphi(q)$ in Eq. (2) is dependent on spatial coordinates. In such situations (or, indeed, for any case involving multiple-coil signal excitation and detection using coils of different size or geometry) the lack of phase coherence between the excitation and reception fields prevents totally coherent signal summation over the sample volume, and leads to partial cancellation of the signal [5, 10]. However, this reduction in signal amplitude is offset by the ability to produce an optimum flip angle (and, hence, maximum M_{xy}) over the entire sample volume.

The signal reduction resulting from spatially-dependent phase mismatch between the $(B_1)_{xy}$ of the transmitter coil and the $(B_1)_{xy}$ of the receiver coil, and its effect on the observed volume of detection, has been described quantitatively for the case of homogeneous excitation, surface-coil reception with a homogeneous sample [5]. The x and y components of $(B_1)_{xy}$ for a surface-coil receiver possess mirror symmetry about the xz plane, as illustrated in Fig. 2(a). Thus, two points that are reflected symmetrically about the xz plane possess identical x components of $(B_1)_{xy}$ but y components that are 180° out of phase with respect to each other. If two identical moments are positioned symmetrically about the xz plane and homogeneously excited, the contributions to the total signal induced in the surface-coil receiver from the x components of $\mathbf{B}_1$ will sum coherently, while the contributions from the y components of $\mathbf{B}_1$ will cancel. The surface-coil reception sensitivity resulting from a homogeneous sample that has been homogeneously excited is, thus, reduced from a dependence on $(B_1)_{xy}$ to a dependence on $(B_1)_x$ alone. (Sample heterogeneity may result in imperfect cancellation of the components due to $(B_1)_y$, altering the volume of detection.)

The signal-amplitude distribution resulting from homogeneous excitation, surface-coil reception of a homogeneous sample is described by Eq. (7), where

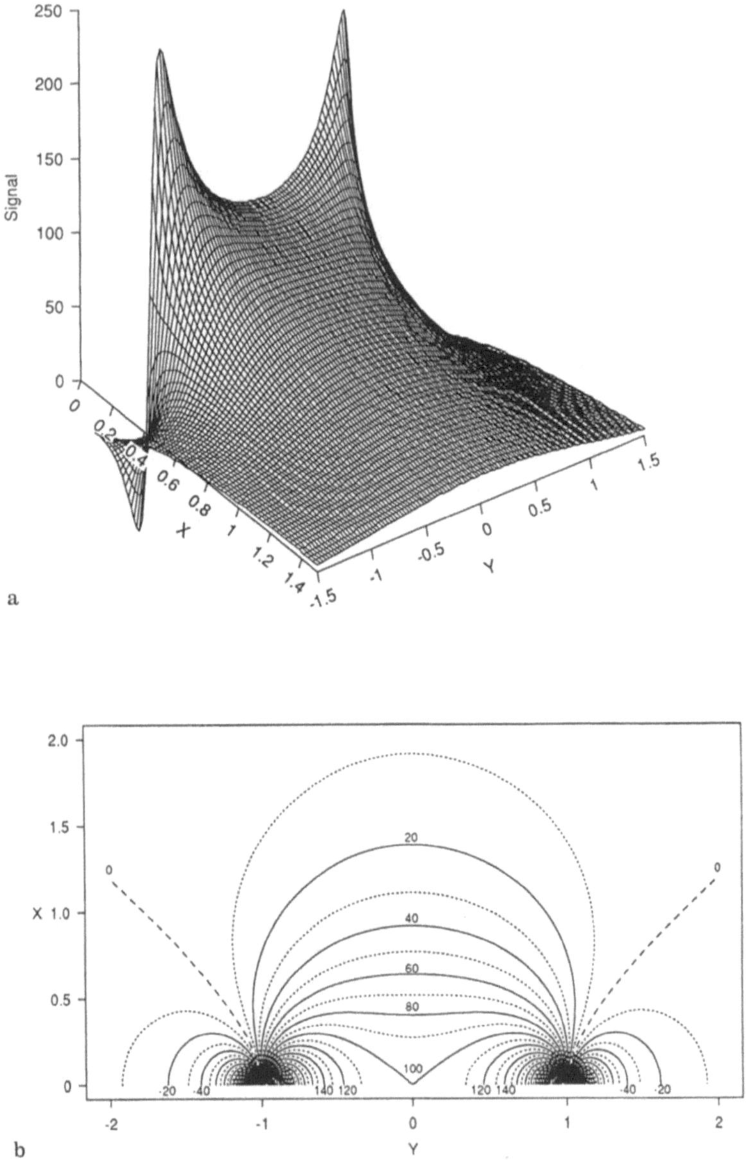

Fig. 7 a, b. Calculated signal-amplitude distribution maps in the *xy* plane for a single homogeneous-excitation pulse with a 1-turn, circular surface coil lying in the *yz* plane used as the NMR receiver. The axes are given in units of coil radii. All signal amplitudes are normalized relative to an amplitude of 100 at the coil center for a homogeneous 90° flip angle. **a** Signal-amplitude surface in the *xy* plane at $z = 0$. **b** Corresponding iso-amplitude contours. The amplitude distribution for the *xz* plane is identical to the one shown in this figure

$\alpha(q) = \alpha$ (e.g. 90°) and the dependence on the (surface-coil) receiver $(B_1)_{xy}(q)$ becomes a dependence on $(B_1)_x(q)$. The calculated signal-amplitude distribution for surface-coil reception of a homogeneous sample that has been homogeneously excited with a single pulse under equilibrium conditions is shown in Fig. 7. The amplitude surface is plotted for the xy plane at $z = 0$ in Fig. 7(a); the corresponding iso-amplitude contours are plotted in Fig. 7(b). Since the signal amplitude depends only on $(B_1)_x$, which is radially symmetric, the signal-amplitude map for the xz plane has the same form as the one for the xy plane. In the maps of Fig. 7, signal amplitudes have been scaled relative to an amplitude of 100 at the coil center.

The effects of the uniform flip angle and transmitter/receiver phase mismatch are seen in comparing Fig. 7 to the single-coil-mode surface-coil excitation and reception signal-amplitude maps in Fig. 5. The large positive peaks near the coil wires correspond to regions of intense $\mathbf{B}_1$ (see Fig. 3). The peaks are positive since the same flip angle is achieved over the entire sample. This is in contrast to the negative peaks in the single-coil excitation and reception signal-amplitude maps where the 90° flip angle achieved in the coil center through surface-coil excitation resulted in flip angles greater than 180° in the regions of intense $\mathbf{B}_1$. The negative signal regions in Fig. 7 which extend laterally (in the y direction) for $|y| > 1$ radius result from reversal in the direction of $(B_1)_x$ relative to that of the homogeneous transmitter. While the uniform flip angle results in increased signal intensity in the region lying within approximately one radius of the x axis for the homogeneous-excitation experiment relative to the single-coil experiment, the dependence on $(B_1)_x$ rather than $(B_1)_{xy}$ results in reduced sensitivity laterally outside this region [5].

The uniform flip angle over the sample volume also results in an increased penetration depth relative to the single-coil experiment. In the homogeneous-excitation experiment, 90% of the total signal is obtained from the sample volume located within 2.8 radii of the coil plane. In comparison, for the single-coil experiment 90% of the signal is obtained within a depth of only one radius [5, 11]. The effects of the uniform flip angles and phase mismatches result in a "narrower" and "deeper" interrogated volume of the surface-coil receiver. However, the total signal intensities for the two experiments are similar for a single pulse at equilibrium [5].

One important application of surface-coil reception with a homogeneously excited sample is found in imaging [8, 12]. This application takes advantage of the high sensitivity over a localized region afforded by the surface-coil receiver, while optimizing experimental conditions for imaging by providing uniform-flip-angle excitation. In order to resolve spatially the NMR signal contributions in the imaging experiment, each volume element in space is assigned a unique frequency and phase. In such an experiment, each volume element is, in effect, detected independently of the others. Thus, assuming that the volume element is small relative to the receiver $(B_1)_{xy}$ inhomogeneities, there is no signal cancellation due to transmitter/receiver phase mismatch [5]. The image signal-intensity distribution of the surface-coil receiver in the presence of homogeneous excitation is, therefore, identical to the $(B_1)_{xy}$ distribution of the surface-coil receiver, as plotted in Fig. 3. Since no signal cancellation results from transmitter/receiver phase mismatch, the signal intensity is a function of $(B_1)_{xy}$ instead of just $(B_1)_x$, and the xy and xz signal distributions are no longer identical.

Comparison of the homogeneous-excitation, surface-coil-reception imaging experiment (equivalent to Fig. 3) with the homogeneous-excitation, surface-coil-reception spectroscopy experiment (Fig. 7) shows that except for the $x=0$ and $y=0$ planes, where the two experiments are equivalent (in magnitude), (a) the interrogated volume of the imaging experiment extends further in every direction than the spectroscopy experiment, and (b) the image signal intensity is greater than the spectroscopy signal at every point over the sample volume. Since the magnitude of the signal is generally displayed in the imaging experiment, no negative signals are produced.

In the above discussion, only homogeneous excitation with surface-coil reception has been considered. However, the transmitter coil need not be homogeneous. For example, coaxial, coplanar surface coils of different diameters have been used as separate transmitter and receiver coils, where the larger diameter coil is used for excitation and the smaller diameter coil for reception [13]. Although a single, optimum flip angle is not achieved over the entire sample, the flip-angle distribution created by excitation through the large coil is broader than that which would be created through excitation by the small coil. In this case signal cancellation due to phase mismatches between the transmitter and receiver still occur, although it is less severe than the homogeneous-excitation experiment [10, 13, 14]. This scheme has found useful applications, particularly in conjunction with "depth" pulse sequences, in enhancing signal localization achievable using the surface coil [6, 10, 13, 15].

2.5 *Effects of Experimental Parameters on Interrogated Volume*

In the previous descriptions of the interrogated volume of the surface coil, only the effects of the inhomogeneous $(B_1)_{xy}$ on excitation and reception were considered; all distributions were calculated from a single pulse or, equivalently, for a single pulse under steady-state pulsing conditions with a pulse repetition time greater than $5T_1$. However, since the interrogated volume of the surface coil is determined not only by the inhomogeneous spatial distribution of $(B_1)_{xy}$, but also by the nonlinear relationship between $(B_1)_{xy}$ and the induced magnetic moment, certain experimental parameters such as pulse width, pulse repetition rate, resonance offset, and sample relaxation times can also significantly affect the shape and location of the interrogated volume. These effects have been considered theoretically by several investigators [7, 11, 16].

Under steady-state pulsing conditions, the longitudinal magnetization (M_z) existing prior to each pulse is related to the equilibrium magnetization M_0 and the experimental pulsing conditions as [17, 18]

$$M_z = \frac{M_0(1 - e^{-T_R/T_1})}{(1 - \cos\alpha e^{-T_R/T_1})}, \tag{9}$$

where T_R is the pulse repetition period (assumed to be greater than $3\,T_2$ so that effects of incomplete transverse relaxation, e.g. spin-echo formation, are neglected), T_1 is the longitudinal relaxation time of the sample, and α is the spin flip angle. The steady-state magnetization produced in the xy plane (M_{xy}) under steady-state pulsing conditions is, thus,

$$M_{xy} = \left[\frac{M_0(1 - e^{-T_R/T_1})}{(1 - \cos\alpha\, e^{-T_R/T_1})} \right] \sin\alpha. \tag{10}$$

In general, for surface-coil excitation α and, hence, M_{xy} will be functions of spatial position. If T_R/T_1 is less than five, M_z will also be a function of position. Finally, if the sample itself is inhomogeneous, M_0 will be a function of position as well.

Substituting the expression for the steady-state magnetization into Eq. (2) and ignoring signal decay due to T_2 processes, the signal induced in a receiver coil by a precessing moment at point q under steady-state conditions can be expressed as

$$S(t)_q = \omega_0 (B_1)_{xy}(q) \left[\frac{M_0(1 - e^{-T_R/T_1})}{(1 - \cos\alpha(q) e^{-T_R/T_1})} \right] \sin\alpha(q) \sin(\omega_0 t + \varphi(q)), \tag{11}$$

where the spatial dependence of α is indicated by $\alpha(q)$. Thus, the steady-state signal-amplitude distribution resulting from surface-coil excitation and detection depends on (a) the ratio T_R/T_1 as well as the excitation $(B_1)_{xy}(q)$ (through the $\sin\alpha(q)$ factor) as a result of the excitation process, and (b) on $(B_1)_{xy}(q)$ (via the receiver $(B_1)_{xy}(q)$) as a result of the detection process.

The dependency of the signal-amplitude spatial distribution on the ratio T_R/T_1 for surface-coil excitation is in distinction to homogeneous-$(B_1)_{xy}$ excitation (with homogeneous or inhomogeneous reception), where the observation volume is independent of pulsing parameters and sample relaxation times because of a constant α over space. The effect of decreasing T_R relative to T_1 for homogeneous excitation is an overall decrease in the signal amplitude (per pulse) due to "partial magnetization saturation" independent of spatial coordinates. For surface-coil excitation, however, the "partial saturation" that occurs with decreasing T_R relative to T_1 is spatially dependent as a result of the $\cos\alpha(q)$ term in Eq. (11). An often unappreciated consequence of this dependence of spatial distribution on relaxation times is that if multiple resonances of a spectrum are characterized by different T_1 times, and if T_R is not three to five times the longest T_1 time, the different resonances will experience different excitation spatial profiles using surface coils. This dependence of the excitation spatial profile on T_1 relaxation time can be removed through use of a saturation-recovery pulse sequence [16].

The effects of pulse repetition rate on interrogated volume shape and relative signal amplitude for on-resonance surface-coil excitation and detection of a homogeneous sample are illustrated in Fig. 8. Signal-amplitude surfaces are plotted for the xy plane for a single pulse under steady-state conditions for three different pulse repetition periods. In each case, the flip angle at the center of the

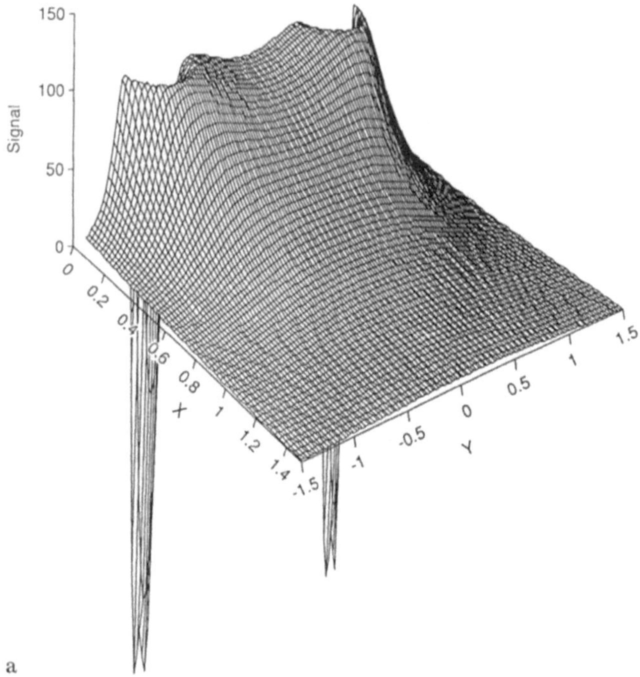

150
100
50
0
Signal
0
0.2
0.4
0.6
0.8
1
1.2
1.4
X
-1.5
-1
-0.5
0
0.5
1
1.5
Y

a

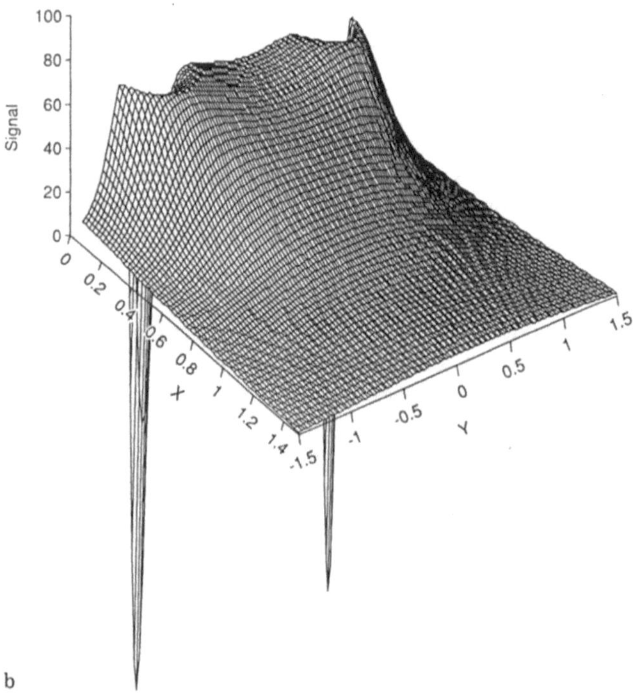

100
80
60
40
20
0
Signal
0
0.2
0.4
0.6
0.8
1
1.2
1.4
X
-1.5
-1
-0.5
0
0.5
1
1.5
Y

b

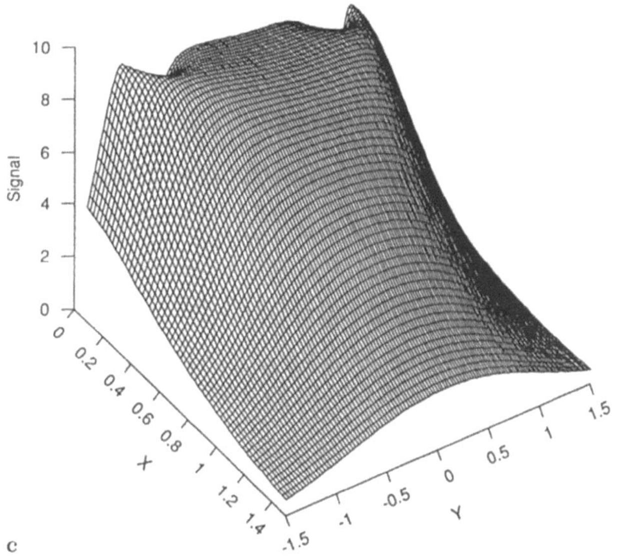

c

Fig. 8 a–c. Calculated signal-amplitude distribution maps for a single pulse from a 1-turn, circular surface coil lying in the *yz* plane used as both transmitter and receiver under steady-state pulsing conditions. The axes are given in units of coil radii. In each case, the flip angle at the center of the coil (α_0) that achieves maximum total sample *S/N* per unit time over a sample separated from the coil by 0.1 radii is used [11]. All signal amplitudes are normalized relative to an amplitude of 100 for a 90° flip angle at the center of the coil for a single pulse at equilibrium. The signal-amplitude surface is plotted for the *xy* plane at $z = 0$ for three different relative pulse repetition periods: **a** $T_R/T_1 = 5, \alpha_0 = 94°$, **b** $T_R/T_1 = 1, \alpha_0 = 85°$, **c** $T_R/T_1 = 0.05, \alpha_0 = 51.5°$. Note the difference in signal scaling in the three maps

coil, α_0, is that which has been determined to produce the maximum signal-to-noise ratio per unit time (*S/N*) over the entire interrogated volume of a large sample separated from the coil by 0.1 radii [11]. The signal amplitudes are normalized relative to an amplitude of 100 for a single pulse at equilibrium conditions ($T_R/T_1 = \infty$) with a flip angle of 90° at the center of the coil. The signal-amplitude map of Fig. 8(a) is calculated for slow pulse-repetition conditions ($T_R/T_1 = 5$); Fig. 8(b) is for intermediate pulse-repetition conditions ($T_R/T_1 = 1$), and Fig. 8(c) is for rapid pulse-repetition conditions ($T_R/T_1 = 0.05$). The negative peaks apparent in the maps for the long and intermediate pulse-repetition conditions (Figs. 8(a) and 8(b)) result from flip angles greater than 180° in the intense-$(B_1)_{xy}$ regions near the coil wires. The lack of negative peaks in the rapid pulse-repetition map (Fig. 8(c)) results from the fact that, as the pulse repetition rate increases the pulse width at the coil center that achieves maximum total signal decreases, producing smaller flip angles near the plane of the coil. Furthermore, as the pulse repetition rate increases, the regions in the strongest $(B_1)_{xy}$ field (i.e. regions experiencing larger flip angles) near the coil will experience greater magnetization saturation than regions farther from the coil, where α is reduced. This is seen in Fig. 8 as a drastic decrease in maximum signal amplitude near the coil as T_R/T_1 decreases and a more moderate fall-off of signal amplitude with distance. (Note different amplitude scales in Fig. 8.) Thus, under rapid pulse-repetition conditions, the regions farther from the coil will contribute proportionately more signal than under slow pulse-repetition conditions, achieving a more uniform sensitivity over the interrogated volume of the surface coil. This has the effect of "spreading out" the interrogated volume of the surface coil and, thus, decreasing the spatial selectivity of the surface coil as the repetition period decreases. This is illustrated for the one-dimensional case in Fig. 9 in which the signal amplitude along the x axis is plotted as a function of distance for several repetition periods. The magnitude of $(B_1)_{xy}$ is also plotted in this figure for comparison.

Although the sensitivity per pulse is much reduced for rapid pulsing conditions in comparison to slow pulsing conditions (Figs. 8 and 9), the sensitivity per unit time increases significantly with repetition rate. Rapid pulsing ($T_R/T_1 = 0.05$) at the optimum pulse width with a surface coil results in a net gain of about 130% in the total S/N per unit time relative to slow pulsing conditions ($T_R/T_1 = 5$) [11]. This is more than the gain achieved with a homogeneous coil under rapid pulsing conditions (approximately 60% [18]), and is a consequence of the proportionately-larger contributions from regions further from the coil in the rapid pulsing surface-coil experiment

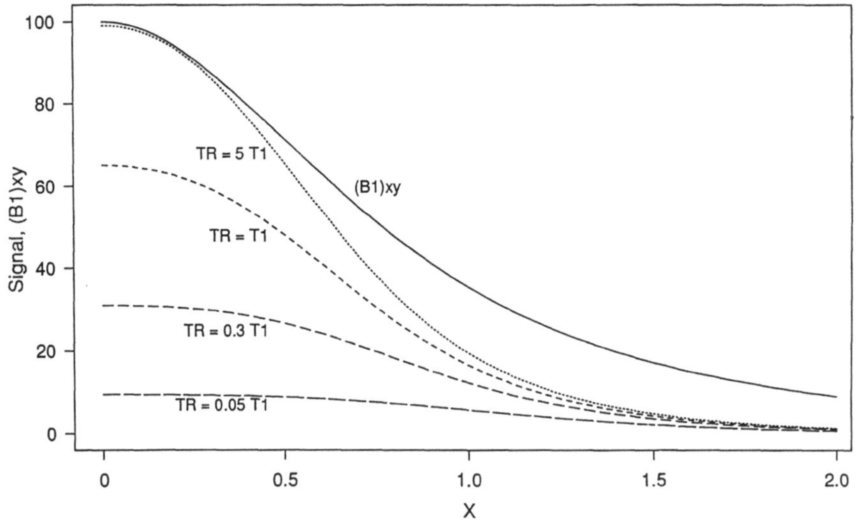

Fig. 9. Comparison of signal amplitude for a one-dimensional x-axis sample as a function of distance along the x axis for a single pulse from a 1-turn, circular surface coil used as both transmitter and receiver under steady-state pulsing conditions for different pulse repetition periods. The x axis is given in units of coil radii. In each case the α_0 that achieves maximum total S/N per unit time was used [11]. The curves are normalized relative to an amplitude of 100 for $\alpha_0 = 90°$ for a single pulse at equilibrium. The relative axial magnitude of $(B_1)_{xy}$ is also plotted. Line assignments are: *solid line*, $(B_1)_{xy}$ magnitude; *dotted line*, $T_R/T_1 = 5$; *short-dashed line*, $T_R/T_1 = 1$; *medium-dashed line*, $T_R/T_1 = 0.3$; *long-dashed line*, $T_R/T_1 = 0.05$

2.6 Sensitivity of the Surface Coil

The sensitivity of the NMR receiver is quantified by the signal-to-noise (S/N) ratio per unit sample volume obtained in the NMR experiment [19]:

$$\frac{S}{N} \propto \frac{\omega^2 (B_1)_{xy}}{\sqrt{R}}, \tag{12}$$

where $(B_1)_{xy}$ is the transverse field produced by unit current in the receiver coil, ω is the Larmor frequency, and R is the equivalent series resistance representing the thermal noise generated in the coil and sample. The noise associated with dielectric and inductive (eddy current) losses in electrically-conductive biological samples usually predominates at high frequencies. This sample noise is coupled to the receiver coil through the electric and magnetic fields of the coil, and degrades the signal-to-noise ratio in the NMR experiment. While dielectric losses can be reduced by using circuits designed to minimize stray electric fields and by using a Faraday shield to screen the electric field of the coil [20], little can be done to eliminate inductive losses. Rigorous analyses of the effects of such loss mechanisms, as well as other factors affecting receiver sensitivity, are treated in detail elsewhere [19–24].

The signal-to-noise ratio obtained for a volume element located within the detection volume of a surface coil is related to the $(B_1)_{xy}$ at that position. Along the axis of a surface coil $(B_1)_{xy}$ is related to the coil radius r as [22]

$$(B_1)_x \propto \frac{r^2}{(r^2 + x^2)^{3/2}}, \tag{13}$$

where x is the axial distance from the coil. $(B_1)_{xy}$ is strongest at the center of the surface coil, where it is inversely proportional to the coil radius, and decreases rapidly with distance from the coil. Sample volume elements close to the surface-coil receiver (in regions of high $(B_1)_{xy}$) contribute strongly to the total signal as well as to the total noise, while volume elements far from the coil contribute markedly less to both. The effective resistance associated with a conductive sample filling the detection volume of a surface coil (i.e. sample-related circuit resistance R representing sample noise) has been determined to vary as the third power of the coil radius [23]. Thus, a small surface coil (small r) will have a larger sensitivity for a volume element at the center of the coil than will a large surface coil. However, the sensitivity of the smaller coil drops off more rapidly with increasing distance from the coil than that of the larger surface coil. Beyond some frequency-dependent depth, the sensitivity of the surface coil becomes nearly independent of coil radius [24]. As a result of their strong $(B_1)_{xy}$, surface coils have much higher sensitivities than large-volume homogeneous coils for regions near the sample surface. However, due to the rapid fall-off of $(B_1)_{xy}$ of the surface coil, large-volume coils outperform surface coils for regions deeper within the sample [22, 23].

3 Addressing Inhomogeneous $\mathbf{B}_1$ Effects

3.1 *Improving Surface Coil Localization*

While the surface coil is well suited for signal detection from a thick, superficial tissue layer, the curved iso-$(B_1)_{xy}$ contours and intense $(B_1)_{xy}$ near the coil make it difficult to detect signals from deeper structures without contamination from the overlying regions. Elimination of signal from superficial tissues is particularly necessary for in-situ spectroscopy of organs such as liver and kidney which lie beneath a thick muscle layer. Selection of a 180° flip angle at the sample surface reduces contributions from the overlying region, but is insufficient to eliminate them (see Fig. 4). Numerous investigators have resorted to surgical exposure of such tissues to isolate signal detection to only the tissue of interest. This compromises the otherwise noninvasive nature of NMR spectroscopy. A number of elegant techniques have been developed and applied to surface-coil spectroscopy to improve the inherent volume localization capabilities of the surface coil. These techniques utilize gradients in either the static $\mathbf{B}_0$ field or the RF $\mathbf{B}_1$ field to delineate a specific sample volume. Many excellent reviews of localized spectroscopic methods have been published (see, for example [13, 25–28]).

Frequency-selective RF pulses applied in the presence of linear $\mathbf{B}_0$ field gradients have been used to select a well-defined slice or volume of interest. Data are acquired in the absence of applied gradients yielding high-resolution spectra. Spins from outside the volume of interest are either not excited, rapidly dephased in the gradient fields, or subtracted out in subsequent data collections. Examples of such techniques include volume-selective excitation (VSE) [29–31], image-selected in vivo spectroscopy (ISIS) [32, 33], and depth-resolved surface-coil spectroscopy (DRESS) [34, 35]. Incremented pulsed field gradients and subsequent multidimensional Fourier transformation have been used to produce spectra from a series of spatially resolved slices [36–38]. In the topical magnetic resonance (TMR) technique [39, 40], non-linear gradients are used to profile the static magnetic field to produce a homogeneous localized region from which a high-resolution spectrum is obtained. Difficulties encountered in the localization schemes employing $\mathbf{B}_0$ gradients include the possible generation of long-lived eddy currents by the switched field gradients and spatial smearing of the localized volumes resulting from chemical shift dispersion (chemical shift artifacts).

The natural $\mathbf{B}_1$ gradients of the surface-coil RF field have been exploited to enhance signal localization. Rotating frame imaging [41–43] and related techniques [44–46] acquire a series of spectra in which the excitation pulse width is incremented, thereby mapping the spatial distribution of the chemical species according to $\mathbf{B}_1$ gradient contours. Composite pulses have been used in depth-selective sequences for enhancing spatial selectivity and reducing off-resonance phase distortion [47–50]. A family of sequences known as "depth pulses" makes use of a series of phase-cycled pulses that result in spatial discrimination based on sensitivity to the flip angle experienced [6, 10, 13, 15]. Cascading of depth-selective pulses further increases spatial selectivity. Figure 10 illustrates the increased spatial discrimination achieved by a simple $\theta; 2\theta[\pm x, \pm y]$ depth pulse sequence (i.e. a pulse producing flip angle θ, in this case equal to 180° at the coil center, followed by a phase-cycled pulse producing flip angle 2θ) over that achieved by a single pulse with a 180° flip angle at the coil center.

Since the shape of the localized volume selected by these sequences is determined by the curved $\mathbf{B}_1$ contours of the surface coil, the signal produced by the selected volume will still be contaminated by the response of superficial tissues. Additionally, alternating positive and negative signal contributions from sample in the high-flux regions close to the coil wire can be significant (Fig. 10(b, c)). Both of these effects can be reduced by using $\mathbf{B}_0$ gradients in addition to the depth selective sequences [6, 34], or by using separate surface-coil transmit and receive coils whose excitation and detection volumes overlap [10, 51]. High-flux signals near the coil have also been suppressed by varying the excitation pulse lengths in the depth-pulse sequence or by including low-flip-angle preparation pulses in the sequence [6, 52–54]. Severe off-resonance effects and lengthy pulse trains can complicate $\mathbf{B}_1$-gradient techniques.

An alternative approach to the elimination of superficial signals that is not plagued by residual eddy currents, curved excitation volumes, or high-flux signal contributions, is the use of a surface-spoiling gradient in which a local (surface) gradient coil is positioned directly over the unwanted tissue region. A highly inhomogeneous $\mathbf{B}_0$ gradient field is produced which rapidly dephases signals within a limited penetration depth [55, 56]. Hence, these sample regions will contribute no net signal in response to subsequent surface-coil interrogation. This technique has been demonstrated in vivo using simple single-pulse sequences and spin echoes [57]. Paramagnetically- and ferromagnetically-generated spoiling gradients have also been used to eliminate superficial signals [58–60].

3.2 Field Insensitive Pulses

While the inhomogeneous $(B_1)_{xy}$ of the surface coil can be advantageous in signal localization schemes, it can be an impediment in other spectroscopic techniques, especially those which depend on uniform signal excitation. Since an optimum flip angle cannot be achieved over the entire sample with surface-coil excitation, the maximum signal-to-noise obtainable in the experiment is limited. To improve sensitivity and minimize spectral distortions caused by the inhomogeneous $(B_1)_{xy}$, phase-cycled composite pulses have been developed which are tolerant of wide variations of $(B_1)_{xy}$ and resonance offsets [61, 62]. In contrast to the composite pulses used in localization schemes, which increase the sensitivity of the signal to variations in $(B_1)_{xy}$, these pulses produce a significantly more uniform excitation profile over a wide range of $(B_1)_{xy}$ values than achievable with simple pulses [47, 63–65]. Adiabatic pulses, in which the amplitude and frequency (or phase) of the pulse are modulated to achieve a more uniform excitation response, also exhibit marked insensitivity to wide variations in the $(B_1)_{xy}$ field [66–68].

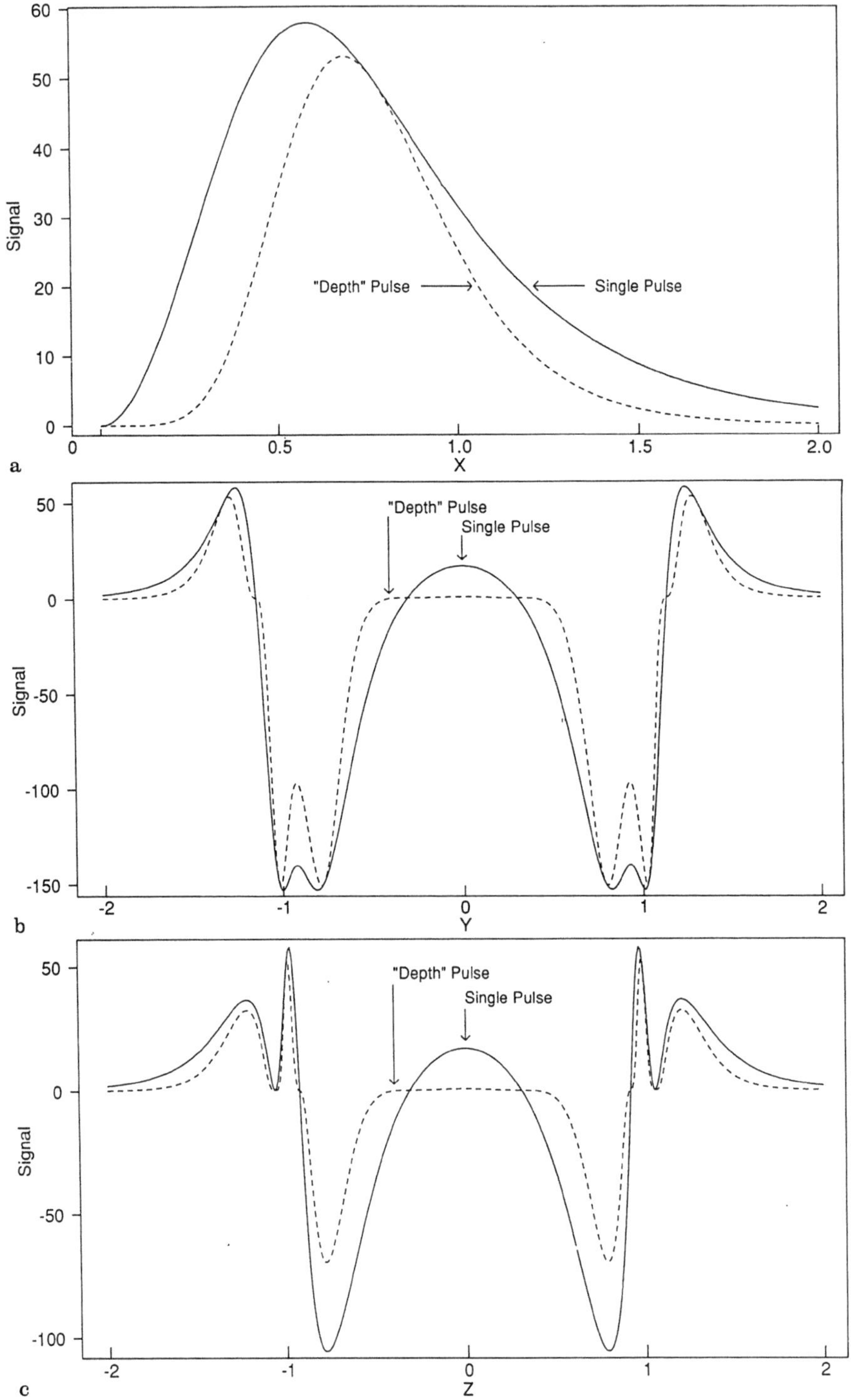

60
50
40
30
20
10
0
Signal
"Depth" Pulse
Single Pulse
0
0.5
1.0
1.5
2.0
X
a
50
0
-50
-100
-150
Signal
"Depth" Pulse
Single Pulse
-2
-1
0
1
2
Y
b
50
0
-50
-100
Signal
"Depth" Pulse
Single Pulse
-2
-1
0
1
2
Z
c

In addition to increasing the signal-to-noise ratio obtainable through more uniform excitation, field-insensitive pulses such as the composite and adiabatic pulses have the added advantage that they can be used with a single surface coil operating as both transmitter and receiver. Hence, the excitation and reception fields will have a constant phase relation over all space. These pulses have been useful for signal excitation, inversion, and refocusing in surface-coil experiments such as spin-echo formation, imaging, localized spectroscopy, and relaxation-time measurements [69–73].

4 Surface Coil Design

The basic surface-coil design consists of a circular loop of wire composed of one or more turns, wound either in a cylindrical or planar fashion (Fig. 1). The coil is tuned to the NMR frequency of interest and matched to the impedance of the driving device, typically 50 Ω. One standard tuning and matching scheme is the capacitively-coupled resonant tank circuit shown in Fig. 11(a) [2]. Degradation in circuit performance resulting from electrical coupling to lossy conductive samples can be reduced by balancing the NMR coil with respect to ground [74–76]. An example of a capacitively-coupled balancing scheme is shown in Fig. 11(b). The single-resonance surface-coil probe has been extended to applications in multi-nuclear NMR spectroscopy through development of multiple-frequency circuit-design technology [77–87]. The use of resonant transmission lines and lumped reactive elements in multiply-tuned circuits is discussed in detail elsewhere [77–82].

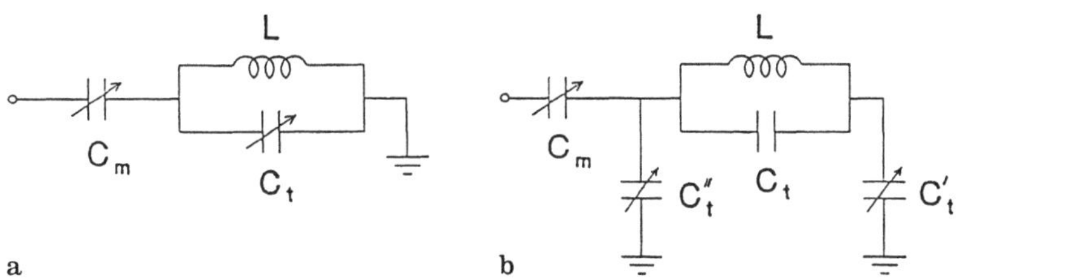

Fig. 11 a–b. Capacitively coupled frequency-tuning and impedance-matching circuits. **a** Tank circuit. **b** Balanced circuit. The inductor, L, is tuned primarily with capacitor C_t and matched primarily with capacitor C_m. Capacitors C_t' and C_t'' are used to tune the circuit and balance the coil

◄

Fig. 10 a–c. Calculated signal amplitude obtained from a 1-turn, circular surface coil lying in the *yz* plane used as both transmitter and receiver for a single pulse on resonance in which a flip angle of 180° is achieved at the coil center (*solid line*), and simple "depth pulse" sequence on resonance with a 180° flip angle at the coil center (*dashed line*). The signal is proportional to $(B_1)_{xy} \sin \alpha(q)$ for the single pulse and $(B_1)_{xy} \sin^3 \alpha(q)$ for the "depth pulse". The signal intensity is plotted **a** along the *x* axis, **b** parallel to the *y* axis $x = 0.2r, z = 0$, and **c** parallel to the *z* axis $x = 0.2r, y = 0$. The *x*, *y*, and *z* axes are given in units of coil radii

Modifications of the circular surface-coil design have been motivated by a desire to better accommodate the sample size or shape, to enhance the signal localization, or to improve the sensitivity of the coil. Anatomically contoured surface coils such as the rectangular spine coil and curved neck coil are useful in imaging applications [88–91]. Surface coils designed for signal detection restricted to very superficial regions include coils with closely-spaced conductors with current flowing in alternating directions such as the parallel-array "zig-zag" coil [92] and multiple-loop counter-rotating-current resonator placed tangential to the sample surface [93].

NMR coils that are balanced with respect to ground (through capacitive coupling [74], inductive coupling [75], or the use of baluns [76, 94]) have been shown to be effective in reducing dielectric losses in conductive samples. Dielectric losses have also been reduced by symmetrically distributing lumped capacitors around the coil to reduce the potential difference between the coil and ground [8, 75]. Explicit grounding at the coil center and electric shielding of the sample from the coil are features of the "crossover" surface coil which minimize dielectric losses in the sample [76].

An important class of coils used for local signal detection is based on the loop-gap resonator design [95]. These low-loss, high-sensitivity coils consist of wide conductor bands that are brought to resonance with capacitive gaps. Two coplanar loops that are laterally displaced and connected by a single gap form the planar-pair loop-gap resonator [96]. This coil detects signals from a slab-shaped region close to the coil and is useful for imaging superficial structures such as the temporomandibular joint. A further advantage of this design when used as an imaging coil is that the two loops are intrinsically isolated from an external homogeneous field of arbitrary orientation, such as a homogeneous transmitter field. This results from the fact that, for the proper circuit resonance condition, an emf induced at the gap by flux linking one of the loops will cancel that induced by flux linking the other loop. The counter-rotating-current coil consists of two coaxial loops that are axially displaced and support current flows in opposite directions [97]. It, too, is intrinsically isolated from an external field. Combinations of the planar-pair and counter-rotating-current coils produce a resonant structure whose elements are intrinsically isolated both from each other and from an external transmitter field [98]. Additionally, the signals induced in the two coil components can be detected in quadrature and combined to produce a net $\sqrt{2}$ improvement in the signal-to-noise ratio [99].

5 The Surface Coil Experiment

5.1 Practical Considerations

Optimization of the static $\mathbf{B}_0$ homogeneity over the interrogated volume of the surface coil is necessary to achieve high-resolution spectra in vivo. This is readily accomplished in biological samples by observing the water-proton free induction decay through the coil, even if the coil is not tuned to the ^1H frequency [100]. Although the sensitivity of the "detuned" coil is far from optimal for proton signal detection, the ability to detect the proton signal is a consequence of the high sensitivity of the ^1H nuclide and the extremely high concentration of water protons in biological tissues. The sensitivity of the coil to the proton signal can be improved by a transmission-line impedance transformation which allows a circuit that is optimized for one frequency (e.g. ^{31}P or ^{13}C) to be remotely tuned and matched to the proton frequency [101].

The tissue-water proton signal can also be useful as an internal chemical shift reference [100]. This is significant for tissues that do not contain a suitable reference peak of the nuclide of interest. Additionally, using a double-tuned surface-coil receiver, the absolute molar concentration of a metabolite can be determined by referencing the metabolite peak area to the water signal [102–105]. This requires correction for coil loading effects at the two frequencies and calibration of signal sensitivities with samples of known concentration. Determination of absolute molar concentrations of a chemical species using the natural-abundance deuterium signal present in tissue water can be accomplished in an analogous manner [106]. These techniques which use the endogenous water signal as an internal reference assume that the tissue is homogeneous and that the water concentration and chemical shift are invariant over normal physiological conditions. Gradient localization techniques have also been used in quantifying metabolite concentrations with surface coils [107, 108].

While the strong water signal detected in biological samples is advantageous for optimizing static field homogeneity and quantifying metabolite concentrations, it presents a severe problem for in vivo ^1H-NMR investigations. Observation of protonated species other than water and fat generally requires suppression of the intense water resonance. Water suppression techniques rely on either (a) avoiding the excitation of the water spins while exciting other species, or (b) selectively saturating (or nulling) the water signal. The applicability of these techniques to surface-coil spectroscopy has been evaluated and demonstrated [109–114]. Hard pulse selective-excitation sequences, such as binomial pulse sequences, have been shown to be effective in suppressing a solvent peak. However, when used with coils possessing inhomogeneous RF fields (e.g. the surface coil), the resulting frequency response is dependent on spatial position. This is problematic for quantitative analysis of resonance area. The use of small flip angles or phase-cycled refocusing pulses, however, largely overcomes this problem [109, 113]. Selective saturation techniques and techniques employing spin echoes are advantageous in that they do not have frequency-selective spatial responses. However, sample heating, saturation transfer to exchangeable protons, or relaxation-time constraints may pose a problem. Water suppression and spectral editing techniques have been used in surface-coil ^1H spectroscopy to monitor lactate levels in the brain in vivo [113, 115, 116].

Knowledge of the spin-lattice relaxation time of a sample is vital not only for the information it provides regarding the spin environment, but also for selection of experimental parameters. Accurate T_1 relaxation times have been shown to be obtainable with a surface coil, in spite of the inhomogeneous surface-coil field, using inversion-recovery and saturation-recovery pulse sequences [117, 118]. In order to apply a perturbation/recovery technique to the surface-coil experiment (or any situation involving inhomogeneous $(B_1)_{xy}$) three criteria must be met at each individual point in the excitation volume of the coil: (a) the longitudinal magnetization must be at the same initial condition prior to the start of each evolution period, (b) the magnetization must then be allowed to return to equilibrium during variable evolution periods, and (c) a constant fraction of the longitudinal magnetization must be sampled at the end of each evolution period. With these three criteria satisfied, and assuming that the relaxation is a first-order process, an accurate T_1 relaxation time can be determined by fitting the recovery curve to a three-parameter exponential function of the form

$$S(\tau) = A + B\exp\left(\frac{-\tau}{T_1}\right), \tag{14}$$

where $S(\tau)$ is the observed signal intensity and τ is the variable evolution period. Use of field-insensitive pulses, while not explicitly required for accurate measurements of T_1 relaxation times, provide more uniform excitation over the sample and, thus, can be used to increase the sensitivity and dynamic range of the T_1 experiment [69, 119].

5.2 *Representative Applications*

The surface coil has been used in a wide variety of multinuclear spectroscopy and imaging applications in vivo. NMR nuclides of obvious biological importance include ^{31}P, ^{13}C, ^{23}Na, ^{15}N, and ^{1}H, although exogenous spin labels such as ^{2}H and ^{19}F have also found use as tracers of metabolism or blood flow. While an exhaustive review of the many surface-coil applications is not attempted here, the following discussions and representative spectra are meant to serve as illustrative examples of the role of the surface coil in in vivo spectroscopy and imaging.

^{31}P NMR has been used most widely in the assessment of tissue energetics. The effects of pathological conditions or physiological insults, such as malignancy, ischemia, or hypoxia, on the high-energy phosphate metabolites present in tissue has been addressed in many ^{31}P surface-coil investigations. The high-energy phosphate metabolites readily observed via surface-coil detection in different tissues are illustrated in Figs. 12–14. Figure 12 is a representative phosphorus spectrum obtained in rat liver in vivo. Signal localization with preservation of maximal sensitivity was achieved via surgical exposure of the liver and placement of the coil directly on the liver (separated by a protective plastic sheet). The major phosphorus metabolites detected include the α-, β-, and γ-phosphate groups of adenosine triphosphate (ATP), inorganic phosphate (P_i), phosphomonoesters (PME), and phosphodiesters (PDE). The ^{31}P spectrum obtained from muscle tissue has an additional, very intense peak at 0 ppm corresponding to phosphocreatine (PCr), as shown in Fig. 13. The relative areas of the PCr, P_i, and ATP resonances in the phosphorus spectrum have been used

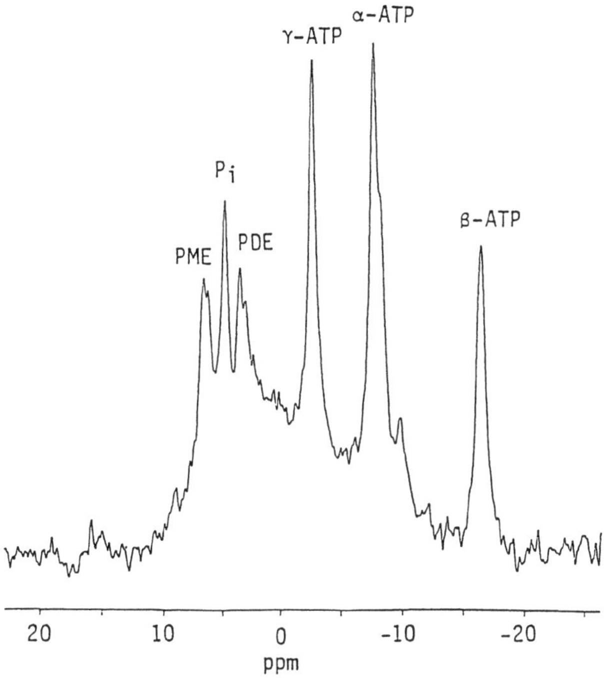

Fig. 12. ^{31}P surface-coil spectrum of rat liver obtained in vivo at 80.98 MHz (4.7 tesla). The liver was surgically exposed to achieve localization. Data consist of 100 transients acquired over five minutes and processed with a 15-Hz exponential filter. Data collection was quantitative for ATP (recycle time greater than $5T_1$) and near-quantitative for P_i (recycle time equal to $3T_1$) [120]. Visible resonances include the α-, β-, and γ-phosphate resonances of adenosine triphosphate (ATP), inorganic phosphate (P_i), and the phosphomonoester (PME) and phosphodiester (PDE) regions. Chemical shifts are referenced relative to phosphocreatine (PCr) at 0 ppm, which is not present in the liver

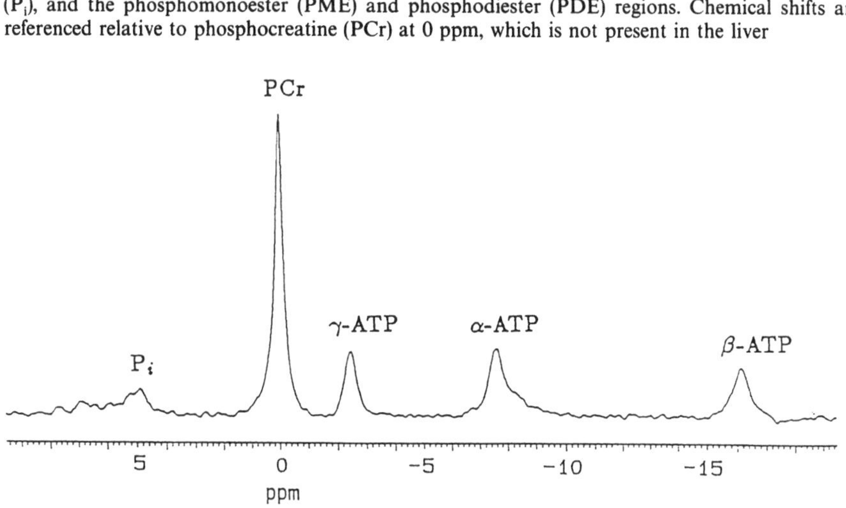

Fig. 13. ^{31}P surface-coil spectrum of rat leg muscle obtained in vivo at 121.42 MHz (7.1 tesla). Data consist of 44 transients acquired under quantitative conditions (20.5 sec recycle time) and processed with a 25-Hz exponential filter. Peak assignments are as indicated in Fig. 12. (Data courtesy of S-K Song)

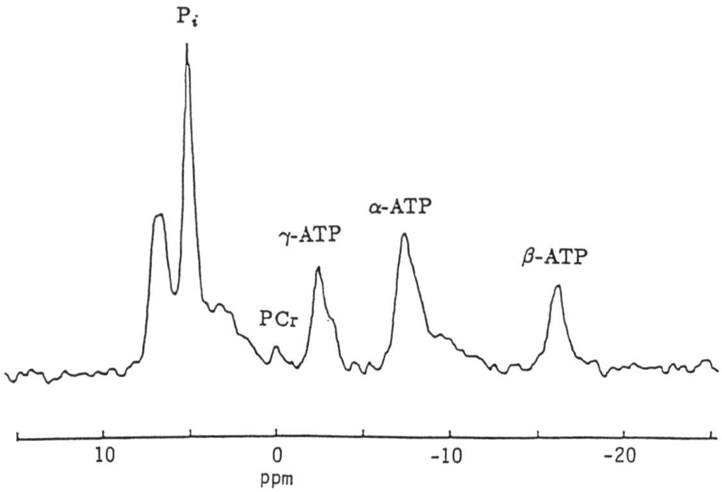

Fig. 14. ^{31}P surface-coil spectrum of a C_3H/Anf mouse RIF-1 tumor in vivo at 80.98 MHz (4.7 tesla). The spectrum consists of five min of data acquisition. Peak assignments are as indicated in Fig. 12. The pH of the tumor, as determined by the chemical shift of P_i, is 6.91. (Figure adapted from Fig. 2 of Ref. [121])

to assess the energy status and viability of the tissue. For example, Fig. 14 is a phosphorus spectrum obtained from a RIF-1 tumor implanted in a C_3H/Anf mouse. Immediately obvious is the greatly increased P_i content, likely indicative of compromised tissue perfusion (i.e. hypoxia/ischemia). Numerous surface-coil ^{31}P-NMR investigations have focused on characterizing tumor tissues and addressing effects of therapeutic interventions [121–128].

A further contribution of ^{31}P NMR to the assessment of the physiological condition of biological tissues is a noninvasive method for measuring intracellular pH, based on the chemical shift of the P_i resonance [129, 130]. Typically, the chemical shift of P_i is determined relative to PCr, whose shift is fairly independent of pH over physiological ranges. In tissues such as liver and kidney, which contain little or no PCr, the chemical shift of P_i is often determined relative to an external reference, or, alternatively, relative to the internal tissue water ^1H signal [100]. The pH of the tumor represented in Fig. 14 is 6.91 [121], as determined from the phosphorus spectrum, which is in contrast to a pH of around 7.1 for healthy muscle tissue.

Magnetization transfer techniques [131, 132] have been used in conjunction with surface-coil NMR to probe chemical reaction kinetics in vivo [69, 119, 133–137]. Reactions studied by ^{31}P NMR include the creatine kinase reaction, the adenylate kinase reaction, and ATP hydrolysis and synthesis. If two species are in chemical exchange, perturbation of the magnetization of one of the species (e.g. saturation or inversion) will result in transfer of the perturbation to the other species. The rate constant governing chemical exchange can be determined from

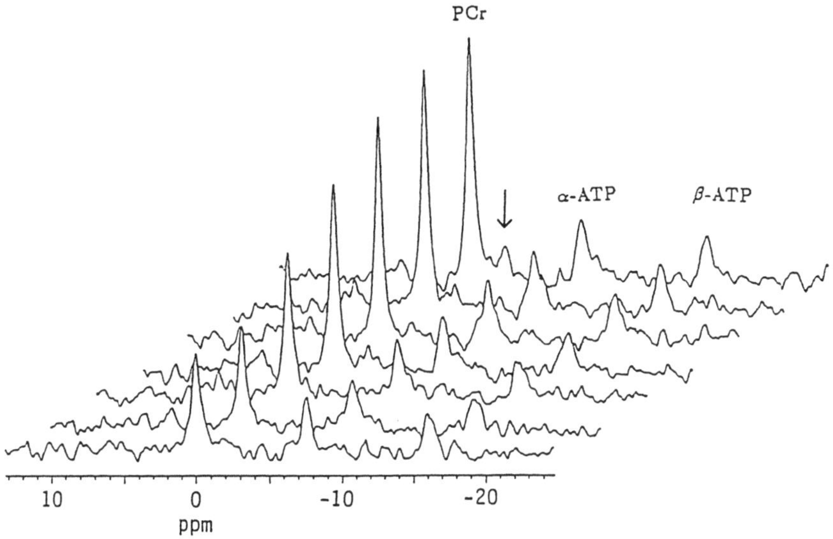

Fig. 15. ^{31}P surface-coil saturation transfer experiment performed on rat muscle in vivo at 121.42 MHz (7.1 tesla). Saturation of the γ-ATP resonance (indicated by the *arrow*) results in a transfer of the saturation to PCr. Chemical exchange between γ-ATP and PCr is evidenced by a decrease in the PCr peak intensity following saturation of the γ-ATP resonance. The length of the saturation pulse is (from top to bottom) 0, 0.5, 1, 2, 5, 10, and 20 sec. (Data courtesy of S-K Song)

the time dependence of the signal amplitude of the second species. An example of saturation transfer between γ-ATP and PCr in rat leg muscle is shown in Fig. 15. Saturation of γ-ATP results in a decrease in the PCr intensity which is dependent upon the length of the saturation time. From the time course of the PCr decrease in Fig. 15, and the T_1 relaxation time of PCr in the absence of exchange, the rate constant for the forward reaction (PCr → ATP) is determined to be 0.26 s^{-1}.

^{13}C NMR provides a valuable tool for the independent and simultaneous detection of multiple metabolic substrates whose fate can be monitored in near-real-time within a single system through a continuous analysis of label flux. The surface coil has been used in ^{13}C-NMR investigations of carbohydrate and fatty acid metabolism in tissues such as fat, liver, brain, and tumor in vivo [138–149]. The low natural abundance of ^{13}C (1.1%) makes ^{13}C-labeled substrates useful as metabolic tracers to elucidate metabolic pathways and substrate utilization. Carbon-containing moieties whose concentrations in the cell are sufficiently high, such as fatty acids and glycogen, allow investigation of metabolic processes from the natural-abundance ^{13}C signal as well [140–143].

The liver has been a primary target for ^{13}C-NMR investigations of the regulation of carbohydrate metabolism because of its central role in the maintenance of glucose homeostasis. Proton-decoupled carbon spectra (^{13}C-$\{^1H\}$)

obtained in rat liver with a surface coil are shown in Fig. 16 [143]. These high-resolution spectra were acquired using a double-coil surface-coil probe tuned to the carbon and proton resonance frequencies [83]. The coils were arranged in a coaxial, coplanar orientation with the larger-diameter coil used for decoupling and for optimizing static field homogeneity. Figure 16(a) is a 10-min acquisition of the natural-abundance ^{13}C-$\{^{1}H\}$ signal observed from a rat fed ad libitum. The natural abundance carbon spectrum is characterized by resonances arising from the methyl and methylene carbons of phospholipids and triacylglycerols in the region around 15–50 ppm, allylic carbons of unsaturated fatty acids around 130 ppm, and carbonyl carbons around 173 ppm [140]. The resonance at 100.4 ppm arises from the C-1 carbon of glycogen, the primary storage form of glucose

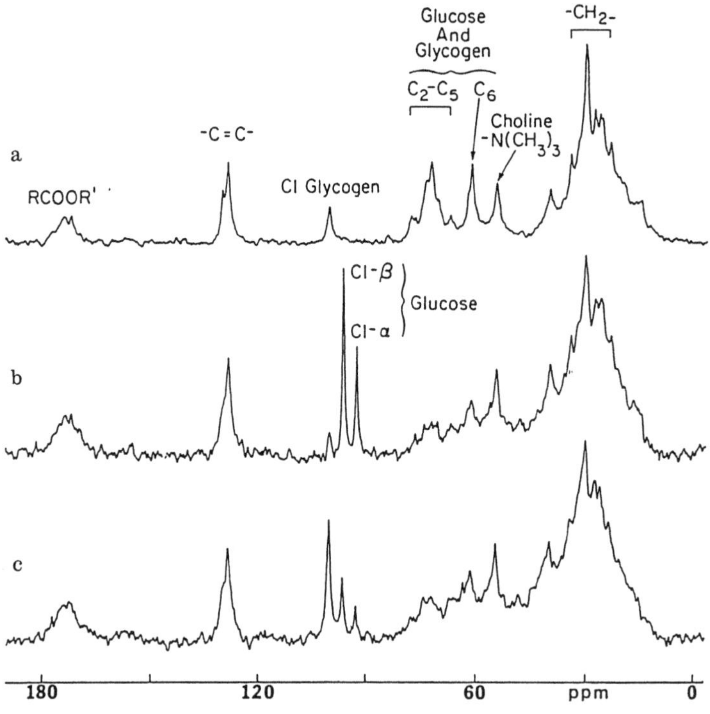

Fig. 16 a–c. Proton-decoupled ^{13}C surface-coil spectra of rat liver in vivo at 90.56 MHz (8.5 tesla). A double-coil, double-frequency surface-coil probe was used. Proton decoupling was achieved with a gated bilevel (5 W/0.5 W) decoupling sequence. The liver was surgically exposed to achieve localization. Data were acquired with a 0.165 sec recycle time, processed with a 40-Hz exponential filter, and baseline corrected. Chemical shifts are referenced relative to TMS via the signal from the methyl carbons of choline at 54.6 ppm. **a** 10-min acquisition of the natural abundance carbon signal from a fed rat. Glycogen resonances occur in the region from 61 to 100 ppm. The C-1 carbon resonance from α-1,4 glycosidic linkage is well-resolved at 100.4 ppm. **b** 3-min acquisition from the liver of a fasted (15 h) rat. Data collection was initiated three min after an intravenous glucose load of 100 mg D-[1-^{13}C] glucose. The resonances at 92.7 and 96.5 ppm arise from the α and β anomers of glucose, respectively. **c** 3-min acquisition initiated 30 min after the glucose load. Glucose incorporation into glycogen is evidenced by label flux to the glycogen resonance. (Figure adapted from Fig. 1 of Ref. [143])

in the liver. The strong resonances between 61 and 78 ppm result from the other carbons of glycogen. The presence of glycogen in the spectrum is indicative of the fed state of the animal.

The ability of ^{13}C-$\{^1H\}$ NMR to monitor evolving metabolism in vivo using surface coils is demonstrated in Figs. 16(b) and 16(c). These spectra were acquired in fasted rat liver using the double-coil probe described above [83] and consist of three minutes of data acquisition [143]. Data collection for Fig. 16(b) was initiated three minutes after an intravenous injection of 100 mg D-[1-^{13}C] glucose dissolved in saline. The high sensitivity afforded by the isotopically enriched substrate is evident in the strong signals detected from the C-1 α and β anomers of glucose (at 92.7 and 96.5 ppm, respectively). Figure 16(c) was acquired 30 min after the glucose administration. Subsequent glycogen formation from the administered glucose load is evidenced by label incorporation into the C-1 carbon of glycogen. The relative contributions from multiple metabolic pathways can be monitored by use of variously-labeled substrates and analysis of label "scrambling" in the products of metabolism [145, 149]. Substrate and hormonal regulation of substrate flux through metabolic pathways such as glycolysis and gluconeogenesis remain areas of active research.

The ^{19}F nuclide has been used as an exogenous spin label for a variety of in vivo applications. Its high sensitivity (0.8 that of proton), 100% natural abundance, and negligible endogenous concentrations make it a sensitive tracer for monitoring metabolism or elimination of fluorinated compounds in vivo. Surface-coil ^{19}F NMR has been used in studies of fluorinated anesthetics [150–154], chemotherapeutic agents [155–157], glucose utilization [158, 159], and blood flow [87, 160, 161]. The rate of metabolism of fluorine-labeled glucose analogues has been used as a measure of cerebral glucose utilization. The surface-coil ^{19}F spectra shown in Fig. 17 illustrate products of metabolism of 2-fluoro-2-deoxy-D-glucose (2FDG) observed in rat brain under proton-coupled and proton-decoupled conditions. The surface coil, tuned to the fluorine frequency, was positioned over the skull of the rat and a Helmholtz pair, tuned to the proton frequency, was positioned orthogonally to the surface coil for proton decoupling. The decoupler was gated on during acquisition of data in Figs. 17(a) and 17(c), and remained off at all times during collection of data in Fig. 17(b). Each spectrum consisted of 10 min of signal averaging and were acquired consecutively in the order presented in the figure. The signal-to-noise ratio and resolution were significantly improved by proton decoupling. These improvements resulted in detection of three additional resonances under proton-decoupled conditions that were not apparent in the proton-coupled case.

The deuterium (^2H, D) nuclide has been used both as an exogenous and endogenous in vivo spin label. In spite of the low natural abundance of ^2H (0.0156%), the enormous concentration of equivalent hydrogen in tissue water results in a natural abundance deuterium signal in tissue water (HOD) that is readily detected. This natural abundance signal has been used in surface-coil spectroscopy to determine the absolute concentrations of other metabolites [106]. The sensitivity of isotopically enriched substances and the low natural

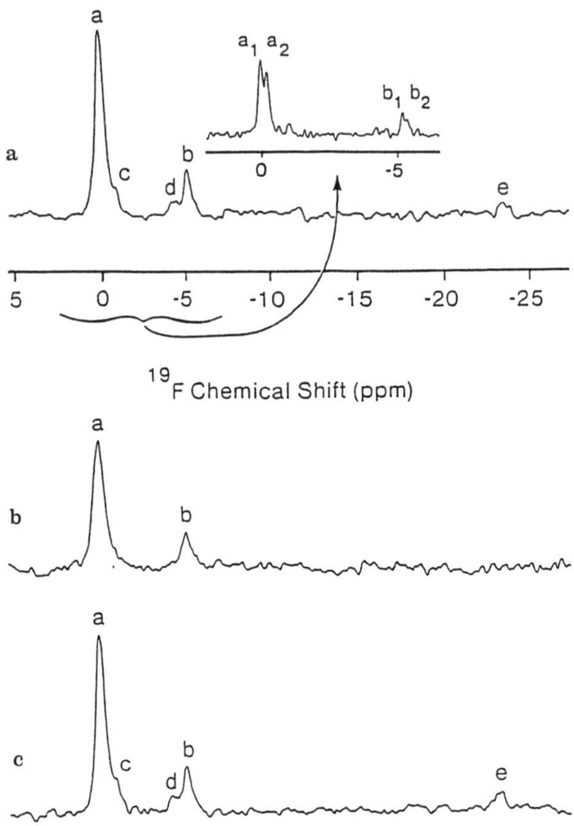

Fig. 17 a–c. ^{19}F surface-coil spectra of rat brain in vivo at 188.154 MHz (4.7 tesla) acquired approximately two hours after an intravenous injection of 50 mg 2FDG. Each spectrum consists of 10 min of data acquisition. Spectra **a**, **b** and **c** were acquired sequentially in consecutive 10-min intervals. The proton decoupler was gated on during data collection in **a** and **c** and turned off throughout collection of **b**. Resonances observed in all three spectra are: a, combination of 2-fluoro-2-deoxy-D-glucose-6-phosphate (2FDG6P) and 2FDG and b, 2-fluoro-2-deoxy-glycerol. Proton decoupling (**a** and **c**) reveals three additional peaks; c, 2-fluoro-2-deoxy-D-sorbitol; d, 2-fluoro-2-deoxy-L-glyceraldehyde; and e, unassigned. Resolution enhancement (insert of **a**) reveals the overlapping β anomers of 2FDG and 2FDG6P (a_1) and the overlapping α anomers of 2FDG and 2FDG6P (a_2). Due to low signal-to-noise ratio, the designation of separate peaks b_1 and b_2 is only tentative. Chemical shifts are arbitrarily referenced relative to the overlapping β-anomers of 2FDG and 2FDG6P (a_1) at 0 ppm. (Figure adapted from Fig. 3 of Ref. [158])

abundance has made the deuterium nuclide useful as an exogenous tracer in studies such as those investigating the metabolism of deuterated amino acids [162].

Surface-coil ^2H NMR has been applied to the quantitative estimation of regional blood flow and tissue perfusion in vivo using D_2O as a freely-diffusible blood-flow tracer [124, 163–166]. In this technique, D_2O is administered to a tissue of interest (e.g. as a bolus input), either via the bloodstream or by direct

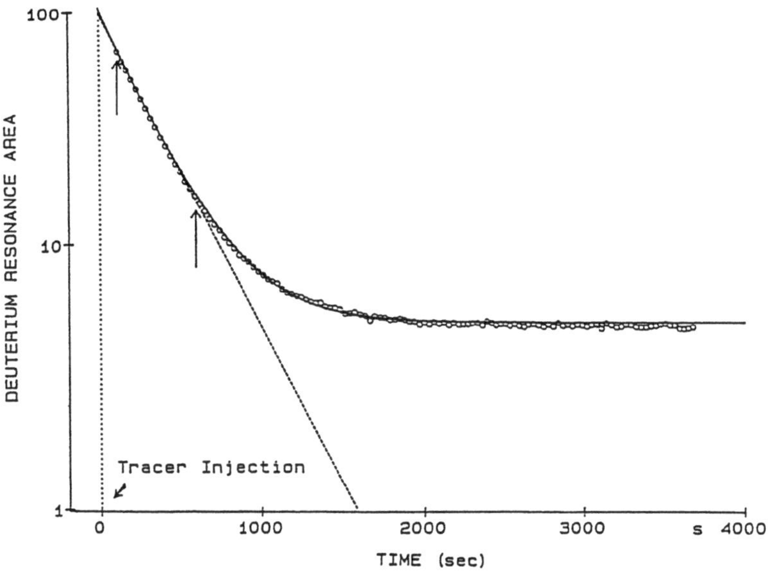

Fig. 18. ^2H washout time course following a 10-µl injection of D_2O saline into a RIF-1 tumor at time $t = 0$, as detected by surface-coil ^2H NMR. Open circles represent experimental data points. The background deuterium signal has been subtracted from the data. Fitting the linear portion of the washout curve (indicated by the *arrows*) to a single exponential function (i.e. using a single-compartment model to describe the tumor blood flow) yielded a tumor blood flow of 17.3 ml/(100g min). The resultant fit is indicated by the *dashed line*. Correcting for tracer recirculation using a two-compartment model, the blood flow rate was estimated to be 18.9 ml/(100g min). The resultant fit is indicated by the *solid line*. (Figure adapted from Fig. 1 of Ref. [165])

injection into the tissue, where rapid proton exchange yields HOD. Surface-coil detection of the HOD signal shows a marked rapid increase in the integrated peak area upon introduction of the label, followed by a gradual decrease in the signal area as the tracer is removed from the tissue via ongoing blood flow. The washout curve asymptotically approaches a new steady-state level as the tracer equilibrates throughout the body. Using mathematical models developed for freely-diffusible radiolabeled tracers [167], the tissue blood flow can be quantitatively estimated from the deuterium washout time course. Single- and multiple-compartment models have been used to describe the HOD washout curves in tissues such as muscle, liver, and tumor [163–166].

An example of the deuterium washout observed in tumor tissue is shown in Fig. 18. In this example, 10 µl of D_2O saline was injected directly into a murine RIF-1 tumor. Using a single-compartment model to describe the washout curve, the blood flow through the tissue is estimated from the exponential time constant of the washout curve and the tissue:blood partition coefficient for the tracer (determined independently for water). The single-compartment model assumes that the tissue is homogeneous in blood flow and that tracer recirculation is negligible. The slope of the linear portion of the semi-log plot of the washout curve in Fig. 18 (i.e. before recirculation becomes significant) is used to estimate blood flow using the single-compartment model. Tracer recirculation can be taken into account by using a two-compartment model and analyzing the entire washout curve to estimate the blood flow rate. Multiple-compartment models have also been developed to describe tissues that are not homogeneous in blood flow [165, 166].

Chronically implanted surface coils have been used to obtain high-sensitivity spectra of deep-lying organs without signal contamination from superficial tissues [74, 168–171]. Such techniques allow multiple investigations over time with high signal sensitivity without the need for repeated invasive surgeries. A minimally invasive technique for localized detection of cardiac muscle has been described in which a small, elliptical catheter coil is inserted directly into the heart via a peripheral blood vessel [172]. High-sensitivity spectra are, thus, obtained without extensive surgical procedures.

The surface coil has become a widely-used tool for imaging superficial tissues such as the orbit, spinal cord, carotid artery, breast, and joints. The high sensitivity of the surface-coil receiver facilitates improved temporal or spatial resolution of the image over that achievable by a large-volume coil. Additionally, matching the surface-coil's detection volume to the sample volume eliminates signal (and noise) detection from unwanted regions. In order to maximize the detected volume in the image, the surface-coil receiver is typically used in concert with a large-volume homogeneous transmitter coil [8, 12, 173, 174]. This requires, however, isolating the two coils to prevent damage to receiver circuitry and distortion of the transmitter field as a result of induced current in the receiver coil during transmission, and to prevent receiver-field distortion and additional noise pickup during reception. Placement of the coils in such a way that their magnetic fields are mutually orthogonal, or use of intrinsically-isolated coils such as the counter-rotating-current coils [93,96–98], is advantageous in that coil interactions are minimized both during excitation and reception. More extensive decoupling schemes involving active or passive detuning networks and resonant transmission-line filters have been developed to isolate multiple coils during transmission [88, 175–179]. Multiple receiver coils and multiple-coil arrays have been used to increase the image signal-to-noise ratio and volume of detection [180–183]. Separate surface-coil transmitter and receiver coils that are positioned such that they have limited overlapping volumes of excitation and detection have been used for depth-selective imaging [184].

The surface coil has also been used in single-coil-mode as both the transmitter and receiver coil in imaging applications [9, 30, 185]. Although this obviates the need for a separate transmit coil and, hence, special coil-to-coil decoupling strategies, the images are acquired at the expense of reduced interrogated volume, difficulty in addressing T_1 effects, and intensity variations resulting from the non-uniform distribution of flip angles produced [13, 173, 185]. The use of $(B_1)_{xy}$-insensitive pulses, such as adiabatic or composite pulses, for more homogeneous excitation and refocusing are helpful in compensating for the deleterious effects of the inhomogeneous $(B_1)_{xy}$ in surface-coil single-coil-mode imaging [186].

Methods to correct for the effects of the inhomogeneous $(B_1)_{xy}$ on image intensity by postprocessing the image data have been developed [9, 187–189]. Surface-coil microscopy of rat eye and ankle using single-coil surface-coil excitation and detection has been demonstrated [190], illustrating the enhanced image signal-to-noise and resolution obtained with a surface coil in comparison to larger-volume coils.

6 Summary

Two issues of paramount concern in in vivo NMR spectroscopy are (a) the extent to which signal detection can be localized to the sample volume of interest, and (b) the signal-to-noise ratio that can be achieved in the experiment. These concerns are especially important for in vivo detection of biologically-significant nuclides which occur in low concentrations and in specific tissue types. These are also important considerations in determining temporal or spatial resolution achievable in magnetic resonance images. The surface coil has found wide-spread applications in in vivo spectroscopy and imaging as a result of its inherently high degree of localization and high sensitivity over regions close to the coil.

The ability to design and position the surface coil such that the sample of interest fills the regions of substantial $(B_1)_{xy}$ (i.e. to optimize the "filling factor") results in high sensitivity per unit volume. This is achieved, however, at the expense of $\mathbf{B}_1$ homogeneity.

The highly inhomogeneous $\mathbf{B}_1$ of the surface coil determines the volume over which signal is excited and, ultimately, detected. When the surface coil is used for both excitation and detection a dual dependence of the signal on $(B_1)_{xy}$ is realized. Although the $(B_1)_{xy}$ of the surface coil is inhomogeneous in amplitude and direction, when the same coil is used to excite and detect signals, the phase angle between the excitation field and the reception field is constant over all space and the signals from the individual spins sum coherently. The theoretically-determined "interrogated" volume of the surface coil is conceptually valuable but may differ quantitatively from the volume observed experimentally since factors such as sample heterogeneity, off-resonance effects, and RF penetration can alter the shape of this volume. Experimental parameters such as pulse repetition period also have pronounced effects on the volume of detection and signal-to-noise ratio achieved.

The intense $(B_1)_{xy}$ near the coil and the curved iso-$(B_1)_{xy}$ lines limit the ability of the surface coil to discriminate against signals originating from superficial tissues. Considerable effort has gone into improving the inherent localization of the surface coil. "Depth pulses" make use of the natural $\mathbf{B}_1$ gradients of the surface coil to maximize signal detection from those regions receiving an optimum flip angle. Use of separate transmitter and receiver coils, with limited overlap of excitation and detection volumes, further enhances signal localization. Non-linear surface-spoiling gradients have also been used to remove signal contributions from overlying tissues. Further definition and shaping of the interrogated volume has been achieved through application of external linear field gradients in conjunction with surface-coil NMR.

The wide distribution of spin flip angles produced upon surface-coil excitation limits the usefulness of the surface coil in applications that require a uniform flip angle over the sample volume. Use of a separate homogeneous transmitter coil with a surface-coil receiver permits homogeneous excitation as well as high-sensitivity reception. A major application of homogeneous excitation with surface-coil reception is found in imaging. $(B_1)_{xy}$-insensitive pulse sequences can be performed in single-coil mode to produce more uniform surface-coil excitation.

Use of the surface coil for in-vivo applications is now well established. Detection of biologically-significant nuclei such as ^{31}P, ^{13}C, and ^{1}H in intact tissue systems has been used to probe dynamic systems at the molecular level, addressing such issues as metabolism, kinetics, flow, and cellular integrity. The high sensitivity and limited volume of detection of the surface coil facilitate collection of high-resolution magnetic resonance images in reasonable time. As a result, surface-coil imaging has become an accepted technique in routine clinical applications.

Acknowledgements: The authors gratefully acknowledge the assistance of Dr. Walter R. Bosch, who produced the surface and contour maps presented herein. Support for this work was provided by NIH grants GM-30331 and CA-40411.

7 References

1. Ackermann JJH, Grove TH, Wong GG, Gadian DG, Radda GK (1980) Nature 283: 167
2. Hoult DI, Richards RE (1976) J Magn Reson 24: 71
3. Hoult DI (1978) Progress in NMR Spect 12: 41
4. Ackerman JJH (1990) Concepts in Magn Reson 2: 33
5. Crowley MG, Evelhoch JL, Ackerman JJH (1985) J Magn Reson 64: 20
6. Bendall MR (1984) J Magn Reson 59: 406
7. Haase A, Hänicke W, Frahm J (1984) J Magn Reson 56: 401
8. Schenck JF, Hart HR, Jr, Foster TH, Edelstein WA, Hussain MA (1986) In: Kressel HY (ed) Magnetic Resonance Annual, Raven Press, New York, p 123
9. Bernardo ML, Jr, Cohen AJ, Lauterbur PC (1982) In: Proceedings of the International Workshop on Physics and Engineering in Medical Imaging, March 15–18, 1982 IEEE Computer Society Press, Silver Spring Maryland, p 277
10. Bendall MR (1983) Chem Phys Lett 99: 310
11. Evelhoch JL, Crowley MG, Ackerman JJH (1984) J Magn Reson 56: 110
12. Kneeland JB, Hyde JS (1989) Radiology 171: 1
13. Bendall MR (1986) Bull Magn Reson 8: 17
14. Garwood M, Robitaille P-M, Uğurbil K (1987) J Magn Reson 75: 244
15. Bendall MR, Gordon RE (1983) J Magn Reson 53: 365
16. Decorps M, Laval M, Confort S, Chaillout J-J (1985) J Magn Reson 61: 418

17. Ernst RR, Anderson WA (1966) Rev Sci Instrum 37:93
18. Becker ED, Ferretti JA, Gambhir PN (1979) Anal Chem 51:1413
19. Hoult DI, Lauterbur PC (1979) J Magn Reson 34:425
20. Gadian DG, Robinson FNH (1979) J Magn Reson 34:449
21. Bottomley PA, Andrew ER (1978) Phys Med Biol 23: 630
22. Hayes CE, Axel L (1985) Med Phys 12:604
23. Harpen MD (1987) Med Phys 14:616
24. Carlson JW (1988) J Magn Reson 78:563
25. Aue WP (1986) Rev Magn Reson Med 1:21
26. Pope JM, Eberl S (1985) Magn Reson Imaging 3:389
27. Pope JM, Eberl S (1988) Magn Reson Imaging 6:157
28. Morris PG (1988) Ann Reports NMR Spect 20:1
29. Aue WP, Müller S, Cross TA, Seelig J (1984) J Magn Reson 56:350
30. Müller S, Aue WP, Seelig J (1985) J Magn Reson 63:530
31. Müller S, Aue WP, Seelig J (1985) J Magn Reson 65:332
32. Ordidge RJ, Connelly A, Lohman JAB (1986) J Magn Reson 66:283
33. Matson GB, Twieg DB, Karczmar GS, Lawry TJ, Gober JR, Valenza M, Boska MD, Weiner MW (1988) Radiology 169:541
34. Bottomley PA, Foster TB, Darrow RD (1984) J Magn Reson 59:338
35. Bottomley PA, Edelstein WA, Foster TH, Adams WA (1985) Proc Natl Acad Sci USA 82: 2148
36. Maudsley AA, Hilal SK, Perman WH, Simon HE (1983) J Magn Reson 51: 147
37. Haselgrove JC, Subramanian VH, Leigh JS, Jr, Gyulai L, Chance B (1983) Science 220: 1170
38. Mareci TH, Brooker HR (1984) J Magn Reson 57:157
39. Gordon RE, Hanley PE, Shaw D, Gadian DG, Radda GK, Styles P, Bore PJ, Chan L (1980) Nature 287:736
40. Balaban RS, Gadian DG, Radda GK (1981) Kidney Int 20: 575
41. Hoult DI (1979) J Magn Reson 33:183
42. Cox SJ, Styles P (1980) J Magn Reson 40:209
43. Haase A, Malloy C, Radda GK (1983) J Magn Reson 55:164
44. Garwood M, Schleich T, Ross BD, Matson GB, Winters WD (1985) J Magn Reson 65:239
45. Peker J, Leigh JS, Jr, Chance B (1985) J Magn Reson 64: 115
46. Metz KR, Briggs RW (1985) J Magn Reson 64:172
47. Shaka AJ, Freeman R (1984) J Magn Reson 59:169
48. Tycko R, Pines A (1984) J Magn Reson 60:156
49. Shaka AJ, Freeman R (1985) J Magn Reson 64:145
50. Shaka AJ, Freeman R (1985) J Magn Reson 63:596
51. Bendall MR, Pegg DT (1985) Magn Reson Med 2:298
52. Bendall MR, Pegg DT (1985) Magn Reson Med 2:91
53. Shaka AJ, Freeman R (1985) J Magn Reson 62:340
54. Bourgeois D, Decorps M, Remy C, Benabid AL (1989) Magn Reson Med 11:275
55. Crowley MG, Ackerman JJH (1985) J Magn Reson 65:522
56. Chen W, Ackerman JJH (1990) NMR Biomed 3: 147
57. Chen W, Ackerman JJH (1990) NMR Biomed 3:158
58. Hennig J, Boesch CH, Gruetter R, Martin E (1987) J Magn Reson 75:179
59. Geoffrion Y, Rydzy M, Butler KW, Smith ICP, Jarrell HC (1988) NMR Biomed 1:107
60. Malloy CR, Lange RA, Klein DL, Jeffrey FM, Peshock RM, Willerson JT, Nunnally RL (1988) J Magn Reson 80:364
61. Levitt MH (1982) J Magn Reson 48:234
62. Freeman R, Kempsell SP, Levitt MH (1980) J Magn Reson 38:453
63. Tycko R, Cho HM, Schneider E, Pines A (1985) J Magn Reson 61:90
64. Hetherington HP, Wishart D, Fitzpatrick SM, Cole P, Shulman RG (1986) J Magn Reson 66:313
65. Hetherington HP, Rothman DL (1985) J Magn Reson 65:348
66. Bendall MR, Pegg DT (1986) J Magn Reson 67:376
67. Bendall MR, Garwood M, Uğurbil K, Pegg DT (1987) Magn Reson Med 4:493
68. Baum J, Tycko R, Pines A (1985) Phys Rev 32:3435
69. Shoubridge EA, Briggs RW, Radda GK (1982) FEBS Lett 140: 288
70. Robitaille P-M, Merkle H, Sublett E, Hendrich K, Lew B, Path G, From AHL, Bache RJ, Garwood M, Uğurbil K (1989) Magn Reson Med 10:14
71. Johnson AJ, Garwood M, Uğurbil K (1989) J Magn Reson 81: 653

72. Norris DG, Haase A (1989) Magn Reson Med 9:435
73. Staewen RS, Johnson AJ, Ross BD, Parrish T, Merkle G, Garwood M (1990) Invest Radiol 25:559
74. Murphy-Boesch J, Koretsky AP (1983) J Magn Reson 54:526
75. Decorps M, Blondet P, Reutenauer H, Albrand JP, Remy C (1985) J Magn Reson 65:100
76. Nagel TL, Stolk JA, Alderman DW, Schoenborn R, Schweizer MP (1990) Magn Reson Med 13:271
77. Styles P, Grathwohl C, Brown FF (1979) J Magn Reson 35:329
78. Cross VR, Hester RK, Waugh JS (1976) Rev Sci Instrum 47:1486
79. Kan S, Fan M, Courtieu J (1980) Rev Sci Instrum 51:887
80. Doty FD, Inners RR, Ellis PD (1981) J Magn Reson 43:399
81. Schnall MD, Subramanian VH, Leigh JS, Jr, Gyulai L, McLaughlin A, Chance B (1985) J Magn Reson 63:401
82. Schnall MD, Subramanian VH, Leigh JS, Jr (1986) J Magn Reson 67:129
83. Reo NV, Ewy CS, Siegfried BA, Ackerman JJH (1984) J Magn Reson 58:76
84. Gonnella NC, Silverman RF (1989) J Magn Reson 85:24
85. Fitzsimmons JR, Brooker HR, Beck B (1987) Magn Reson Med 5:471
86. Grist TM, Jesmanowicz A, Kneeland B, Froncisz W, Hyde JS (1988) Magn Reson Med 6:253
87. Eleff SM, Schnall MD, Ligetti L, Osbakken M, Subramanian VH, Chance B, Leigh JS, Jr (1988) Magn Reson Med 7:412
88. Boskamp EB (1985) Radiology 157:449
89. Doornbos J, Grimbergen HAA, Booijen PE, te Strake L, Bloem JL, Vielvoye GJ, Boskamp E (1986) Magn Reson Med 3:270
90. Yamanashi WS, Lester PD, Letcher JH (1989) Magn Reson Imaging 7:591
91. Lufkin RB, Votruba J, Reicher M, Bassett L, Smith SD, Hanafee WN (1986) AM J Roentgen 146:409
92. Nakada T, Kwee IL, Miyazaki T, Iriguchi N, Maki T (1987) Magn Reson Med 5:449
93. Hyde JS, Jesmanowicz A, Kneeland JB (1987) Magn Reson Med 5:456
94. Sank VJ, Chen C-N, Hoult DI (1986) J Magn Reson 69:236
95. Grist TM, Hyde JS (1985) J Magn Reson 61:571
96. Hyde JS, Froncisz W, Jesmanowicz A, Kneeland JB (1986) Med Phys 13:1
97. Froncisz W, Jesmanowicz A, Kneeland JB, Hyde JS (1986) Magn Reson Med 3:590
98. Hyde JS, Jesmanowicz A, Grist TM, Froncisz W, Kneeland JB (1987) Magn Reson Med 4:179
99. Chen C-N, Hoult DI, Sank VJ (1983) J Magn Reson 54:324
100. Ackerman JJH, Gadian DG, Radda GK, Wong GG (1981) J Magn Reson 42:498
101. Gordon RE, Timms WE (1982) J Magn Reson 46:322
102. Thulborn KR, Ackerman JJH (1983) J Magn Reson 55:357
103. Tofts PS, Wray S (1988) Magn Reson Med 6:84
104. Wray S, Tofts PS (1986) Biochim Biophys Acta 886:399
105. Cady EB, Azzopardi D (1989) NMR Biomed 2:305
106. Song S-K, Hotchkiss RS, Ackerman JJH (1992) Magn Reson Med (in press)
107. Roth K, Hubesch B, Meyerhoff DJ, Naruse S, Gober JR, Lawry TJ, Boska MD, Matson GB, Weiner MW (1989) J Magn Reson 81:299
108. Meyerhoff DJ, Karczmar GS, Matson GB, Boska MD, Weiner MW (1990) NMR Biomed 3:17
109. Blondet P, Decorps M, Confort S, Albrand JP (1987) J Magn Reson 75:434
110. Alger JR, Brunetti A, Nagashima G, Hossmann K-A (1989) Magn Reson Med 11:73
111. De Graaf AA, Bovee WMMJ, Deutz NEP, Chamuleau RAFM (1988) Magn Reson Imaging 6:255
112. Chang L-H, Pereira BM, Weinstein PR, Keniry MA, Murphy-Boesch J, Litt L, James TL (1987) Magn Reson Med 4:575
113. Hetherington HP, Avison MJ, Shulman RG (1985) Proc Natl Acad Sci USA 82:3115
114. Behar KL, den Hollander JA, Stromski ME, Ogino T, Shulman RG, Petroff OAC, Prichard JW (1983) Proc Natl Acad Sci USA 80:4945
115. Gadian DG, Frackowiak RSJ, Crockard HA, Proctor E, Allen K, Williams SR, Russell RWR (1987) J Cereb Blood Flow Metab 7:199
116. Hanstock CC, Rothman DL, Prichard JW, Jue T, Shulman RG (1988) Proc Natl Acad Sci USA 85:1821
117. Evelhoch JL, Ackerman JJH (1983) J Magn Reson 53:52
118. Matson GB, Schleich T, Serdahl C, Acosta G, Willis JA (1984) J Magn Reson 56:200
119. Clarke GD, Rehr RB (1990) Magn Reson Med 14:522

120. Evelhoch JL, Ewy CS, Siegfried BA, Ackerman JJH (1985) Magn Reson Med 2:410
121. Evelhoch JL, Sapareto SA, Nussbaum GH, Ackerman JJH (1986) Rad Res 106:122
122. Evelhoch JL, Sapareto SA, Jick DE, Ackerman JJH (1984) Proc Natl Acad Sci USA 81:6496
123. Ng TC, Evanochko WT, Hiramoto RN, Ghanta VK, Lilly MB, Lawson AJ, Corbett TH, Durant JR, Glickson JD (1982) J Magn Reson 49:271
124. Ross BD, Mitchell SL, Merkle H, Garwood M (1989) Magn Reson Med 12:219
125. Stubbs M, Rodrigues LM, Griffiths JR (1989) Br J Cancer 60:701
126. Griffiths JR, Stevens AN, Iles RA, Gordon RE, Shaw D (1981) Biosci Rep 1:319
127. Lilly MB, Ng TC, Evanochko WT, Katholi CR, Kumar NG, Elgavish GA, Durant JR, Hiramoto R, Ghanta V, Glickson JD (1984) Cancer Res 44:633
128. Negendank WG, Crowley MG, Ryan JR, Keller NA, Evelhoch JL (1989) Radiology 173:181
129. Moon RB, Richards JH (1973) J Biol Chem 248:7276
130. Petroff OAC, Prichard JW, Behar KL, Alger JR, den Hollander JA, Shulman RG (1985) Neurology 35:781
131. Forsén S, Hoffman RA (1963) J Chem Phys 39:2892
132. Brown TR (1980) Phil Trans R Soc Lond B 289:441
133. Rees D, Smith MB, Harley J, Radda GK (1989) Magn Reson Med 9:39
134. Ogawa S, Lee T-M, Glynn P (1986) Arch Biochem Biophys 248:43
135. Meyer RA, Brown TR, Krilowicz BL, Kushmerick MJ (1986) Am J Physiol 250:C264
136. Degani H, Alger JR, Shulman RG, Petroff OAC, Prichard JW (1987) Magn Reson Med 5: 1
137. Gonzalez-Mendez R, Moseley ME, Murphy-Boesch J, Chew WM, Litt L, James TL (1985) J Magn Reson 65:516
138. Kotyk JJ, Rust RS, Ackerman JJH, Deuel RK (1989) J Neurochem 53:1620
139. Ross BD, Higgins RJ, Boggan JE, Willis JA, Knittel B, Unger SW (1988) NMR Biomed 1: 20
140. Canioni P, Alger JR, Shulman RG (1983) Biochem 22:4974
141. Alger JR, Behar KL, Rothman DL, Shulman RG (1984) J Magn Reson 56:334
142. Jue T, Rothman DL, Tavitian BA, Shulman RG (1989) Proc Natl Acad Sci USA 86:1439
143. Reo NV, Siegfried BA, Ackerman JJH (1984) J Biol Chem 259:13664
144. Alger JR, Sillerud LO, Behar KL, Gillies RJ, Shulman RG, Gordon RE, Shaw D, Hanley PE (1981) Science 214:660
145. Pahl-Wostl C, Seelig J (1986) Biochem 25:6799
146. Cross TA, Pahl C, Oberhänsli R, Aue WP, Keller U, Seelig J (1984) Biochem 23:6398
147. Siegfried BA, Reo NV, Ewy CS, Shalwitz RA, Ackerman JJH, McDonald JM (1985) J Biol Chem 260:16137
148. Shalwitz RA, Reo NV, Becker NN, Ackerman JJH (1987) Magn Reson Med 5:462
149. Shalwitz RA, Reo NV, Becker NN, Hill AC, Ewy CS, Ackerman JJH (1989) J Biol Chem 264:3930
150. Wyrwicz AM, Pszenny MH, Schofield JC, Tillman PC, Gordon RE, Martin PA (1983) Science 222:428
151. Litt L, González-Méndez R, James TL, Sessler DI, Mills P, Chew W, Moseley M, Pereira B, Severinghaus JW, Hamilton WK (1987) Anesthesiology 67:161
152. Mills P, Sessler DI, Moseley M, Chew W, Pereira B, James TL, Litt L (1987) Anesthesiology 67:169
153. Selinsky BS, Perlman ME, London RE (1988) Mol Pharmacol 33:559
154. Selinsky BS, Perlman ME, London RE (1988) Mol Pharmacol 33:567
155. Stevens AN, Morris PG, Iles RA, Sheldon PW, Griffiths JR (1984) Br J Cancer 50:113
156. Wolf W, Albright MJ, Silver MS, Weber H, Reichardt U, Sauer R (1987) Magn Reson Imaging 5:165
157. Semmler W, Bachert-Baumann P, Gückel F, Ermark F, Schlag P, Lorenz WJ, van Kaick G (1990) Radiology 174: 141
158. Berkowitz BA, Ackerman JJH (1987) Biophys J 151: 681
159. Nakada T, Kwee IL, Conboy CB (1986) J Neurochem 46:198
160. Rudin M, Sauter A (1989) NMR Biomed 2:98
161. Barranco D, Sutton LN, Florin S, Greenberg J, Sinnwell T, Ligeti L, McLaughlin AC (1989) J Cereb Blood Flow Metab 9:886
162. London RE, Gabel SA, Funk A (1987) Biochem 26:7166
163. Ackerman JJH, Ewy CS, Kim S-G, Shalwitz RA (1987) Ann NY Acad Sci 508:89
164. Ackerman JJH, Ewy CS, Becker NN, Shalwitz RA (1987) Proc Natl Acad Sci USA 84:4099
165. Kim S-G, Ackerman JJH (1988) Cancer Res 48:3449
166. Kim S-G, Ackerman JJH (1988) Magn Reson Med 8:410

167. Kety SS (1951) Pharmacol Rev 3:1
168. Koretsky AP, Wang S, Murphy-Boesch J, Klein MP, James TL, Weiner MW (1983) Proc Natl Acad Sci USA 80:7491
169. Schnall MD, Barlow C, Subramanian VH, Leigh JS, Jr (1986) J Magn Reson 68:161
170. Lotito S, Blondet P, Francois A, von Kienlin M, Rémy C, Albrand JP, Décorps M, Benabid AL (1989) Neurosci Lett 97:91
171. Malloy CR, Cunningham CC, Radda GK (1986) Biochim Biophys Acta 885:1
172. Kantor HL, Briggs RW, Balaban RS (1984) Circ Res 55:261
173. Axel L (1984) J Comput Assist Tomogr 8:381
174. Schenck JF, Hart HR, Jr, Foster TH, Edelstein WA, Bottomley PA, Redington RW, Hardy CJ, Zimmerman RA, Bilaniuk LT (1985) Am J Roentgenol 144:1033
175. Edelstein WA, Hardy CJ, Mueller OM (1986) J Magn Reson 67:156
176. Bendall MR, Connelly A, McKendry JM (1986) Magn Reson Med 3:157
177. Bendall MR (1988) Magn Reson Med 8:380
178. Haase A (1985) J Magn Reson 61:130
179. Styles P, Smith MB, Briggs RW, Radda GK (1985) J Magn Reson 62:397
180. Hyde JS, Froncisz W, Jesmanowicz A, Kneeland JB (1988) Magn Reson Med 6:235
181. Hardy CJ, Katzberg RW, Frey RL, Szumowski J, Totterman S, Mueller OM (1988) Radiology 167:835
182. Hyde JS, Jesmanowicz A, Froncisz W, Kneeland JB, Grist TM (1986) J Magn Reson 70:512
183. Requardt H, Offermann J, Erhard P (1990) Magn Reson Med 13:385
184. Carlson JW, Arakawa M, Kaufman L, McCarten BM, George C (1987) Radiology 165:251
185. Fisher MR, Barker B, Amparo EG, Brandt G, Brant-Zawadzki M, Hricak H, Higgins CB (1985) Radiology 157:443
186. Garwood M, Uğurbil K, Rath AR, Bendall MR, Ross BD, Mitchell SL, Merkle H (1989) Magn Reson Med 9:25
187. Axel L, Costantini J, Listerud J (1987) Am J Roentgenol 148:418
188. Lufkin RB, Sharpless T, Flannigan B, Hanafee W (1986) Am J Roentgenol 147:379
189. Brey WW, Narayana PA (1988) Med Phys 15:241
190. Rudin M (1987) Magn Reson Med 5:443

Rotating Frame Spectroscopy and Spectroscopic Imaging

P. Styles

MRC Biochemical and Clinical Magnetic Resonance Unit, John Radcliffe Hospital, Headington, Oxford OX3 9DU, U.K.

Table of Contents

Rotating frame spectroscopy and spectroscopic imaging methods have been used to obtain high-resolution spectra from a wide range of tissues in vivo. The localisation is effected by detecting the response of the nuclear spins in a spatially dependent B_1 transmitter field gradient. Both single volume and 1 D imaging protocols are available. The basic experiment may be extended to make relaxation time and magnetisation transfer measurements. The design of the NMR probe is an important factor in the experiment, and a variety of solutions are described. When the sensitivity profile of the receiver coil provides inadequate lateral localisation, rotating frame methods may be combined with other techniques such as ISIS to improve the certainty of the localisation. In common with other localisation methods, rotating frame spectroscopy has both merits and disadvantages. These are discussed together with an overview of investigations of both animals and man which have been performed.

1 Introduction

Rotating frame spectroscopic imaging is one of a family of techniques which are used to obtain spatially resolved 'maps' of, for example, metabolites in intact biological systems. Most other methods employ gradients in the B_0 magnetic field, but rotating frame spectroscopy employs a gradient B_1 field for the purpose of encoding position. Some techniques are able to provide resolution in up to three spatial dimensions, but, to date, rotating frame spectroscopic imaging has only been used in one dimension. (The use of the term 'imaging' may seem inappropriate, but it has gained acceptance in experiments where data is collected from a large sample, and later decoded into smaller voxels). A reduction of the imaging experiment in which signal is collected from only one part of the sample offers an alternative method of rotating frame localisation.

2 The Development of Rotating Frame Spectroscopic Imaging

The principle of using the response of nuclear spins in a gradient B_1 transmitter field as a means of spatial encoding was first proposed by Hoult in 1979 [1]. His paper, entitled 'Rotating frame zeugmatography', describes how a proton image may be obtained by using a static B_0 gradient in combination with an orthogonal gradient in the transmitted B_1 field. It was quickly appreciated that the spatial dimension that Hoult had encoded with his static B_0 gradient could be replaced by a chemical shift dimension. Thus the first one dimensional rotating frame spectroscopic 'image' was obtained using (for reasons of practical convenience) the deuterium resonance from ^2H_2O and $C_6{}^2H_6$ [2].

At that time, the prospect of performing routine biochemical imaging seemed somewhat optimistic. Lauterbur had proposed localised chemical shift spectroscopy [3, 4] by using selective pulses in the presence of a B_0 gradient. The use of different gradient directions and a projection reconstruction algorithm allowed for a two dimensional image to be obtained. Unfortunately, the method didn't work well with extensively overlapping spectral lines, and was therefore of limited generality. It was appreciated that the Fourier imaging method [5] could be adapted to preserve chemical shift but it would be two years before such an experiment would be demonstrated, first on phantoms [6] and later on an intact animal [7]. The rotating frame approach was also beset with difficulties, not least being the problem of generating the necessary B_1 gradient across any large intact biological sample, and providing sufficient r.f. power to keep off-resonance effects to manageable proportions. Hoult's transmitter coil had been constructed as a saddle-shaped coil—one turn on one side, and three turns wound in the opposite sense on the other. An orthogonal coil was used for the receiver. This arrangement didn't lend itself to the particular demands of human spectroscopy, namely high sensitivity and realistic power requirements.

The first step towards a more reasonable coil arrangement was the use of a simple surface coil which performed both transmit and receive functions [8, 9]. Such a probe will certainly produce a B_1 gradient, but planes of equal field strength are curved, and the gradient is not linear. However, by using small bulb phantoms placed along the axis of the coil, good 'images' of phosphocreatine, ATP and inorganic phosphate were obtained. Later, the single surface coil was applied to biological samples, but the problem of field curvature meant that studies were initially restricted to objects that did not extend laterally much beyond the bounds of the coil. These included the bovine eye [10], and the exteriorised rat kidney [11]. Recently, hybrid localisation (in which B_0 methods are used for lateral localisation) has extended the scope of the single surface coil probe [12–14].

An alternative solution to the probe problem is to use separate surface coils for transmit and receive, and this arrangement was employed to first demonstrate that spatially resolved metabolite maps of human organs could indeed be obtained [15]. Subsequently these probes have been used to study a wide range of anatomies in both animal models and man.

Regarding the rotating frame method itself, two modifications are of interest. First, a reduction of the 'imaging' protocol is possible where only one volume of the sample is interrogated. This has variously been termed "harmonically analysed sensitivity profile (HASP)" [16], "Fourier series window (FSW)" [17–19], and "Fourier depth selection", [20, 21]. Secondly, early rotating frame experiments used amplitude modulation of the signal intensity to encode position. Hoult appreciated that this method wasted sensitivity, and proposed that phase modulation could overcome the problem [1, 22]. His idea has now been implemented [23, 24], and is to be preferred over the amplitude modulated experiment.

3 The Principles of Rotating Frame Spectroscopy

We may consider the rotating frame experiment as a general method of determining the value of B_1 field in a sample. A routine task in one-dimensional spectroscopy is the measurement of the length of the 90° pulse. Usually, one performs several 'pulse and collect' experiments, varying the length of the pulse, and observing the consequential changes in signal intensity which will be approximately sinusoidal for a good probe. Putting this another way, the B_1 field strength determines the *rate of change* of signal amplitude with respect to pulse duration. In the case of a homogeneous B_1 field, this rate of change is readily ascertained by direct observation of the signals, but if the B_1 field is non-uniform, we must apply more formal analysis to decode the information. Having first collected the several one-dimensional spectra, a second Fourier transformation can be performed to determine the frequency components that are present in the observed rate of change of signal strength with respect to pulse duration.

The Amplitude Modulated Rotating Frame Imaging Sequence

This experiment, which is illustrated in Fig. 1, simply requires the acquisition of a series of 'pulse and collect' f.i.d's where the excitation pulse is incremented for each collection. A two-dimensional Fourier transform creates a data matrix which has chemical shift on one axis, and B_1 field strength on the other. Assuming that there is a linear gradient in the B_1 field, this is equivalent to a plot of chemical shift vs. linear distance. We may notate this protocol thus:

$$n\theta_x - \text{acquire} - \text{relaxation delay} \qquad (1)$$

where $n=0$ to $(NI-1)$, (NI being the total number of pulse increments and hence spectra which are acquired). A zero f.i.d. is helpful in that it avoids the

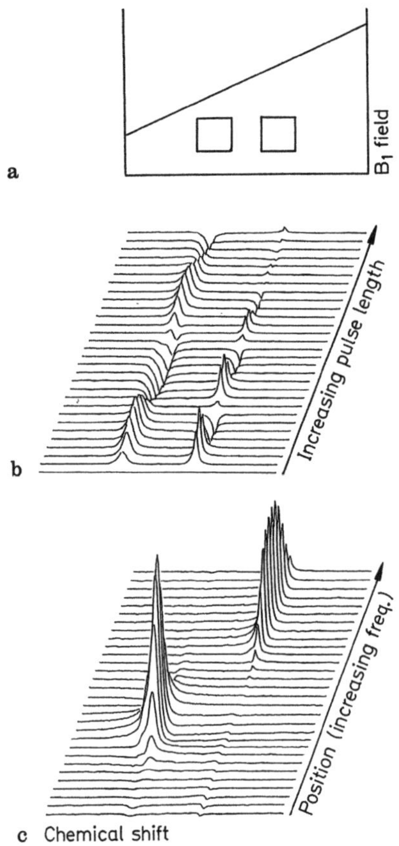

Fig. 1 a–c. The acquisition of a rotating frame spectroscopic 'image'. **a** Schematic representation of two-compartment phantom placed in a gradient B_1 transmitter field. **b** A set of 'pulse and collect' spectra obtained from the phantom illustrated in **a**. Note that the signal strength modulates at a rate which is dependent on the position of the phantom in the gradient B_1 field. **c** A data matrix derived by performing a second Fourier transformation on the set of spectra shown in **b**. The vertical axis represents the frequency with which the signal modulates, and hence the position of the phantom in the gradient B_1 field.

need for phase correction after the second Fourier transformation. The appropriate transform in the second (spatial) dimension is a real sine transform performed on only the absorption mode spectra.

The Phase Modulated Rotating Frame Imaging Sequence

The principle disadvantage of the amplitude modulated experiment is that the maximum signal will only be collected when the pulse happens to be a 90° (or 270° etc.) pulse for a particular part of the sample. This drawback is avoided by following the incremental pulse by a $90°_y$ pulse, giving the sequence:

$$n\theta_x - 90°_y - \text{acquire} - \text{relaxation delay} \tag{2}$$

The effect of this sequence is that the magnetisation is tipped into the xy plane by the $90°_y$ pulse, and the variation with respect to the $n\theta$ pulse is now encoded in the phase of the signal. (By direct analogy to a one dimensional experiment, this manoeuvre creates quadrature data, with the consequential sensitivity improvement of $\sqrt{2}$). This is illustrated in Fig. 2. It will be seen that now we always obtain the maximum signal irrespective of the tip angle imposed by the incremented pulse. Accurate implementation of this sequence requires a second tansmitter coil able to generate a homogeneous $90°_y$ pulse over the whole sample. However, it is possible to use the gradient transmitter for both pulses, although some loss of signal occurs in regions where the phase encoding pulse is not 90°.

One drawback of phase encoded data sets, common to many 2DFT experiments, is that the two-dimensional Fourier transformation gives line-shapes which contain dispersion modes. This familiar phase twist would render the experiment useless. The solution [1, 22] is to collect a second set of spectra, negating the phase of one of the pulses. The effect is to reverse the direction in which the magnetisation rotates in the xy plane. After the separate transform-

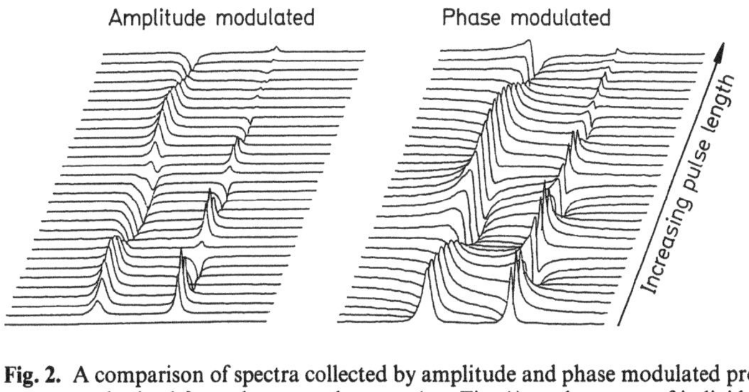

Fig. 2. A comparison of spectra collected by amplitude and phase modulated protocols. These data sets were obtained from the same phantom (see Fig. 1), and are sets of individual spectra prior to the second Fourier transformation. Note that the full signal strength is obtained in the phase modulated experiment.

ation of each data set, one is reversed along the spatial dimension to compensate for the reversal imposed by the phase change. Subsequent addition of the two data sets gives cancellation of the unwanted dispersion signals, leaving the required absorption mode spectra. Thus the usual phase-modulated sequence is:

$$n\theta_x - \lambda_{(+\text{ or }-)_y} - \text{acquire} - \text{relaxation delay} \tag{3}$$

where $n = 0 \rightarrow (NI-1)$, and λ is a nominal 90° pulse. The $n\theta$ pulse is termed the "incremental pulse", and the λ pulse the "phase-encoding pulse". The number of pulse increments, NI, will be for each data set, and so an experiment employing 16 increments will involve a total of 32 separate data collections.

An alternative method for avoiding the sensitivity penalty in the amplitude-modulated method is to collect two f.i.d.'s sequentially after each incremental pulse [25, 26]. This sequence is:

$$n\theta - \text{acquire}(1) - 90° - \text{acquire}(2) - \text{relaxation delay} \tag{4}$$

Each set of amplitude modulated acquisitions are processed separately before addition of the two data sets. In essence, this method is identical to the phase modulated experiment in that the z component of magnetisation, which is lost in the amplitude modulated protocol, is recovered by the 90° pulse. One advantage of the above scheme is that the 90° rotation can be performed with a composite pulse, thus reducing the problems associated with the inhomogeneous transmitter field. However, any T_1 relaxation between the two pulses will give a loss in sensitivity.

The idea of collecting a string of f.i.d.'s has also been suggested by Pekar and Leigh [27], but their 'very fast rotating frame zeugmatography' scheme has not been developed for routine use.

Single Volume Rotating Frame Sequences

The experiments that have been described collect signal from throughout the sensitive region of the probe, and decode these signals to form a multi-voxel data set. If the position of the region of interest is known, then it is possible to perform a modified rotating frame experiment such that only signals from that region are acquired. Several groups independently suggested similar protocols [16–21].

Consider the effect of applying a single transmitter pulse in a gradient B_1 field. The induced transverse magnetisation will vary sinusoidally with the B_1 field. The periodicity and position of this variation will depend on the length of the transmitter pulse, the B_1 strength and gradient. By applying different pulse lengths, the variation of magnetisation throughout the sample can be adjusted. If, therefore, we require a particular excitation shape (in space), we need to apply different pulse lengths such that the several sinusoidal magnetisation profiles form a Fourier series whose addition creates the desired shape. Taking a simple example (Fig. 3), apply (in separate pulse and collect experiments)

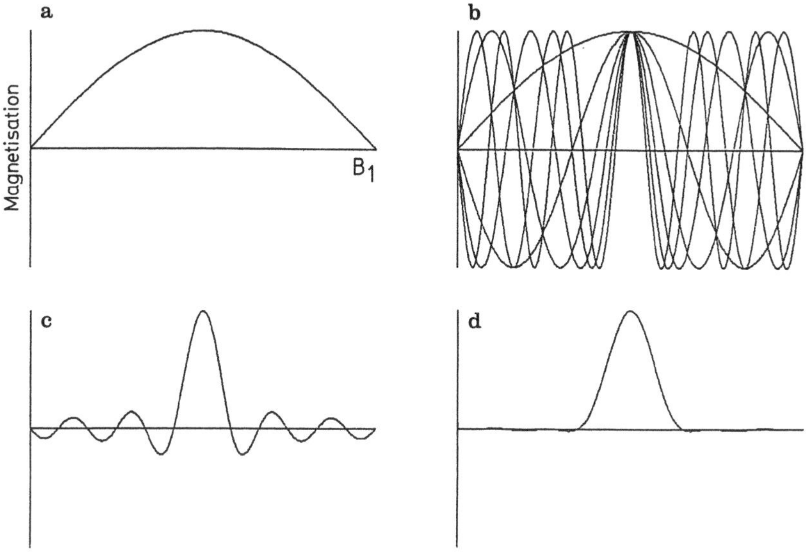

Fig. 3 a–d. A schematic representation of the principles of Fourier series rotating frame localisation. **a.** The transverse magnetisation produced in a sample placed in a gradient B_1 transmitter field. The transmitter pulse (θ) is chosen to be a 180° pulse at the maximum B_1 field. **b.** Superimposed plots of the magnetisation resulting from pulse lengths of $+\theta$, -3θ, $+5\theta$, -7θ, $+9\theta$, -11θ. Note that the magnetisation is always positive in the region where θ is a 90° pulse. **c.** The effect of summing experiments performed as in **b**. **d.** As for **c**, except that different number of data acquisitions are employed for each pulse length. These coefficients have been chosen to approximate to a Gaussian envelope, being 20, 18, 14, 9, 5 and 2 at θ, 3θ, 5θ, 7θ, 9θ and 11θ, respectively

pulses that generate 90°, 270°, 430°,... tip angles at a particular place in the sample. Alternate addition and subtraction (to allow for the sign change between 90° and 270° pulses, etc.) of the spectra will give coherent accumulation of signals from that point. The shape of the sensitive region will be the Fourier equivalent of the various sinusoids—in our example ω, 3ω, 5ω, 7ω,... will result in a sinc function response in space (Fig. 3c). A more useful sequence would adjust the contribution from each pulse width, either by multiplication of the acquired data or varying the number of accumulated f.i.d.'s, to avoid the undesirable side lobes in the sinc function (Fig. 3d). This leads to the general expression for a Fourier series protocol of:

$$\sum C_n(-1)^n[(2n+1)\theta] \tag{5}$$

where C_n is the coefficient describing the relative contribution of signal from each value of incremental pulse. A variation of this sequence, which avoids certain phase and off-resonance effects, is used by Garwood and colleagues [28]. The incremental $n\theta$ pulse is replaced by a phase-alternated $2n\theta$ pulse which generates spatially dependent longitudinal magnetisation. A subsequent readout pulse generates the f.i.d.'s which are added together as before.

One particular difficulty that may be encountered, particularly when using simple surface coil probes, is that of harmonic response. In the earlier example, the sensitive region is centred around the position where the first pulse is 90°, but if regions of the sample experience 3 (or 5, 7,...) times the B_1 field in the required position, these will also give rise to coherent signal accumulation. It is, of course, possible to devise suitable Fourier series that have no such harmonic response, but in this case a loss of sensitivity is incurred in the desired region.

The Fourier series approach has the advantage of great simplicity, and avoids the signal loss inherent in the amplitude-modulated experiment. However, by comparison with the phase-modulated imaging protocol, it is a somewhat wasteful method because information from the remainder of the sample is lost.

4 Probes for Rotating Frame Experiments

The ideal probe for rotating frame spectroscopy would consist of three separate components—a transmitter coil to generate the linear B_1 gradient, a second transmitter to produce a homogeneous phase-encoding pulse, and a receiver coil to obtain maximum sensitivity from the region of interest. In practice, these ideals are difficult to realise and compromise solutions are required (some possibilities are shown in Fig. 4). As mentioned earlier, single surface coil probes can be used in some circumstances, but the curved iso-flux contours are unsuitable for most applications unless some other form of lateral localisation is employed. An alternative is the double surface coil probe where a smaller receiver coil collects signals from close to the axis of a larger transmitter [15, 29–32]. The linearity of the transmitter gradient is improved by offsetting the transmitter a small distance behind the receiver [15, 32].

With two coils closely coupled by mutual inductance, it is necessary to electrically isolate transmitter and receiver. Possible circuits for performing this function may be found in Refs. [15, 29–34]. In applications where the sample can be surrounded with a receiver coil, then several coil arrangements are possible. Moonen et al. [35] use a 'Dadok' type of receiver coil together with a single loop transmitter for perfused heart studies, and animal studies have been performed using a double-eliptical receiver coil in front of a surface coil transmitter [24, 36, 37].

Certain applications require extra coils to be added to the probe. It may be useful to obtain 'scout' images from the region adjacent to the probe, and this has been achieved by incorporating a 'butterfly coil' for the proton frequency [38]. This arrangement has the advantage that there is little interaction between the conventional surface coils and the butterfly coil, and so no extra isolation circuits are needed.

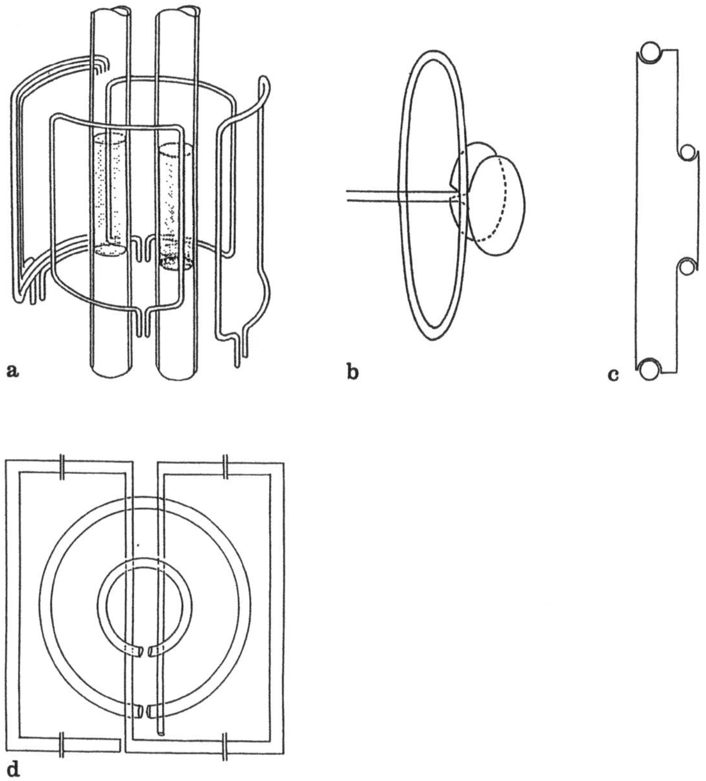

Fig. 4a–d. Schematic representation of several probe designs which have been used in the Oxford laboratory for rotating frame spectroscopy. **a.** The probe used for the original spectroscopic rotating frame experiment [2]. The transmitter coil is Hoult's design [1]. The three turns on the left are wound in the opposite sense to the one on the right. **b.** A single turn surface coil transmitter is combined with a more homogeneous receiver coil. This arrangement has been used in studies of animal models [37, 81]. **c.** The double offset surface coil probe, shown here in cross section, which is used for human spectroscopic imaging. **d.** The incorporation of a butterfly proton coil into an offset surface coil probe. Series capacitors are incorporated into the proton coil to facilitate tuning at the higher frequency. The proton coil may be used to obtain scout images and an agarose sample is incorporated into the probe to identify its position in the image

5 The Use of Adiabatic Pulses in Rotating Frame Experiments

As pointed out in the previous section, accurate implementation of rotating frame methods is hindered by practical considerations with regard to probe design. Some of the problems can be alleviated by employing composite or adiabatic pulses to perform prescribed spin manipulations under conditions of inhomogeneous B_1 fields. These methods are extensively reviewed elsewhere in this volume [39], but of particular interest are plane rotation pulses [40, 41] which can be used to implement phase encoding, and adiabatic inversion and detection pulses [42, 43], which have been employed in hybrid Fourier window protocols.

6 Extensions to Rotating Frame Experiments

Spin Lattice Relaxation Time Measurements

Saturating pulses may be incorporated in a rotating frame imaging experiment to achieve localised T_1 measurements [9, 44]. Several images are collected with different delay times, and the signal amplitudes analysed in accordance with the usual saturation recovery equation. Blackledge [45] used inversion recovery and fast inversion recovery in conjunction with a Fourier series protocol to measure T_1's and obtain a degree of spectral editing for ^{31}P human liver data. When using the Fourier window method, the sensitive region is defined by the pulse width, and so there is a direct correlation between inversion and excitation pulses. In the imaging protocol, a similar sequence will give a wide range of inversion over the data matrix. However, as the spatial axis of the matrix is defined by the B_1 transmitter field strength, then the tip angle produced by the nominal inversion pulse can be accurately determined, and correction applied in fitting the data to the exponential function [46, 47].

Spin Echo Experiments

The rotating frame pulse sequence can be extended with a refocussing pulse to generate a localised spin echo experiment. Ideally, this extra pulse should be generated by a homogeneous coil, but acceptable results may be obtained if the usual transmitter coil is used [48]. Provided that 'EXORCYCLE' phase cycling is employed [49], the principle error is a reduction in signal amplitude away from the centre (in the spatial dimension) of the data matrix. Such sequences are valuable for in-vivo investigations both as a means of determining T_2 relaxation times [50] and for spectral editing [48].

Magnetisation Transfer Experiments

Saturation transfer measurements have been made in conjunction with rotating frame localisation [46, 51]. Cadoux-Hudson et al. [51] have studied the flux through creatine kinase in the human brain, and demonstrated different rates in superficial (mainly grey matter) and deep (mainly white matter) tissue. In order to obtain even saturation, the usual double surface coil probe was augmented with an orthogonal untuned Helmholtz coil. The lack of tuning avoided many of the problems of coil interaction, but was adequate for producing the low level of irradiation necessary for saturation.

Proton Spectroscopy

The problem of water suppression in proton spectroscopy in vivo has been addressed. Both the incremental and phase-encoding pulses may be replaced by binomial (1331) pulses to avoid excitation of the water resonance [52]. The addition of binomial spin echo pulses have also been used [53], offering the possibility of additional T_2 water suppression and J-modulated spectral editing.

^{13}C Spectroscopy

Rotating frame methods are not well suited to the direct observation of the carbon resonances due to the extensive chemical shift range of this nucleus. However, the heteronuclear coupling with hydrogen can be exploited for indirect ^{13}C observation. Böttcher et al. [54] employed a double tuned surface coil probe to implement a polarisation transfer experiment, using adiabatic pulses to avoid problems created by the inhomogeneous field. An alternative approach is to use difference spectroscopy—inversion of the carbon spins in a spin echo experiment and subsequent subtraction of data with and without this inversion, gives proton ‘images’ containing only signals from protons coupled to ^{13}C nuclei [53].

Absolute Quantification

Most spectroscopic investigations of biological systems in-vivo measure only concentration ratios. The well-known difficulties of obtaining absolute concentrations by NMR are compounded by the ubiquitous use of the surface coil receiver with its inhomogeneous sensitivity. If, however, the spatial distribution of the sample within this nonuniform detection field is known, then it becomes easier to determine these concentrations. By including a small phantom within the rotating frame probe, two calibrations may be made. First, any coil loading and tuning effects which effect the spatial calibration are manifest in the position in the data matrix of the signal from the phantom. Secondly, once this correction has been made, the phantom signal can be used as a concentration reference. Usually a probe is calibrated empirically using an extensive standard sample, and determining calibration factors with a variety of incremental pulses to give an appropriate range of positions of the reference phantom within the matrix. This method has been successfully applied in human studies of brain [51] and liver [55].

7 Artifacts in Rotating Frame Spectroscopy

Imperfections in the rotating frame experiment have been analysed by Blackledge [21, 56, 57] and Garwood [28]. In many instances, there is a direct equivalence between the imaging and Fourier series protocols, and one or other may provide the simplest way of understanding a particular phenomenon. The following is a nonmathematical description of the most important effects.

Partial Saturation Effects

In the amplitude modulated imaging protocol and the equivalent Fourier series experiment, distortions occur if the magnetisation does not fully return to equilibrium between acquisitions. Under conditions of partial saturation, the signal amplitude will depend on the residual z magnetisation, which itself varies with the tip angle. Analysis of this effect shows that the expected sinusoidal variation of signal with pulse length contains higher harmonics. In an imaging experiment, these harmonics give rise to spurious signals at multiples of the correct spatial frequency, and in the Fourier series experiment, unwanted regions of the sample are rendered sensitive. One solution is to follow each data acquisition with a saturation pulse or sequence, thus ensuring that the magnetisation always recovers from the same starting point irrespective of the incremental pulse duration [10]. The artifact is avoided in the phase encoded protocol because the phase encoding pulse performs the same function as the saturating pulse previously described [1]. However, when the phase encoding pulse is not an ideal 90° pulse, then T_1 harmonics will again have to be considered.

Off-Resonance Effects

The nuclear magnetisation precesses around a vector $B_{1\,(\mathrm{eff})}$ which is given by the vectorial addition:

$$B_{1\,(\mathrm{eff})} = B_1 + \frac{(\omega_0 - \omega_1)}{\gamma} \tag{6}$$

where $(\omega_0 - \omega_1)$ is the off-resonance frequency, γ is the gyromagnetic ratio, and the $(\omega_0 - \omega_1)/\lambda$ vector is in the direction of the static B_0 field. Thus, away from resonance, the magnetisation nutates at a higher frequency than on resonance, and in a plane which is tilted away from the $y'z'$ plane. This has three main implications in the rotating frame experiments:

i. Off-resonance signals appear in the wrong place in the matrix—at a position equivalent to a higher B_1 field than would be correct. This spatial distortion varies across the matrix, becoming more acute away from resonance and in regions of low B_1 field.

ii. Off-resonance signals have a phase shift which is dependent on both the frequency offset and B_1 strength.

iii. There is an x' component of magnetisation which gives rise to a narrow dispersion signal at the origin (zero-frequency) in the spatial dimension. If zero filling but no windowing is employed prior to the second Fourier transformation, this narrow origin spike will give rise to truncation 'sinc-wiggles' across the data matrix.

Imperfect Pulse Effects

When the phase-modulated experiment is performed using the same transmitter coil for both pulses, errors will result from the imperfect phase encoding pulse. The signal amplitude is modulated by the function $(1 + \sin\lambda)/2$, where λ is the tip angle of the phase-encoding pulse. In addition, dispersive components are introduced into signals off-resonance. These effects have been extensively modelled [56], and the conclusions reached are: "The effect of pulse angle imperfection is small—less than 15% of the signal is lost due to a phase-encoding pulse of $\pi/4$ or $3\pi/4$ instead of $\pi/2$. Generally, the introduction of dispersion mode terms into the matrix is significant when the off resonance angle is greater than 15°".

Probe Imperfections

Most of the probes used for in-vivo studies are a compromise between realistic r.f. power requirements, good sensitivity and acceptable performance in terms of B_1 profile. Thus the offset double surface coil probe gives acceptable transmitter linearity, and the planes of equal field are reasonably flat. However, correction should be made for the B_1 gradient if accurate depth calibration is required. In principle, the calculated r.f. field profile could be modified by eddy currents in a conducting sample, but experiments using phantoms and static B_0 gradients have failed to detect significant errors at the frequencies and sample sizes that we have used in this laboratory [38, 58].

A potentially more serious difficulty arises as a consequence of using a surface coil receiver in the presence of an essentially unidirectional transmitter field. In this situation, the principle of reciprocity does not apply, in that phase differences occur due to variations in the relative orientation of transmitter and receiver flux lines throughout space [59, 60]. This effect gives rise to out-of-phase signals being generated away from the axis of the probe. If the sample and B_0 field are both homogeneous over a resolved slice, then dispersion signals from opposite sides have opposite phases and hence cancel. If these criteria are not met, then spurious dispersion signals will be present, and the lateral localisation will also suffer. This topic has been extensively discussed by Garwood et al. [28] who use a longitudinal Fourier window experiment to circumvent the problem. (Readers should, however, be warned of an unfortunate error in this paper. The authors model the phase modulated rotating frame pulse sequence in the context of a Fourier series experiment, and predict quite unacceptable results. Their finding is related to the phase twist problem in the 2DFT experiment which is eliminated by the collection of two data sets as has previously been described. This is clarified in Ref. [57].) In situations where the probe phase problem is important, the best solution for the imaging experiment is to use some other form of lateral localisation (see Sect. 9 below).

8 Optimal Data Collection and Processing

In his original rotating frame paper [1], Hoult considers questions of sensitivity and resolution, and gives valuable advice on the choice of data collection parameters and processing methods. His analysis uses analogies from optics including the Rayleigh criterion for minimum resolution. This approach, however, has limitations in spectroscopy where considerations about resolution must include an assessment of the contamination of any voxel by signal from outside that region. No formal treatment of this problem has been undertaken for rotating frame spectroscopy, but some general advice is available [1, 58].

A common feature of all Fourier localisation methods is that the true spatial response is convoluted by the window through which the data is collected. This gives rise to the "point spread function" which defines the way in which signal from a small discrete region is spread across the transformed data matrix. In the case of a rectangular window (i.e. equal weighting to signals collected at each pulse increment), the point spread function is of the form $\sin(x)/x$ (a sinc function), which has characteristic side lobes away from the centre, and a width which is inversely proportional to the duration of the data collection. The actual presentation of this phenomenon depends on the form of the sample (i.e. the way in which concentrations vary through space), the length of the data collection window, and the method of processing (in particular, whether zero-filling is employed).

In the case of a non-point source sample, the signal tends to decrease as more pulse increments are collected because signals from different parts of the sample destructively interfere at long pulse lengths. Correspondingly, the sinc-wiggles in the transformed data set become less evident because side lobes from every individual point in the sample also overlap. However, as the resolution is increased by increasing the collection window, so the sensitivity decreases as less signal is collected in successive accumulations. The dilemma, therefore, is how to best compromise between spatial resolution, sensitivity and truncation artifacts.

An absolute requirement in in-vivo spectroscopy is to obtain the best possible signal to noise, and this dictates that the number of pulse increments be the minimum needed to resolve the tissue of interest. Zero filling [61] will help to optimise the resolution, but sinc distortions, not visible in non-zero-filled data, will now become visible [62].

In order to avoid truncation effects, the data can be windowed by a decaying function so that the sinc wiggles are reduced to less than the noise. There are several suitable functions (i.e. the Hanning window, the Hamming window), and we employ a Gaussian decaying to ~5% for the last data collection. This windowing may be implemented in one of two ways. Either the data is apodised prior to the second Fourier transformation or differrent numbers of scans can be collected for each pulse increment. An analogous situation exists in the Fourier series protocol, where the shape of the spatial response is governed by the same criteria. Pekar [16] showed that tailored data acquisition is more efficient than multiplication of a uniformly acquired data set by the desired function. Considering a window function $f(n)$ (where the collected f.i.d.'s are numbered from 1 to n), then it can be shown that the relative sensitivities of the two methods are [58]:

$$\Delta \frac{\mathrm{S:N}}{\sqrt{t}} = \frac{\sum f(n)}{(n_{\max} \cdot \sum f(n)^2)^{1/2}} \tag{2}$$

where the summations are from 1 to n, and $\Delta \mathrm{S:N}/\sqrt{t}$ is the relative sensitivity of the apodisation method compared to the tailored data acquisition technique. Taking as an example a Gaussian window truncated at 5% of its maximum value, the apodisation of equal acquisitions reduces the sensitivity to about 83% of that obtained by tailoring the number of scans.

One other possible approach to the data processing problem is to use a non-Fourier method for decoding the data. Hore and Daniel [63] have presented some very impressive results using maximum entropy. The processing avoids difficulties with truncation artifacts, and also corrects for off-resonance effects. If this technique can be made to work with broad lines (in both chemical shift and spatial dimensions) and indifferent signal to noise, then some of the problems discussed above may be alleviated.

9 Localisation in Two and Three Dimensions

The rotating frame techniques that have been described are only able to provide one-dimensional localisation. In many instances, the lateral response of a surface coil receiver provides sufficient localisation in the other two dimensions, but sometimes additional methodology is needed to obtain better control of the sensitive region. Hoult has discussed the use of additional B_1 gradients and phase-modulated B_1 pulses which are able to perform slice selection in a second rotating frame [22]. A less elegant, but perhaps more practical approach is to employ hybrid methods where both B_1 and B_0 gradients are used.

Many of the single volume methods involving B_0 gradients may be implemented in conjunction with rotating frame protocols. The Fourier series experiment has been combined with two-dimensional ISIS [64] to form the IDESS (improved depth selective single surface coil spectroscopy) method [12], similar in concept to the later FLAX-ISIS method [14, 65]. There is also a phase modulated version (RAPP-ISIS) which employs adiabatic plane-rotation pulses to achieve increased sensitivity [65].

Böttcher et al. [13] investigated the use of prepulses to selectively saturate all but a column of sample on the axis of a single surface coil probe. This presaturation was achieved by applying noise pulses [66] or hyperbolic secant pulses [42] during the application of gradients perpendicular to the surface coil axis. Experimental data confirming the efficacy of the noise pulse method may be found in Refs. [38, 58].

10 The Problems and Benefits of Rotating Frame Spectroscopy

There is no doubt that rotating frame methods lack the versatility of the more usual B_0 localisation techniques. Equally, with the possible exception of echo methods for proton spectroscopy, there are significant technical and theoretical difficulties with all methods. It is inappropriate here to attempt a critical analysis of each protocol, but rather to allude to those features of the rotating frame experiments which may offer certain advantages for particular applications.

Disadvantages of Rotating Frame Methods

Most rotating frame probes are designed to provide acceptable performance at realistic *r.f.* power levels. Consequentially, the field gradient is not highly linear, neither are the planes of equal field absolutely flat. In use, these imperfections are not too significant because tissue anatomy doesn't conform to precise geometric patterns either. There are also localisation errors due to off-resonance effects—a problem not unique to rotating frame methods—but these are in principle correctable in the imaging experiment by suitable data processing.

More serious is the lack of good lateral discrimination, particularly when investigating localised lesions such as tumours. The use of B_0 techniques (such as 2D ISIS or noise pulses) to avoid these problems might seem to negate the benefits of the B_1 approach, but such hybrid techniques can offer an effective way of 'imaging' short T_2 metabolites such as ATP and the phosphodiesters.

In B_0 localisation methods, direct correlation between the localised spectra an a proton image is easily achieved. For rotating frame imaging, this is less straightforward, and calibration relies on independent determination of the B_1 field profile.

Advantages of Rotating Frame Methods

The most significant advantage of rotating frame methods is that no switched B_0 gradients are needed. This is a particular benefit for nuclei with short relaxation times. Considering, for example, the Fourier CSI method, one must either collect signal immediately following the gradient which demands an impeccable gradient set and produces phase correction difficulties, or generate a spin echo which is useless for studying short T_2 species. State of the art gradients are alleviating these difficulties, but most routine instruments are presently unable to achieve the quality of data that a few specialist groups have demonstrated.

As both the imaging and Fourier series rotating frame experiments select slices, usually parallel to the surface of the subject, then these techniques will be most useful when this matches the anatomy of interest. Obvious examples include the liver and heart, both situated behind a relatively flat region containing intercostal muscle. One important feature is that the localisation is relative to the probe itself, and this can be important if the subject is moving. Breast tumours, for example, may be localised by attaching the probe over the lesion and, provided that magnet homogeneity is good, small breathing movements which raise and lower the chest will have minimal effect on the measurement.

11 Applications of Rotating Frame Spectroscopy

A wide variety of tissues and organs in both animals and man have been investigated by rotating frame methods. Most have involved the phosphorus nucleus, but sodium spectra have been obtained from the perfused rabbit heart [35] and the rat kidney [11]. In studies of the rat brain, ^{19}F rotating frame imaging has been used to investigate the distribution of fluorine labelled metabolites [67–69] and halothane [70]. Early experiments on biological samples did little more than demonstrate that the methodology was able to detect localised tissue heterogeneity, but later, the studies have addressed more significant questions of biochemical and clinical interest. The following discussion is intended to give a brief overview of the scope of rotating frame spectroscopy.

The Liver

The liver has become the "in-vivo phantom" for phosphorus localisation techniques on account of it being a large and homogeneous organ, situated near to surface, and distinguishable from the overlying intercostal muscle as it contains no creatine phosphate (Fig. 5). For these reasons, the first rotating frame spectra from humans were of the liver [15], and more recently, quantitative images have demonstrated cell necrosis as a consequence of Paracetamol poisoning [55]. The ease with which ^{31}P spectra of liver can be obtained is demonstrated by the efficacy of the simplest of all Fourier series protocols for such investigations. Using the offset double surface coil probe, Angus et al. [71] added spectra collected with pulse angles of θ and 3θ, the relative number of scans being 256:90 respectively. This two step Fourier series is quite adequate to suppress the signal from the surface muscle and produce pure liver spectra—in this case from patients with alcoholic hepatic disease.

The Brain

Rotating frame imaging has been used to study both focal and generalised pathology on the human brain. If the anatomy of interest is situated in the

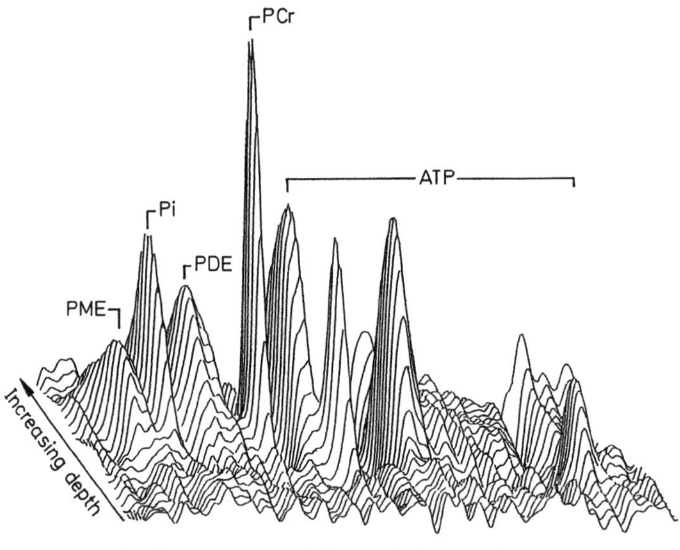

Fig. 5. Rotating frame spectroscopic image obtained by placing a double surface coil probe against the lower right hand side of the chest of a healthy volunteer. Spectra from the overlying intercostal muscle are readily distinguished from the signals which come from liver. Adenosine triphosphate (ATP) is present in both tissues, but creatine phosphate (PCr) is present only in the muscle, and inorganic phosphate (P_i), phospho-monoesters (PME) and -diesters (PDE) arise almost exclusively from the liver. (Spectra by courtesy of Dr. Ruth Dixon)

temporal region, then slice type of localisation is well suited to separating signals from brain and the overlying temporalis muscle. Within the brain, regional differences between white and grey matter may be observed—phosphodiesters are raised in white matter [51], pH changes due to a hypercapnic load are greater in white matter [72], and creatine phosphate flux higher in grey matter [51]. Localised T_2 measurements have also been made to investigate the composition of the phosphodiester peak [50]. Brain injury has been studied in both the adult [73] and infant [74]. The method is less ideal for evaluating deep sited tumours, but by selecting superficial lesions, Cadoux-Hudson et al. [75] have investigated meningiomas and astrocytomas.

Skeletal Muscle

Most studies of muscle in vivo rely on simple surface coil localisation, but rotating frame imaging has been used to detect metabolic changes in the overlying gastrocnemius and soleus muscle. In the human, data have been composition of the phosphodiester peak [50]. Brain injury has been studied in both the adult [73] and infant [74]. The method is less ideal for evaluating deep sited tumours, but by selecting superficial lesions, Cadoux-Hudson et al. [75] have investigated meningiomas and astrocytomas.

The Heart

There are many factors which, in combination, make it difficult to obtain high quality spectra from the heart in vivo. Although the organ is large, the cardiac muscle is moving and occupies only a fraction of the volume of the heart. The phosphorus spectrum of heart muscle is very similar to the overlying intercostal muscle. Finally, there is a large quantity of blood in the heart, both in the chambers and in the cardiac vascular system. Signals from the blood will obscure the P_i signal from heart (which is overlapped by the 2–3 DPG resonances) and contribute to the ATP peaks. Despite these problems, good spectra have been obtained from the human myocardium [23]. In order to better assess diseases of the heart, Conway et al. [76] have devised an exercise device which can be used in the magnet, and enables the subject's heart to be stressed during the NMR investigation. Rotating frame images have been obtained both at rest and during exercise, and in healthy volunteers, these show no significant differences. Patients with hypertrophic cardiomyopathy have also been investigated, but only at rest [77].

Rotating frame methods have also been applied with great success in investigating metabolic heterogeneity across the myocardium in open chested animal experiments. The important benefit here is that the rotating frame probe can be attached directly to the ventricular wall, and localisation is then with respect to the probe, and hence the outer surface of the heart muscle. One approach has been to use FLAX-ISIS and a single surface coil probe. Robitaille and colleagues have studied transmural metabolite gradients in both the normal and ischaemic heart [14, 78, 79]. Alternatively, the imaging experiment (using a double surface coil probe) may be performed, and Rajagopalan et al. [80] have performed gated experiments which incorporate reference phantoms on both sides of the ventricle wall to confirm the spatial calibration.

The Kidney

Investigations of the kidney by rotating frame have been restricted to animal models and perfused organs. Bogusky et al. [11] used both phosphorus and sodium spectroscopy to detect regional differences in both the normal and hypokalemic rat kidney, and Moonen et al. have investigated post hypotensive damage in a similar preparation [81].

Tumours.

Rotating frame imaging has been used in humans to localise the signals from breast [82] and brain [75] tumours, and detect heterogeneity within implanted tumours in animals [26]. Infiltrating lymphoma within the human liver has also been studied using Fourier depth selection [83], and therapeutic response monitored.

12 Summary

Rotating frame spectroscopy can be applied in a variety of guises to the study of metabolism in both animal models and man. By comparison with the more usual B_0 methods of localisation, rotating frame spectroscopy has been employed in only a few centres. However, choices are often made on the basis of available instrumentation and expertise. Improvements in gradient coil technology have undoubtedly alleviated many of the problems associated with 2, 3 and 4DFT Fourier CSI methods, but it will be some time before production instruments can routinely perform spectroscopic imaging on short relaxation time species. In this article I have attempted to give an even-handed overview of rotating frame spectroscopy so that the reader can independently assess its place amongst the various strategies for investigating biochemistry by in-vivo NMR.

Acknowledgements: I would like to thank the many colleagues who have helped with the development and implementation of rotating frame methods in the Oxford laboratory. Although I hesitate to single out particular individuals, I am especially indebted to Martin Blackledge, Tom Cadoux-Hudson and Jonathan Allis. This Medical Research Council Unit is directed by Prof. George Radda, and enjoys financial support from the British Heart Foundation and the Department of Health.

13 References

1. Hoult DI (1979) J Magn Reson 33:183
2. Cox SJ, Styles P (1980) J Magn Reson 40:209
3. Lauterbur PC, Kramer DM, House WV, Chen CN (1975) J Amer Chem Soc 97:6866
4. Bendel P, Lai CM, Lauterbur PC (1980) J Magn Reson 38:343
5. Kumar A, Welti D, Ernst RR (1975) J Magn Reson 18:69
6. Brown TR, Kincaid BM, Ugurbil K (1982) Proc Natl Acad Sci USA 79:3523
7. Haselgrove JC, Subramanian VH, Leigh JS, Gyulai L, Chance B (1983) Science 220:1170
8. Haase A, Malloy C, Radda GK (1983) J Magn Reson 55:164
9. Haase A, Malloy C, Haenicke W (1983) Ann Int Super Sanita 19:65
10. Garwood M, Schleich T, Matson GB, Acosta G (1984) J Magn Reson 60:268
11. Bogusky RT, Garwood M, Matson GB, Acosta G, Cowgill LD, Schleich T (1986) Magn Reson Med 3:251
12. Segebarth C, Luyten PR, Den Hollander JA (1987) J Magn Reson 75:345
13. Böttcher U, Haase A, Norris D, Leibfritz D (1988) Soc Magn Reson Med Abstracts, p 933
14. Robitaille P-M, Merkle H, Sublett E, Hendrich K, Lew B, Path G, From AHL, Bache RJ, Garwood M, Ugurbil K (1989) Magn Reson Med 10:14
15. Styles P, Scott CA, Radda GK (1987) Magn Reson Med 2:402
16. Pekar J, Leigh JS, Chance B (1985) J Magn Reson 64:115
17. Garwood M, Schleich T, Ross BD, Matson GB, Winters WD (1985) J Magn Reson 65:239
18. Garwood M, Schleich T, Bendall MR, Pegg DT (1985) J Magn Reson 65:510
19. Metz KR, Briggs RW (1985) J Magn Reson 64:172
20. Styles P, Galloway G, Blackledge M, Radda G (1985) Soc Magn Reson Med Abstracts p 422
21. Blackledge MJ, Styles P, Radda GK (1987) J Magn Reson 71:246
22. Hoult DI, Chen C-N, Hedges LK (1987) Ann NY Acad Sci 508:366
23. Blackledge MJ, Rajagopalan B, Oberhaensli RD, Bolas NM, Styles P, Radda GK (1987) Proc Natl Acad Sci (USA) 84:4283
24. Styles P, Blackledge MJ, Moonen CTW, Radda GK (1987) Ann NY Acad of Sci 508:349
25. Karczmar GS, Weiner MW, Matson GB (1987) J Magn Reson 71:360
26. Karczmar GS, Shine N, Lawry TJ, Weiner MW, Matson GB (1989) NMR Biomed 1:159
27. Pekar J, Leigh JS (1985) Soc Magn Reson Med Abstracts, p 178
28. Garwood M, Robitaille PM, Ugurbil K (1987) J Magn Reson 75:244
29. Bendall MR (1983) Chem Phys Lett 99:310
30. Garwood M, Ugurbil K, Schleich T, Petein M, Sublet E, From AHL, Bache RJ (1986) J Magn Reson 69:576
31. Garwood M, Schleich T, Ugurbil K (1987) Ann NY Acad Sci 508:512
32. Styles P (1988) NMR Biomed 1:61
33. Bendall MR, Connelly A, McKendry JM (1986) Magn Reson Med 3:157
34. Edelstein WA, Hardy CJ, Mueller OM (1986) J Magn Reson 67:156
35. Moonen CTW, Anderson SE, Unger S (1987) Magn Reson Med 5:296
36. Blackledge MJ, Hayes DJ, Challiss RAJ, Radda GK (1986) J Magn Reson 69:331
37. Challiss RAJ, Blackledge MJ, Radda GK (1988) Am J Physiol 254:C417
38. Styles P, Cadoux-Hudson T, Hogan G Magn Reson Med Biol (in press)
39. Ugurbil K, Garwood M (this volume)
40. Ugurbil K, Garwood M, Bendall MR (1987) J Magn Reson 72:177
41. Garwood M, Ugurbil K (1987) Anal NY Acad Sci 508:527

42. Silver MS, Joseph RI, Hoult DI (1984) J Magn Reson 59:347
43. Bendall MR, Pegg DT (1986) J Magn Reson 67:376
44. Boehmer JP, Metz KR, Briggs RW (1985) J Magn Reson 62:322
45. Blackledge MJ, Oberhaensli RD, Styles P, Radda GK (1987) J Magn Reson 71:331
46. Blackledge MJ (1989) Magn Reson Med Biol 2: 159
47. Blackledge MJ, Styles P (1989) J Magn Reson 83: 390
48. Allis JL, Styles P, Cadoux-Hudson TAD, Rajagopalan B (1991) J Magn Reson 92: 158
49. Bodenhausen G, Freeman R, Turner DL (1977) J Magn Reson 27: 511
50. Kilby P, Allis J, Radda GK (1990) FEBS Letts 272: 163
51. Cadoux-Hudson TAD, Blackledge MJ, Radda GK (1989) FASEB J 3:2660
52. Böttcher U, Norris D, Leibfritz D (1989) Soc Magn Reson Med Abstracts, p 1111
53. Mitsumori F, Bolas NM (1990) XIV Int Conf Magnetic Resonance in Biological Systems (Warwick) Abstracts No. P15–24.
54. Böttcher U, Norris D, Leibfritz D (1989) Soc Magn Reson Med Abstracts, p 1113
55. Dixon RM, Angus PW, Rajagopalan B, Radda GK (1990) Europ Congr of NMR in Med and Biol (Strasbourg) Abstract No. 67.
56. Blackledge MJ, Styles P (1988) J Magn Reson 77:203
57. Blackledge MJ, Styles P, Radda GK (1988) J Magn Reson 79:176
58. Styles P (1991) Magn Reson Med 17: 3
59. Styles P, Smith MB, Briggs RW, Radda GK (1985) J Magn Reson 62:397
60. Crowley MG, Evelhoch JL, Ackerman JJH (1985) J Magn Reson 64:20
61. Bartholdi E, Ernst RR (1973) J Magn Reson 11:9
62. Becker ED, Ferretti JA, Gambhir PN (1979) Annalyt Chem 51:1413
63. Hore PJ, Daniell GJ (1986) J Magn Reson 69:386
64. Ordidge RJ, Connelly A, Lohman JAB (1986) J Magn Reson 66:283
65. Hendrich K, Merkle H, Weisdorf S, Garwood M, Ugurbil K (1989) Soc Magn Reson Med Abstracts p 627
66. Ordidge RJ (1987) Magn Reson Med 5:93
67. Nakada T, Kwee IL (1987) Magn Reson Med 4:366
68. Nakada T, Kwee IL (1987) Magn Reson Imag 5:259
69. Nakada T, Lee IL (1989) Magn Reson Imag 7:543
70. Wyrwicz AM, Conboy CB (1989) Magn Reson Med 9:219
71. Angus PW, Dixon RM, Rajagopalan B, Ryley NG, Simpson KJ, Peters TJ, Jewell DP, Radda GK (1990) Clin Sci 78:33
72. Cadoux-Hudson TAD, Rajagopalan B, Ledingham JGG, Radda GK (1990) Clin Sci 79:1
73. Cadoux-Hudson TAD, Wade D, Taylor DJ, Rajagopalan B, Ledingham JGG, Briggs M, Radda GK (1990) Acta Neurochir (Wien) 104:1
74. Moorcraft J, Bolas NM Ives NK, Sutton P, Blackledge MJ, Rajagopalan B, Hope PL, Radda GK (1991) Paediatrics 87: 273
75. Cadoux-Hudson TAD, Blackledge MJ, Rajagopalan B, Taylor DJ, Radda GK (1989) Br J Cancer 60:430
76. Conway MA, Bristow JD, Blackledge MJ, Rajagopalan B, Radda GK (1991) Br Heart J 65: 25
77. Rajagopalan B, Blackledge MJ, McKenna WJ, Bolas NM, Radda GK (1987) Ann NY Acad Sci 508:321
78. Robitaille PM, Lew B, Merkle H, Sublett E, Lindstrom P, From AHL, Garwood M, Bache RJ, Ugurbil K (1989) Magn Reson Med 10:108
79. Ugurbil K, Merkle H, Robitaille PM, Hendrich K, Yoshiyama M, Path G, Zhang J, Trisiani M, From AHL, Bache RJ, Garwood M (1989) NMR Biomed 2: 317
80. Rajagopalan B, Bristow JD, Radda GK (1989) Cardiovascular Res 23:1015
81. Moonen CTW, Blackledge MJ, Ratcliffe PJ, Radda GK (1985) Soc Magn Reson Med Abstracts, p 508
82. Kalra R, Hands L, Styles P, Wade K, Greenall M, Smith K, Harris A, Radda GK (1990) Soc Magn Reson Med Abstracts, p 1217
83. Dixon RM, Angus PW, Rajagopalan B, Radda GK (1989) Soc Magn Reson Med Abstracts, p 69

Depth Resolved Surface Coil Spectroscopy DRESS

Paul A. Bottomley

GE Research and Development Center, PO Box 8 Schenectady NY 12301, USA

Table of Contents

Depth resolved surface coil NMR spectroscopy (DRESS) employs a slice selective excitation in conjunction with surface coil detection to localize NMR signals to a plane of preselected thickness and location. How and why DRESS came about is discussed in the first section. The two factors affecting the definition of the plane, the surface coil and the slice selection, are analyzed in detail in the description section, where the problems of shimming the magnetic field homogeneity,

NMR Basic Principles and Progress, Vol. 27

distortions introduced by delayed acquisition, and chemical shift artefact are also identified and addressed. Extensions and improvements of the technique, including point-resolved (PRESS), slice-interleaved (SLIT DRESS), point-resolved rotating gradient (PROGRESS), and hybrid adaptations to improve spatial characteristics are then explored, along with the incorporation of DRESS into sequences for measuring relaxation times and reaction kinetics via saturation transfer, and performing decoupling and solvent suppression. The scientific findings from ^{31}P and ^{1}H DRESS studies of the human brain and heart, including stroke, dementia, and myocardial infarction are summarized in the results section. It is concluded that many of the problems faced by DRESS are endemic to contemporary spectral localization techniques, and that identifying useful applications is key to the future of the technology in general.

1 Historical Perspective

Depth resolved surface coil spectroscopy (DRESS) was arguably the first spatial localization technique employing NMR imaging magnetic field gradients to be applied to in vivo spectroscopy and successfully acquire localized ^{31}P and ^{1}H spectra from human subjects [1]. Its publication in 1984 was the culmination of years of experiments with a variety of other approaches to spectral localization that met with serious practical defects when applied to the practical problem of acquiring useful in vivo ^{31}P spectra.

Why localized spectroscopy? Why the ^{31}P *test?* Work by Hoult et al. in 1974 [2] and Jacobus et al. in 1977 [3], had shown that ^{31}P NMR could measure and monitor high-energy phosphate metabolism in living tissue, most notably, adenosine triphosphate (ATP), the body's fundamental energy currency. Ischemia and hypoxia wrought characteristic, dramatic, and almost instantaneous changes in ^{31}P spectra from muscle and heart, changes not reflected in ^{1}H relaxation times nor observable by ^{1}H-NMR imaging. Thus the information provided by ^{31}P NMR was unavailable to imaging and highly relevant to ischemic disease. Since ischemic disease is a major cause of mortality and morbidity, the additional functional information provided by ^{31}P NMR may ultimately be of significant value to patient management. Just as the work of Damadian showing elevated ^{1}H relaxation times in cancer [4] provided an early impetus to the development of ^{1}H-NMR imaging, the detection of energy metabolism via ^{31}P NMR provided the first *raison d'etre* for human localized spectroscopy.

What were the problems? The fundamental problem with using imaging gradients to localize the spectral information is that the effect on the NMR frequency of a gradient applied to an unknown inhomogeneous object is the same as that produced by the presumably unknown chemical shift variations throughout that object. Thus chemical shift differences cannot normally be distinguished from anatomical variations in an object of unknown structure and chemical composition by use of imaging gradients. If an object comprised of multiple chemical moieties is imaged by conventional NMR methods, the resultant picture is a superposition of images showing the distribution of each moiety displaced in space (in the NMR frequency domain) by its chemical shift. For ^{1}H-NMR imaging of the body where the 2 dominant moieties are H_2O and —CH_2—, the ghost image from —CH_2— is known as *chemical shift artefact* [5–7]. The direction of displacement of the ghost in the image is always in the direction of the imaging gradient field that is held constant during data acquisition (the *readout* gradient), as exemplified in Fig. 1.

We first observed this effect in 1977 during ^{19}F-NMR imaging experiments of perfluorinated artificial blood substitutes, which have a large chemical shift dispersion. We used a technique that employed two continuously applied orthogonal oscillating magnetic field gradients and a static gradient in the third-dimension to localize NMR signals to a sensitive line and produce an image by sequentially scanning the line [5]. Our surprise in diagnosing the

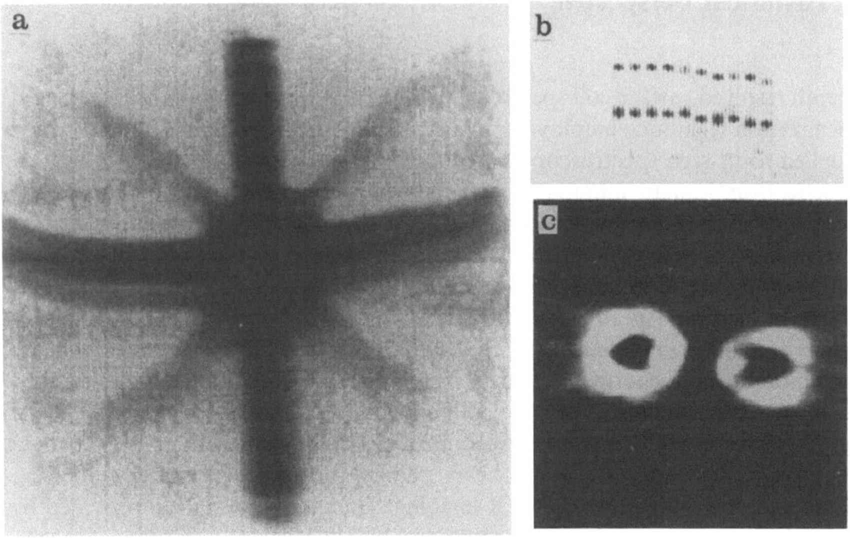

Fig. 1 a–c. Ghosts from the past [5, 7]. (**a**) is a ^{19}F-sensitive line image of a star-shaped phantom of perfluorotributylamine, a component of artificial blood, obtained in 1977 showing vertically displaced chemical shift ghost artefact in the direction of the static gradient, resulting from the 2 principal moieties separated by 45 ppm. (**b**) is a ^{1}H-sensitive line image of a tube containing n-hexane and benzene, recorded in the absence of the static gradient showing the 2 peaks chemically shifted by 6 ppm. (**c**) is a ^{31}P image of 2 tubes containing unbuffered ATP, reconstructed from projections in 1979. Chemical shift artefact from the two phosphate resonances renders the tubes as annuli as the static gradient is reoriented through 180°

source of the artefact was that the continuous oscillating gradient had not destroyed the chemical shift information: the chemical shift spectrum had remained encoded in the signal in a recoverable fashion up to the point that it was rendered into artefact by the readout gradient.

The experiment in which the readout gradient was replaced by a third oscillating gradient was performed in 1979 in Baltimore [7] on apparatus designed to perform both imaging and localized spectroscopy, an early prototype to the strategy of acquiring a conventional NMR image, designating a region of interest and then acquiring therefrom a localized spectrum [5]. Unfortunately, several major flaws in the oscillating gradient method were revealed when we tried to acquire localized ^{31}P spectra from an isolated regionally ischemic heart, most notably, significant broadening of resonances and artefacts analogous to spinning sidebands [7]. Moreover, using a volume coil, the acquisition time for a recognizable ^{31}P spectrum was around 80 min., precluding a dynamic study since the average survival time of our perfused (Langendorf) hearts was then only about 3 h. The obvious lesson was that small localized volumes generate proportionately small NMR signals.

Happily, a solution to the sensitivity/scan-time problem arrived without delay in the form of surface coils, first published by Ackerman et al. in 1980

[8]. Thus early 1980 found us, too, winding 1-cm surface coils which provided access to regional high-energy phosphate metabolism in perfused rabbit hearts, and permitted monitoring of the effect of drug therapy on regional ischemia [9]. In this preparation, the surface coil alone provided adequate localizaton to the anterior left ventricle and apex of the heart, so additional localization methods were not obligatory (Fig. 2). But the problem of defining and localizing in vivo spectra with versatility and in a practical manner was only postponed.

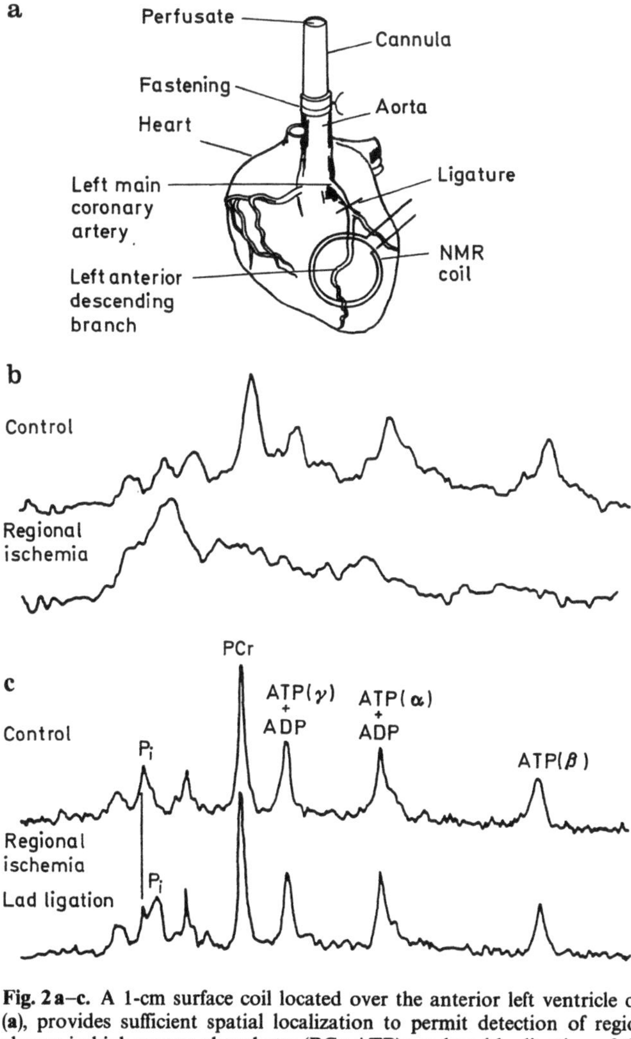

Fig. 2a–c. A 1-cm surface coil located over the anterior left ventricle of an isolated rabbit heart (**a**), provides sufficient spatial localization to permit detection of regionally ischemic metabolic change in high-energy phosphates (PCr, ATP) produced by ligation of the left anterior descending coronary artery (**b**). Such changes were not detectable when the same experiment was performed with a conventional volume coil around the heart (**c**) [9]

By 1981 then, we had concluded that surface coils may be necessary to enhance sensitivity sufficiently to detect low-abundance moieties such as phosphate metabolites, and that the key to gradient localized spectroscopy was to arrange that *all* of the imaging gradients were turned off during signal acquisition. A number of methods that met the latter criterion by spatially encoding the NMR signal prior to signal acquisition were explored in Schenectady, and patents filed a year later [10, 11]. These included several slice-selective phase-encoding, and phase-encoding spin-echo sequences [11], and the basic point resolved spectroscopy sequence (PRESS) comprised of a series of selective 90°, 180°, and 180° NMR pulses in the 3 orthogonal dimensions ([10]; Fig. 3), with additional sustained (*crusher*) gradients applied prior to signal acquisition to dephase unwanted magnetization [12].

We set about the design and construction of a whole-body system capable of performing both imaging and localized spectroscopy, based on a 1-m bore 1.5 T superconducting magnet. The feasibility of body ^{1}H imaging and human ^{31}P surface-coil spectroscopy on the same instrument was demonstrated late 1982 [13]. Brown et al. published their key paper on spectroscopic imaging with phase-encoding gradients [14], and in 1983 we attempted ^{31}P spectroscopic imaging of the brain at 1.5 T with a volume receiver coil, phase-encoding gradients and 27 cm^3 volume elements [15]. We could detect no recognizable ^{31}P brain spectra in scan times of over 1/2 h, times that we regarded as clinically

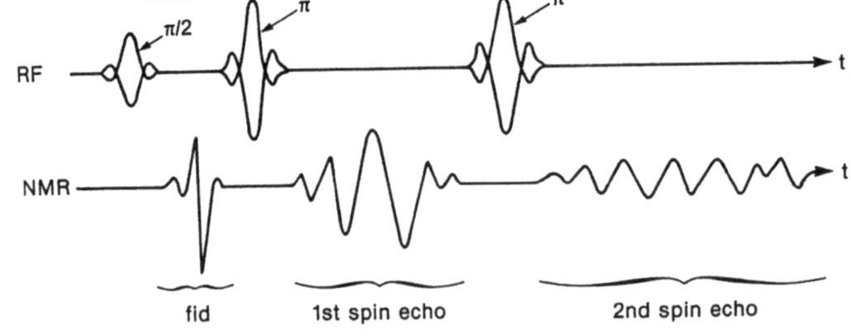

Fig. 3. Gradient and NMR pulse sequence timing diagram for PRESS employs spatially selective 90°, 180°, and 180° modulated RF pulses applied in the presence of the 3 orthogonal magnetic field gradients, G_X, G_Y, and G_Z, to localize a spectrum to a volume at the intersection of the 3 selected planes [10, 50]. The NMR signal from the sensitive volume is in the form of a spin-echo centered at time $t = 4\tau$, where τ is the time between 90° and the first 180° pulses. Gradients applied during 180° pulses are left on for short but equal periods before and after the 180° pulses to dephase unwanted signals, serving as *crusher* pulses [12]

useful, even though the technique worked fine with ^{1}H-NMR on artificial phantoms.

Annoyed or desperate, we tried simplicity. All of the gradients were removed except for a single slice selection to provide depth resolution (Fig. 4), and a surface coil oriented parallel to the slice was deployed to maximize sensitivity of the detected NMR signals. This was DRESS. Success arrived at last in the form of the first ^{31}P spectrum localized to the human brain early in 1984 [1]. A ^{31}P DRESS brain spectrum could be acquired in 5 to 20 min at 1.5 T, depending on depth (Fig. 5), allowing time for prior conventional ^{1}H-NMR imaging of the patient during the same exam to identify the region of interest for metabolic analysis.

Our goals were not technique development *per se*, but rather their application to real problems in medical diagnosis, evaluation, and research. Indeed, the five years through 1989 were mainly spent exploring applications to patients with stroke [16], senile dementia [17], tumors, and the development of cardiac ^{31}P spectroscopy in humans [18], from the detection of regional ischemia in dog models [19, 20] to patients with recent myocardial infarction [21]. Inevitably, the applications research prompted a number of enhancements, especially slice-interleaved DRESS (SLIT DRESS) for acquisition of spectra from multiple slices in the same time as a single DRESS acquisition [22], solvent-suppressed DRESS for suppressing the water resonance and acquiring metabolite spectra in vivo by ^{1}H spectroscopy [23], Nuclear Overhauser enhanced DRESS (NO DRESS) for in vivo ^{13}C spectroscopy [24, 23], and DRESS combined with two-dimensional spatially selective NMR pulses (PROGRESS) to provide full three-dimensional gradient controlled localized spectroscopy [25].

Today, DRESS remains a useful work-horse for examining the spectral response from a selected slice, for calibration, for shimming the magnetic field homogeneity over a coarse volume prior to finer spectroscopic imaging of the

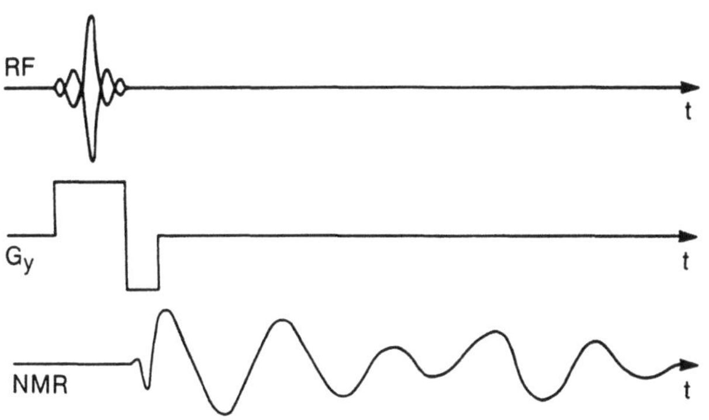

Fig. 4. DRESS pulse sequence [1]. A modulated RF excitation pulse applied in the presence of a magnetic field gradient G_Y directed normal to a surface detection coil excites NMR signals a plane parallel to the coil, and signal acquired as soon as the signal is rephased by the negative G_Y lobe at time T_D

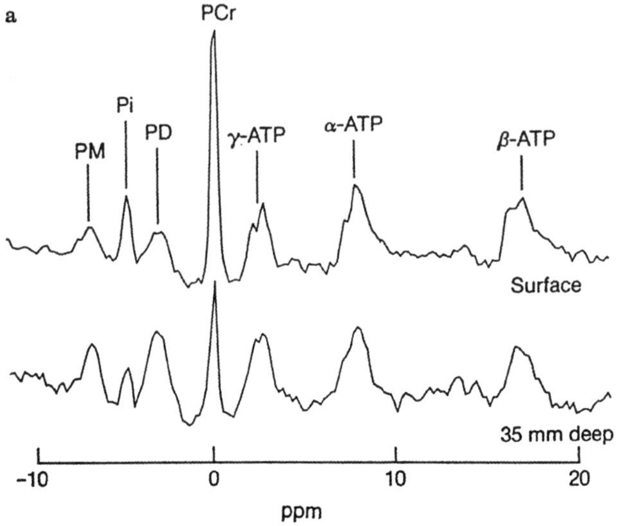

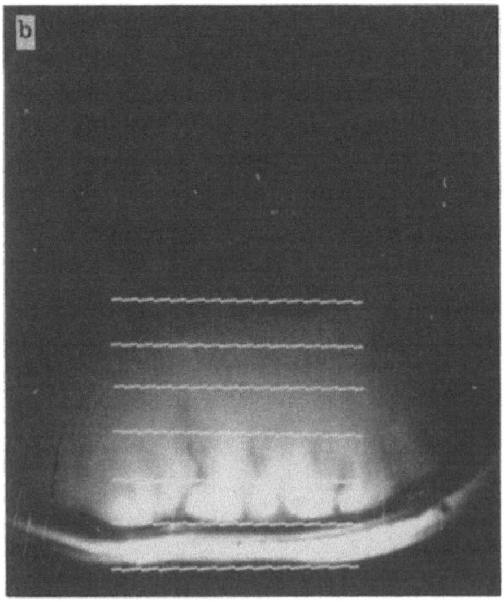

Fig. 5 a, b. ^{31}P-DRESS spectra from 1-cm thick slices at the surface of the human head (*top*), and at a depth of 3.5 cm (*lower*) in the brain (**a**), with corresponding surface coil ^{1}H-NMR image showing approximate location of the slice through the brain (**b**). The surface coil was located above the temple, and the spectral acquisition times were 10 and 20 min for the shallow and deep spectra. T_R was 1 s, and a 3-Hz line-broadening filter was employed, but no baseline flattening. (PCr = phosphocreatine; α-, β-, γ-ATP = α-, β-, γ-phosphates of adenosine triphosphate; PD = phosphodiesters; PM = phosphomonoesters; Pi = inorganic phosphate)

region, or for measuring spectral distortion due to partial saturation, measuring relaxation times, or adjusting radiofrequency (RF) pulse power as part of a spectroscopy exam that may well involve other localization methods with better localization properties. PRESS is finding new applications in water-suppressed ^{1}H spectroscopy of the brain. The strategy of using conventional NMR imaging to provide a reliable anatomic reference during a spectroscopy exam is still a key element of successful clinical spectroscopy research: the sensitivities and concentrations of NMR signals from energy metabolites in vivo are too puny to provide sufficient spatial resolution for anatomic reference in their own right [26]. It seems that spectroscopy must prove its worth on the basis of the unique functional, rather than anatomic, information that it provides.

2 Description

The spatial localization properties of DRESS depend critically upon the combined spatial selection characteristics of the NMR selective excitation procedure and the NMR surface coil used for detection.

2.1 *Surface Coil Effects*

A typical surface coil is comprised of one or more coplanar coaxial circular loops of conductor oriented with cylindrical axis perpendicular to the direction of the main magnetic field designated as the z-axis, and tuned to the NMR frequency of interest with capacitive elements. Their spatial response characteristics are well-documented [8, 27–30]. Surface coils improve the NMR signal-to-noise ratio because they can often be located closer to the NMR sources of interest and because their restricted spatial sensitivity significantly reduces noise from the bulk of the sample. For DRESS, the sensitivity loss as a function of distance from the cylindrical axis of the surface coil in planes parallel to the surface coil also provides the basis for spatial localization in the two dimensions perpendicular to the coil's cylindrical axis [1].

By the principle of reciprocity, the NMR sensitivity profile is given by the component of the magnetic field produced by the surface coil perpendicular to the z-axis (the transverse field), when energized with unit current [8], which can be calculated numerically from the Biot–Savart or related Laws [8, 26–30]. Figure 6 shows contour plots of the amplitude and relative phase of the transverse field in the xy- and yz-planes, where the y-axis is coincident with the coil axis [31]. The sensitivity falls rapidly with depth in proportion to $(a^2 + y^2)^{-3/2}$ on axis, and is 1/2 of the value at the center at a depth of only about $0.8\,a$, where a is the coil radius (Fig. 7a). However, this decrease in sensitivity with depth should not be confused with the width of the sensitivity profile in planes parallel to the coil. The sensitivity within xz-planes falls to 1/2 of the value that occurs at the intersection of the plane with the y-axis at about $z = \pm 1.3\,a(\pm 0.1\,a)$ virtually independent of depth for $y \leqq 2\,a$. Thus by this measure, the DRESS volume is about $1.7\,\pi a^2 \Delta y$, approximating the volume to a disk, where Δy is the thickness of the selected slice.

Unfortunately, a volume estimate based on the width of the sensitivity profile at half-maximum amplitude is not in general a good measure of the volume

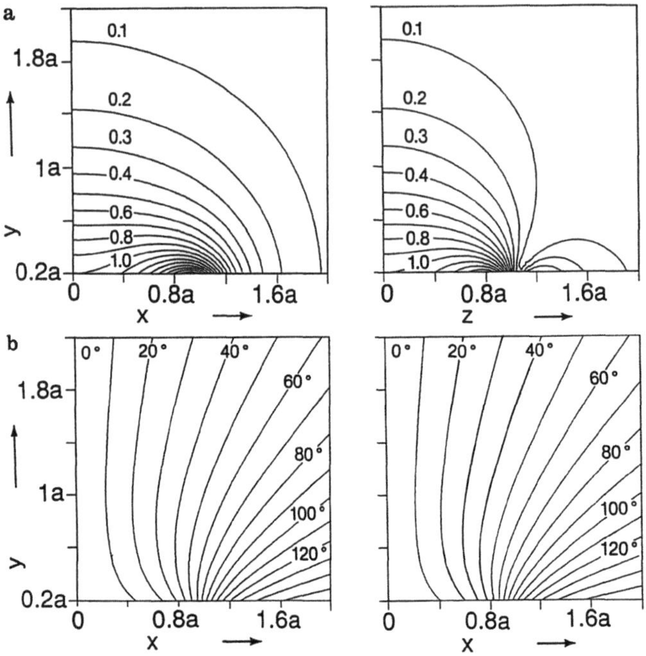

Fig. 6 a, b. Computed contour plots [31] of the amplitude of the transverse magnetic field (or magnetization) of a surface coil of radius a in the xy (*left*) and zy (*right*) planes, where the coil is oriented coplanar with the xz plane with origin at the centre, and the z-axis parallel to the main field of the NMR magnet (**a**). The phase variation of the transverse field in the xy plane at $z=0$ (*left*) and $z=0.4\,a$ (*right*), relative to the phase in the $x=0$ plane where it is constant throughout, is plotted in (**b**). The plots are symmetric about the $x=0$ and $z=0$ planes, and offset to $y=0.2a$.

actually responsible for the major portion of an observed spectrum, which is comprised of the integrated signals from all of the excitable and detectable NMR sources in the sample [32]. A better measure therefore obtains by integrating the transverse field in planes parallel to the surface coil over the range of the useful main magnetic field. This should represent the total signal detected by the coil assuming a homogeneous sample. The size of the DRESS volume can then be defined as that volume which is responsible for, say, 1/2 of the observed signal, measured by noting the radius of the sensitive disk when the integrated field corresponds to 1/2 of the total integral. By this measure the sensitive volume size in planes parallel to the surface-coil is no longer constant with depth because the dipolar field profile of the surface-coil flattens out as distance from the coil increases. If only the amplitude of the field is integrated, the total integral in a plane at constant depth rapidly and monotonically approaches an asymptote as the radius of the sensitive disk increases. The radius of the sensitive disk contributing 1/2 of the total signal then increases essentially linearly with depth (Fig. 7B). Although we have not accounted for the effect of

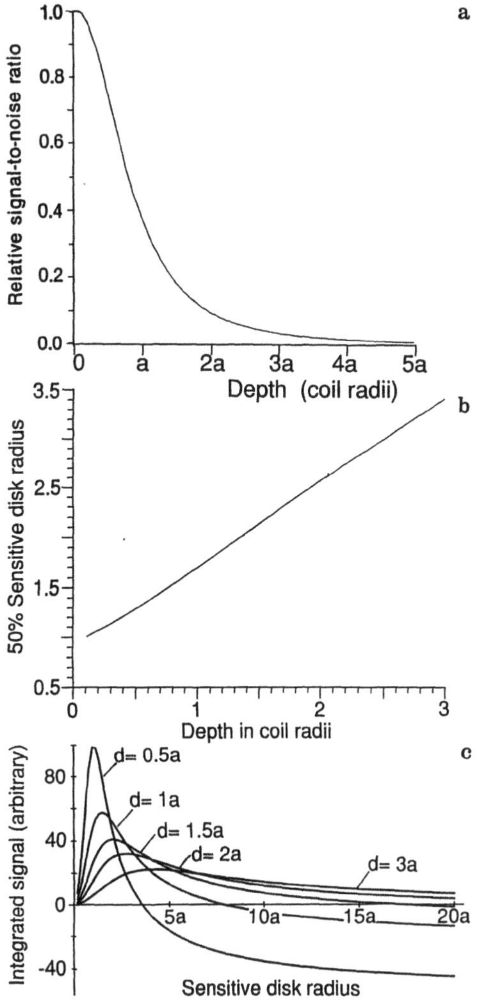

Fig. 7 a. Relative NMR sensitivity (or amplitude of the transverse magnetic field), assuming a uniform excitation field, of a surface coil of radius a along its axis as a function of depth in coil radii [50]. **(b)** Radius of the sensitive disk representing 50% of the total integrated signal in planes parallel to the surface coil, assuming a homogeneous sample and neglecting phase cancellaton due to phase variations in the detected transverse field [32]. **(c)** The integrated signal from a disk of radius r measured in coil radii from the cylindrical axis of the coil (abscissae), at different depths, d, after accounting for phase variations in the detection field [32]

nonuniformity in the transmitter field which will tend to further reduce the sensitive disk size [33], this situation approximates DRESS performed using a surface-coil for both detection and excitation [18].

For the case where a uniform excitation field is employed[1], phase differences between the excitation and detection fields must be accounted for [32]. Figure 6 shows that the field is entirely in the y-direction on the surface coil axis, whereas off-axis, the x-component first increases and the y-component decreases until eventually the field direction far off-axis is completely reversed from that at the center. Thus, if the signal from planes perpendicular to the coil were integrated taking this phase difference into account, the integral as a function of sensitive disk radius would no longer be monotonic: first the integral would increase as positive signals are added, but then at large distances it must decrease since outer volumes would add negative (reversed) field contributions, giving rise to situations where the signal from the center of the sensitive disk could be cancelled by signals from outside the center (Fig. 7c).

Fortunately, experiments with phantoms and in vivo show that this disturbing problem is not usually applicable to real situations. The reasons are three-fold. First and foremost, we have neglected the nonuniformity in the main magnetic field which increases dramatically away from the region of interest over which the field has been optimized for spectroscopy. This nonuniformity causes phase cancellation and signal loss from regions away from the coil axis, significantly diminishing their effect on the real integrated response. Second, practical excitation fields are also limited in extent, with phase and amplitude nonuniformity being greatest further from the coil axis, again diminishing the effect

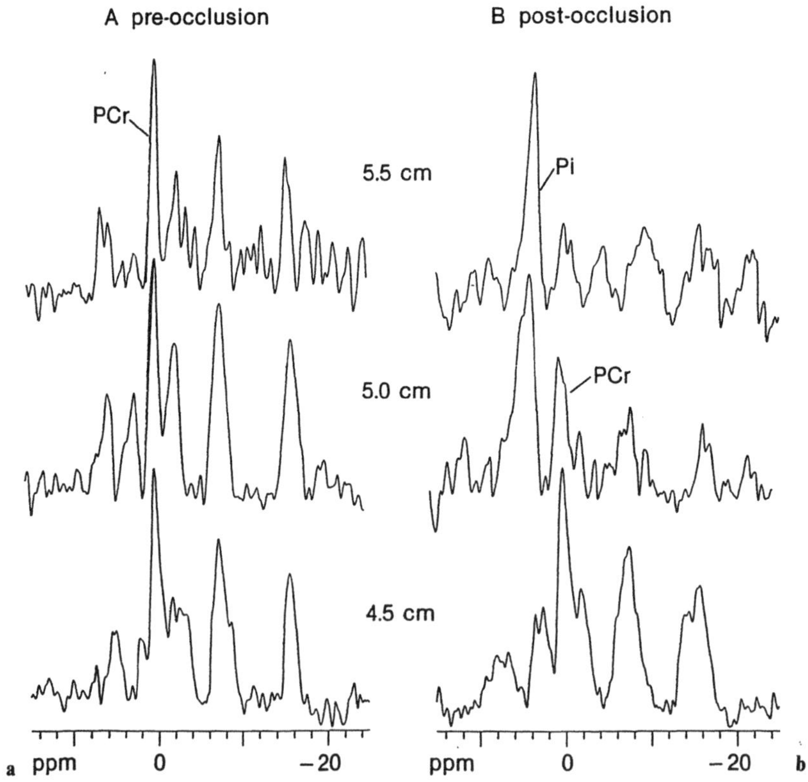

Fig. 8 a, b. in vivo ^{31}P DRESS spectra recorded noninvasively at 0.5-cm intervals through the anterior left ventrical of a dog heart (**a**) before, and 50–70 min after occlusion of the left anterior descending coronary artery (**b**) [19]. Postmortem staining of the excised heart revealed at 14-g endocardial infarction. Total depletion of high-energy phosphates in the postocclusion endocardial DRESS spectrum at 5.5-cm is consistent with an effective volume size of $\leqq 14\ cm^3$

of signals from the outer volume. Third, samples themselves are usually heterogeneous in both their source distribution and their magnetic susceptibility over such extended volumes, again altering their integrated response. An in vivo example demonstrating that such cancellation effects can effectively restrict localized volumes to regions close to the surface coil axis is shown in Fig. 8: the ^{31}P spectrum from a 1-cm thick slice through a 14-g myocardial infarction is uncontaminated by signals from surrounding normal tissue [19].

All this leaves us with the unsatisfactory conclusion that we do not have a good analytic way of precisely measuring the extent of selected volumes parallel to the plane of the surface coil in terms of their relative contribution to the observed signal. For quantitative studies requiring volumetric information such as is needed for absolute metabolite concentrations, therefore, careful calibration with phantoms precisely simulating the in vivo protocol would be necessary. However, given the likelihood of significant systematic error in such procedures [34], it is dubious whether reliable concentration measurements could be made practical for clinical studies without additional localization strategies to restrict NMR sensitivity within planes parallel to the surface coil.

It is important to note that these problems are intrinsic to surface detection coils and hence will arise in all localization protocols that employ surface coils and single or multiple slice selection such as the rotating frame or one-dimensional (1-D) phase-encode methods [35, 36], or indeed surface coils alone.

2.2 *Selective Excitation*

Happily, the spatial response of a sample to an amplitude and/or frequency modulated NMR pulse applied in the presence of a linear magnetic field gradient (Fig. 4) is well-defined from NMR imaging lore [37–41]: for low pulse flip-angles of around 90° and less, it approximates the Fourier transform (FT) of the modulation function of the NMR pulse [37]. Thus, a Gaussian-modulated 90° NMR pulse produces an approximately Gaussian-shaped localization profile [39], a $\sin(t)/t$ modulation produces a rectangular sensitivity profile [33, 40], and so on (Fig. 9a). Of course the $\sin(t)/t$ function, which extends indefinitely, must be truncated in a practical NMR experiment, thereby generating some ringing in the spatial response [41]. This effect is generally ameliorated by truncating at the nodes of the $\sin(t)/t$ function and windowing (multiplying) it with a Hanning, or other suitable window function [40, 38]. Relative to the crude localization provided by surface coils in the other two dimensions, all of these modulation functions provide excellent spatial selection of slices. Moreover, unlike some techniques employing gradients in the NMR excitation (RF) field such as is produced with surface coils, the slices are flat because the imaging gradients are linear over the entire sample.

When DRESS is performed with a homogeneous excitation field, the distribution of excited signal is obviously the same as for a conventional NMR imaging experiment. The nonuniform surface coil sensitivity results in a decrease

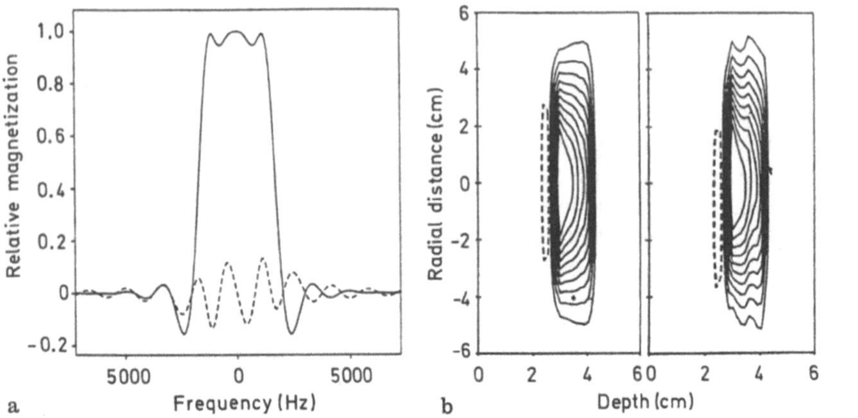

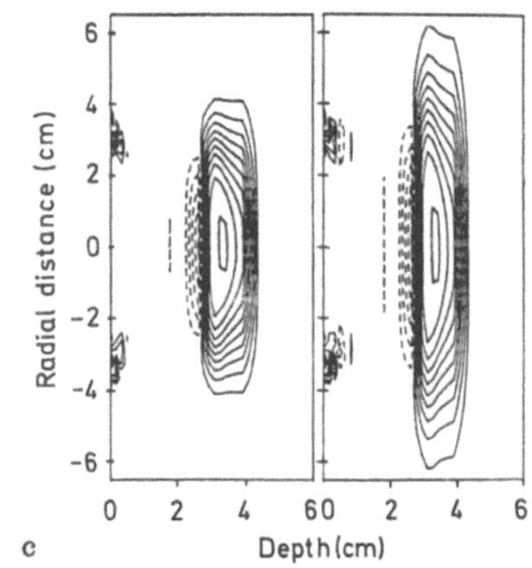

Fig. 9 a. Excitation profile of transverse in-phase (*solid-curve*) and quadrature (*dashed curve*) magnetization produced by a 1.2-ms sin(*t*)/*t* ($p = 2$) modulated 90° ^{31}P selective excitation pulse from a homogeneous transmitter coil applied in the presence of a 14-mT/m gradient pulse, and a $T_D = 3$ ms refocusing delay. (**b**) Sensitivity profile for the homogeneous excite DRESS experiment with a 6.5-cm surface detection coil and a 1-cm thick slice at depth $y = 3.5$-cm, in the *zy* plane (*left*) and the xy plane (*right*). Dashed lines represent negative contours. (**c**) Sensitivity profile for the surface coil excite/detect DRESS experiment in the *zy* (*left*) and *xy* (*right*) planes with the NMR flip-angle adjusted to 90° at the center of the slice. (Reproduced with permission from Garwood et al. [33])

in signal amplitude from the anterior to the posterior edge of the slice, depending on the slice thickness relative to the field profile (Fig. 6), weighting the response towards the front of the slice (Fig. 9b).

When the DRESS experiment utilizes the surface coil for both excitation and detection, the amplitude of the excitation field must be varied to optimize the NMR flip-angle according to the depth of the slice [18, 22]. When multiple DRESS spectra are acquired as a function of depth, *y*, the flip-angle is held constant by adjusting the amplitude of the excitation field to compensate for the $(a^2 + y^2)^{-3/2}$ decrease in the surface coil's field with depth at each slice (Fig. 7a). Thus the flip-angle need only be adjusted once, at a single depth, during a study. The principal effect of the nonuniform excitation relative to the homogeneous excitation case then, is to reduce the lateral extent of the excited volume about 20% on average, and to introduce some small regions of NMR sensitivity immediately adjacent to the coil, outside of selected slices [33]. The latter artefacts arise because as the field of the surface coil is increased to optimize the flip-angles at deeper and deeper slices, fields adjacent to the coil can greatly exceed 90° (or 180°), exciting signals even though those regions are rendered far off-resonance by the gradient that is simultaneously applied.

Computed sensitivity plots for DRESS experiments performed with homogeneous excitation and surface coil excitation using a 1.2-ms duration $\sin(t)/t$ modulated excitation, a 1-cm thick slice, and a 6.5-cm diameter surface coil, from Garwood et al. [33] are plotted in Fig. 9.

The location of the DRESS slice relative to the center of the gradient field is given by $y = 2\pi f/\gamma G$, where f is the difference in frequency between the NMR pulse and the NMR center frequency, f_0, in the absence of the gradient, G is the strength of the gradient applied during the RF pulse, and γ is the gyromagnetic ratio of the observed nucleus [22]. Thus, depth is varied by changing the offset frequency f in a single side-band transmitter.

For thin slices (that is, neglecting the effect of the decrease in surface coil sensitivity across the width of the slice), slice thickness is given by $\Delta y = 2\pi\Delta f/\gamma G$, where Δf measures the spectral width of the excitation pulse [22]. Note that both slice location and thickness vary with γ, so switching from say ^1H to ^{31}P NMR necessitates decreasing f to $[f\gamma(^{31}P)/\gamma(^1H)]$ to maintain the same slice location, and increasing G to $[G\gamma(^1H)/\gamma(^{31}P)]$ to maintain the same slice thickness (decreasing Δf usually being less desirable).

2.3 Problems

2.3.1 Eddy Currents

The earliest problem encountered in DRESS was inhomogeneity in the main magnetic field during data acquisiton caused by eddy currents induced in the magnet by the time-dependent slice-selective gradient. This was solved by shimming directly on the NMR signal from the DRESS slice by adjusting the static magnetic field shim coils. It was found that adjustment of only the linear x-, y-, and z-gradients sufficed, and that the transient nature of the induced fields was sufficiently slow relative to the free induction decay, that a matched transient compensation [42], though desirable, was not always necessary.

For low natural abundance nuclei like ^{31}P or ^{13}C, the shimming procedure is best performed in vivo using the ubiquitous ^1H-NMR signal from a DRESS slice at the identical depth [1], but with a thinner slice thickness corresponding to the same absolute value of G (e.g. $\Delta y(^1H) = \Delta y[^{31}P]\gamma[^{31}P]/\gamma[^1H]$). G is held constant in order to preserve the same eddy current environment during the ^{31}P (or ^{13}C) study that was present during ^1H shimming. Today, the use of self-shielded gradient coils [43] has reduced the fields induced in the magnet by the gradient pulses by about two orders of magnitude, substantially eliminating the eddy current problem. However, shimming on the volume of interest is still critical to ensure good spectral resolution and signal-to-noise ratio in a clinical environment where magnet homogeneity and patient magnetic susceptibility are apt to vary.

2.3.2 Acquisition Delay

The short delay, T_D, of a few milliseconds between excitation and detection necessitated by the gradient refocusing lobe (Fig. 4) is a source of several types of spectral artefacts, also shared by phase-encoding gradient localization methods [14, 36]. First, the spin–spin relaxation time (T_2) of many moieties of interest are comparable to T_D, so some of their signal is lost. For example, in the human brain, we compared phosphate concentrations measured in 15 normal adults at $T_D = 0.4$ ms with values measured at $T_D = 2.2$ ms during the same exam and found a 6% decrease in the apparent concentration of ATP and a 55% decrease in the phosphodiester (PD) concentration due to relaxation occurring during the latter delay relative to the former [34]. According to the Bloch equations, the signal decays as $\exp(-T_D/T_2)$. Thus, to quantify the signal lost via T_2 relaxation, either the signal decay or the metabolite T_2 values must be measured directly and on an individual basis, or values thereof must be assumed from prior in vivo work.

Second, since different chemically shifted moieties resonate at slightly different NMR frequencies relative to the center frequency of the NMR pulse, each resonance will dephase by a different amount during T_D, giving rise to a significant first-order (linear) phase variation in the acquired spectrum. The difference in phase of a resonance chemically shifted by δ Hz from the centre frequency is just $\phi(\delta)^\circ = 360\, T_D\delta$, with T_D in seconds, so the first-order phase correction can be done automatically (bearing in mind that instrumental sources, such as receiver filters, may contribute additional first-order phase variation to a spectrum, independent of T_D). It is interesting to note also that the purpose of the gradient refocusing lobe applied during T_D in Fig. 4 is to overcome just such a first-order phase variation, $\phi(y)^\circ = 360\, t(\gamma G)$, across the slice imparted during selective excitation [39].

Third, the loss of data during T_D can cause significant baseline artefacts in a spectrum produced by FT of the raw signal. Setting the missing data to zero during T_D is equivalent to multiplying the true complete time domain signal with a step function that is zero from $0 \leqq t \leqq T_D$, and 1 for $t \geqq T_D$. The observed FT spectrum is then the convolution of the true spectrum with the FT of the step function, which is a $\sin(t)/t$ function [44, 45]. The effect is to depress the baseline around each peak, and propagate the characteristic $\sin(t)/t$ wiggles along the baseline. When peaks are nearby or overlap, their wiggles may interfere and alter the true intensities of the resonances thereby affecting absolute quantification [34], although relative measures from spectra recorded with the same timing parameters should be comparable. The missing data problem often becomes entangled with the chemical shift dephasing problem when solutions are attempted [44].

Allman et al.'s [45] solution to the missing data problem is to generate copies of a $\sin(t)/t$ function corresponding to the FT of a step function of width T_D which are shifted to the frequencies of, and scaled to the amplitudes of, the major peaks in the spectrum. The shifted and scaled $\sin(t)/t$ functions are added

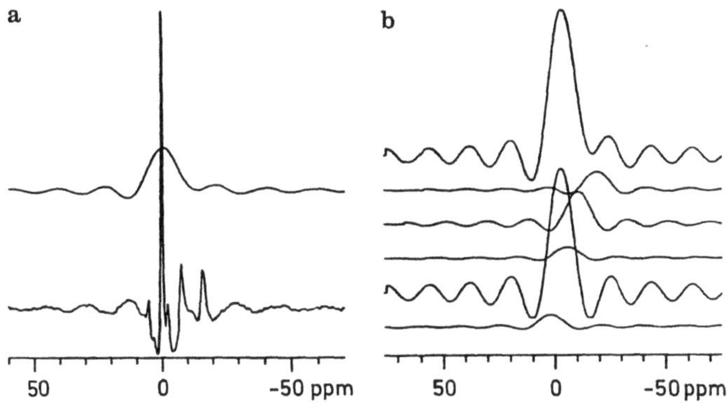

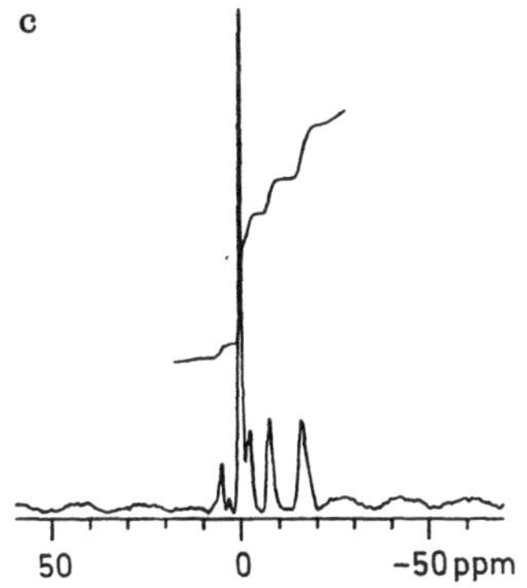

Fig. 10 a–c. Rolling baseline artefacts in this ^{31}P-DRESS spectrum from the human calf muscle (**a**, *bottom spectrum*) are produced by FT of an acquisition in which the first $T_D = 2.9$ ms of data was lost due to gradient application. The artefacts are deconvolved by calculating a magnitude spectrum and measuring the intensities and shifts of the tallest peaks. A $\sin(t)/t$ function is generated by FT of a step function of width 2.9 ms, and copies thereof shifted and scaled to each peak (**b**, *bottom 5 traces*), and added (**b**, *top trace*). This new function is scaled to the baseline artefact of the original spectrum (**a**, *top trace*), and added to the original spectrum to minimize the distortions (**c**). (Reproduced with permission from Allman et al. [45])

to produce a net spectrum whose intensity is scaled to minimize the baseline artefact, and then added to the convoluted spectrum, as exemplified in Fig. 10. Other solutions to this problem include fitting the time domain data directly, for example by the least squares method [46] or using maximum likelihood statistical estimation theory [47], maximum entropy reconstruction of spectra [48], and statistical spectral estimation theory [49]. In overcoming some of the limitations of the FT, these approaches usually also alleviate problems with phasing spectra.

2.3.3 *Chemical Shift Artefact*

Finally, DRESS is not immune from chemical shift artefact. The selective excitation procedure maps an isochromatic NMR signal source to a single location in space that satisfies the Larmor equation ($2\pi[f + f_0] = \gamma[B_o + Gy]$) in the presence of the gradient field. In a spectrum comprising multiple chemically shifted peaks at offset frequencies f_i, the Larmor equation is satisfied at different locations y_i, in direct proportionality. That is, different peaks derive from different parts of the sample [1, 22]. To ensure that peaks in a DRESS spectrum derive substantially from the same slice of a sample, the selective excitation pulse must be adjusted so that the slices corresponding to each peak are of sufficient thickness to substantially overlap those of all the other peaks, that is, $\Delta y > (y_i - y_{-i})$ [1]. This occurs when $\Delta f > (f_i - f_{-i})$, the bracketed term representing the chemical shift range in the sample that maps to the spatial range $(y_i - y_{-i})$. Since increasing Δf requires increasing G to hold Δy constant, the largest bandwidth excitation pulse compatible with the available gradient strength and desired slice thickness should be employed.

In practice, pulse bandwidth requirements can be difficult to meet for nuclei other than ^1H such as ^{13}C with a very large in vivo chemical shift dispersion and low γ, or where thin slices (e.g. $\Delta y < 1$ cm, $G \leqq 10$ mT/m) are desired. This does not mean that such experiments are invalid or should not be performed. Since the relationship between location and chemical shift is linear and all parameters are known, the location of each excited resonance in space is precisely determinable, as exemplified in Fig. 11 [24]. The acquisition of a series of spectra from different depths can ensure that measurements of multiple resonances of interest corresponding to the same depth are obtainable. In fact, a stack of

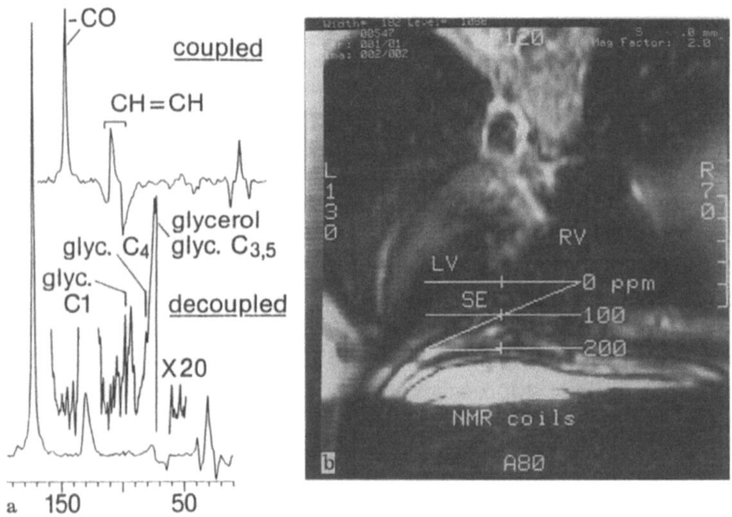

Fig. 11 a, b. Coupled (top) and broadband ^1H-decoupled cardiac-gated (twice per beat) ^{13}C DRESS spectrum acquired in 7 min from a 2-cm thick slice in the human heart (**a**), and corresponding 2 min annotated transaxial ^1H image, obtained immediately prior to spectroscopy, all at 1.5 T [24]. The center of the spectra at 100 ppm derive from the anterior myocardial wall at a depth of 5 cm, chemical shifts of 0 ppm and 200 ppm are respectively posteriorly and anteriorly displaced in direct proportion to chemical shift, as indicated in the image. Image artefacts derive from electrical interactions between the imaging coil and the surface coil used for decoupling. *Insert in bottom spectrum* (**a**) shows glycogen resonances at 100 ppm with gain increased 20-fold. The enhancement of —CO due to decoupling is 2.5 ± 0.1, but decoupling efficacy decreases with depth and hence chemical shift due to the nonuniformity of the decoupling field, generated here by a surface coil

DRESS spectra from different depths possessing significant chemical shift artefact can be replotted into an artefact-free stack of spectra simply by interpolation: in a stacked plot with y-dimension vertical and chemical shift horizontal, the effect of the artefact is to reorient the chemical shift axis by an angle $\alpha = \arctan([y_i - y_{-i}]/[f_i - f_{-i}])$ [22].

3 Improvements and Modifications

Enhancements to DRESS are of two types: those that improve its spatial localization properties, and those required to perform more advanced localized in vivo NMR spectroscopy experiments.

3.1 *Localization Improvements*

3.1.1 PRESS

The PRESS localization sequence [10, 50] was introduced to extend to all three dimensions the spatial localization provided by and under the control of the NMR imaging gradients, by addition of two sequential selective excitation 180° NMR pulses in the two dimensions orthogonal to the initial selective excitation pulse, as shown in Fig. 3. Because each 180° pulse generates a spin-echo signal from a whole line of the sample that intersects the excited plane, whereas signal from only the small selected volume at the intersection of all the planes is desired, and because imperfections in the 180° pulses may produce spurious transverse magnetization, it is advantageous to add large gradient crusher pulses [12] to eliminate signals from outside the selected volume. To avoid destroying the desired signal from the sensitive volume, the crusher pulses must be balanced each side of a 180° pulse, but they can be placed on any gradient axis [12, 51]. A spin-echo signal from the selected volume occurs at a time $t = 4\tau$, where τ is the time between the 90° and 180° pulses, and the two 180° pulses are spaced by 2τ.

For ^{31}P and ^{13}C in vivo spectroscopy, the additional spin-echo pulses and delays in PRESS have the disadvantage of increasing the time between excitation and acquisition to a point where a serious loss of metabolite signals due to T_2 processes may become unavoidable [50]. Accordingly, the technique has found few applications for these nuclei so far. However, for ^{1}H spectroscopy where T_2 values are much longer, useful applications are emerging in the brain when combined with water-suppression methods [51]. PRESS has a two-fold sensitivity advantage over a selective 90°–90°–90° stimulated-echo localization pulse sequence [51].

3.1.2 SLIT DRESS

The SLIT DRESS sequence resulted from a dire need to acquire simultaneously spectra from multiple localized volumes in clinical studies comparing diseased and adjacent normal brain tissue in order to limit the patient exam time [16], and in dynamic studies monitoring metabolic changes during myocardial isochemia and infarction in animals where the precise location of the jeopardized myocardium was unpredictable a priori [19, 20].

The idea is to interleave DRESS experiments exciting some N slices, each at different locations during the normal DRESS sequence repetition period T_R [22]. The pulse sequence is the same as for the conventional DRESS experiment (Fig. 4), except that it is repeated N times faster at intervals of T_R/N with each successive application exciting a new slice $y_j = 2\pi f_j/\gamma\ G$ by offsetting the transmitter frequency f_j in the single side-band transmitter. Ideally, since any one slice is only excited at a period of T_R, there is no increase in spectral distortion due to partial saturation effects. However, in practice the edges of adjacent slices may partially saturate due to overlapping slice profiles. Consequently, it is prudent to order the offset frequencies so that adjacent slices are nonsequentially excited [22]. For example, a stack of six slices might be excited in the order 1, 4, 2, 5, 3, 6. Shimming is more difficult than in DRESS since a reasonable homogeneity must be obtained over all of the slices of interest. A successful strategy is to shim first on the central slice and then make minor adjustments so that the two extreme slices exhibit reasonable linewidths. When using a single surface coil for both excitation and detection, the RF pulse amplitude must be varied to maintain a constant NMR flip-angle as each of the interleaved slices is scanned, as described in Sect. 2.2.

SLIT DRESS has several advantages over 1-D phase-encoding techniques [36]. First it is less susceptible to the eddy current problem because the amplitude of the gradient pulses are fixed, not varied, during the course of the NMR experiment. Induced eddy currents that are the same from pulse to pulse are more easily countered by a static shim coil set and a single adjustment. Second, the localization profiles are generally sharper and relatively independent of the effects of large anatomical variations in metabolites that might occur within slices. Because of the discrete nature of the FT and the way space is sampled by phase-encoding gradients, large signal changes within a slice can produce extensive ringing artefacts in a 1-D data set [32]. Disadvantages of SLIT DRESS relative to 1-D phase-encoding are that the chemical shift artefact problem must be dealt with [24], interleaving in gated heart studies means that spectra from different slices correspond to different phases of the cardiac cycle, and T_D is significantly longer for a given slice thickness and maximum gradient strength because selective pulses take longer than nonselective ones and less gradient is required for localization in the absence of an NMR pulse than must be applied during one. An example of a SLIT DRESS data from the brain of a normal subject and a patient with a seizure disorder acquired in 10 min is shown in Fig. 12.

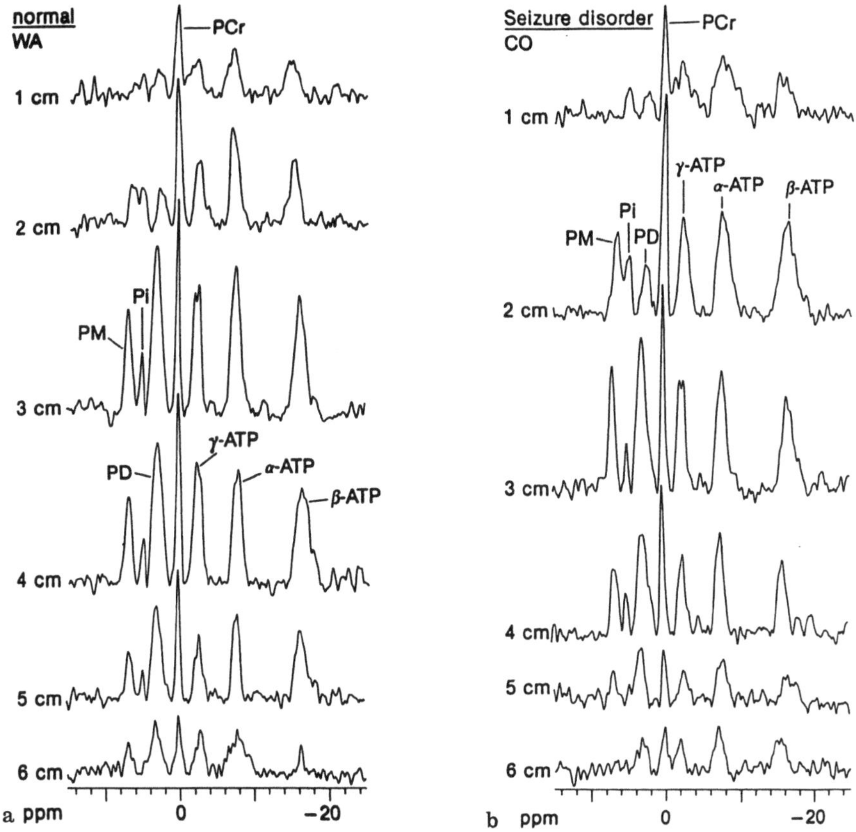

Fig. 12 a, b. ^{31}P SLIT DRESS sets of spectra from 1-cm thick slices as a function of depth at 1-cm intervals relative to the surface, recorded in 10 min from a normal human brain (**a**), and from a patient with a seizure disorder (**b**) [22]. The surface coil was 6.5-cm in diameter and positioned superiorly to the temple with the head lying sideways. The surface coil was used for both excitation and detection. Spectra from depths of 3 cm and greater derive predominantly from the cortex. The patient did not undergo a seizure episode during the study

3.1.3 PROGRESS

Point-resolved rotating gradient surface coil spectroscopy (PROGRESS) is the most recent DRESS variant [25]. Like PRESS, it addresses the problem of fuzzy localization within the plane of the surface coil by providing additional selective excitation with imaging gradients. Unlike PRESS it does so without spin-echoes or increased T_D, which makes it suitable for spectroscopy of moieties with short T_2, such as occur with in vivo ^{31}P NMR. The sequence is based on the invention of a new type of selective excitation pulse which provides 2-D localization in a single application by applying a gradient that reorients through those two dimensions during application of a modulated NMR pulse [52, 53]. In PROGRESS, the 2-D pulse is a 180° selective inversion that precedes the DRESS excitation on alternate applications of the sequence. The NMR signal acquired with the inversion is subtracted from that acquired without the inversion pulse to yield a spectrum from a 3-D localized volume [25].

One version of a 2-D pulse that is practical for ^{31}P spectroscopy employs a sine and a two-cosine waveform for the z- and x-gradients respectively, corresponding to a Fig. 8 trajectory of the gradient which is applied during a Gaussian modulated 180° pulse. This limits the surface coil sensitivity to a disk in the $z-x$ plane, as illustrated in Fig. 13 [25]. For this pulse, the half-height widths of the central lobe of the sensitivity profile are $\Delta x = 1.3 \times 10^4/(\gamma G_x)$ and $\Delta z = 1.8 \times 10^4/(\gamma G_z)$, in meters, where subscripts x and z denote the respective gradient amplitudes, and the G's are in T/m. Thus changing the x- and z-gradient

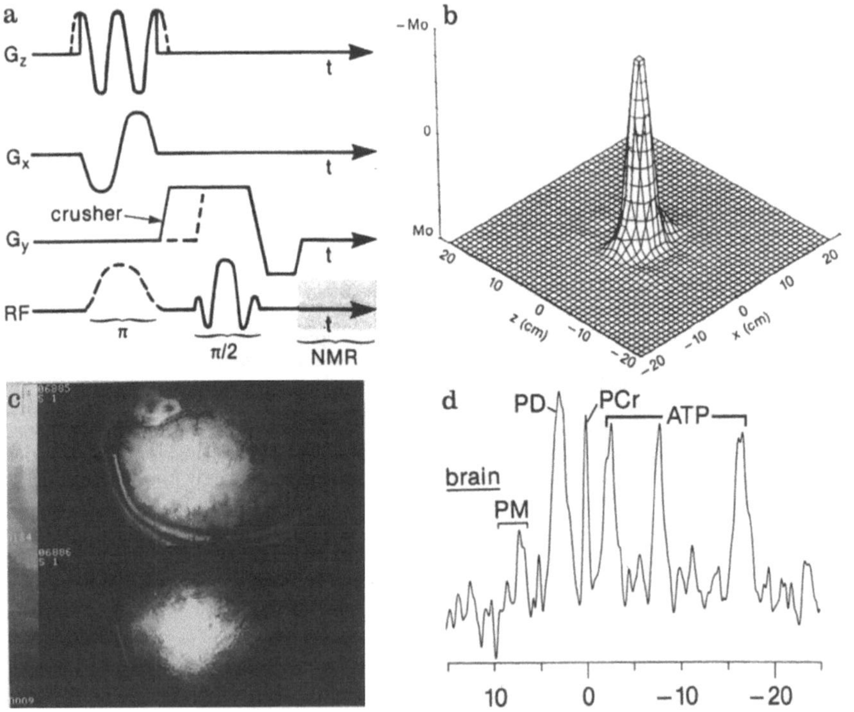

Fig. 13 a. PROGRESS 3-D localization sequence with a 2-D selective 180° (π) inversion pulse in the xz plane followed by a DRESS y-slice selection [25]. The inversion pulse is applied and not applied in 2 sequence applications, and the resulting signals subtracted. (**b**) The sensitivity profile for a 2-ms Gaussian 180° ^{1}H-NMR pulse (*fwhm* = 1 ms) with maximum G_X and G_Y amplitudes of 1 mT/m, and a 6-cm diameter detection coil. (**c**) ^{1}H surface coil image from the brain at a depth of about 5-cm without (*top*) and with (*bottom*) the 2-D pulse demonstrating improved localization. (**d**) ^{31}P-PROGRESS spectrum corresponding to the lower image in (**c**) acquired in 10 min with 3 mT/m gradient amplitudes during the 2-D pulse, and $T_R = 2$ s amplitudes does provide some control over selected volume size. Unfortunately, control is incomplete as this pulse does excite some signal from sources residing outside of the central lobe. These contributions are eliminated when the pulses are used in conjunction with a surface detection coil (Fig. 13; [25, 52, 53]). Gradient crusher pulses are again deployed during the interval between the 2-D pulse and the DRESS portion of the sequence to destroy unwanted transverse magnetization.

There are many different waveforms or trajectories that can be used for 2-D pulses [52–56], including spiral trajectories which can significantly reduce contributions from outside the central lobe [54, 55]. Unfortunately, the pulse length of those spiral gradient trajectories that can achieve good outer-volume suppression with gradient strengths and slew rates that are practical on body NMR imaging systems, can be so long that the spectral bandwidth excited by the 2-D pulse is too narrow for spectroscopy applications. One solution is to break the spiral into short *pin-wheel* segments which are applied in consecutive applications of the sequence, and the resultant spectra added [56]: we have obtained useful 3-D localized ^{31}P spectra from the human heart by such means.

For the Gaussian pulse described above with a full-width at half-maximum duration of 1-ms, an off-resonance shift of 10 ppm distorts the sensitivity profile into an ellipse with major and minor axes about 30% larger and 30% smaller than the on-resonance profile, with major axis parallel to the x-axis [25]. The location of the sensitive disk can be moved within the plane selected by DRESS either by changing the ratio of currents in separate halves of the x- and z-gradients [52], or by modulating the NMR excitation pulse to coincide with the new location [57].

3.1.4 CRISIS and Other Hybrids

Other spectroscopy localization techniques that accomplish 3-D spatial localization prior to applying a conventional (hard) excitation pulse, nominally 90°, could benefit from the replacement of the hard pulse with DRESS slice selection, and a reduction by one dimension of the preceding localization subsequence [58, 25]. Examples of such sequences from which DRESS hybrids could be derived include: volume selective excitation (VSE: [59]), spatially resolved spectroscopy (SPARS; [60]), image selected in vivo spectroscopy (ISIS; [61]), and spatial and chemical-shift-encoded excitation (SPACE; [62, 63]). The advantage to the DRESS experiment is again to provide ways of improving localization within selected planes. In addition, a DRESS hybrid has the advantage relative to the original versions of substantially eliminating any localization artefacts in the form of spurious signals from outside the region of interest that may occur in the direction parallel to the slice selection gradient.

Perhaps the most useful hybrid combines DRESS and ISIS (CRISIS; [50, 58, 25]). ISIS spatially localizes signals to a small volume in the sample by a procedure of successive subtraction of the large signals from the entire sample in a sequence cycle comprised of at least 8 T_R intervals [61]. Since small changes in position may cause large changes in the differences, the technique is susceptible to motion artefacts, and may be at a disadvantage for cardiac studies [25]. The CRISIS sequence, with *x*- and *z*-selective 180° pulses preceding the 90° slice selection pulse (Fig. 14; [50]), will largely eliminate artefactual signals from motions in the y-direction because only the NMR signal in the slice is ever excited. Also, the cycle time is reduced to 4 T_R (Table 1), thereby halving the period over which slow motions could be deleterious. Since the usual cardiac protocol has the patient in a prone or supine orientation with the *y*-axis vertical, which happens to correspond to the dimension where the greatest displacement due to physiological motions occurs, this hybrid approach should ameliorate concerns about whether an ISIS spectrum is significantly compromised by motion artefact.

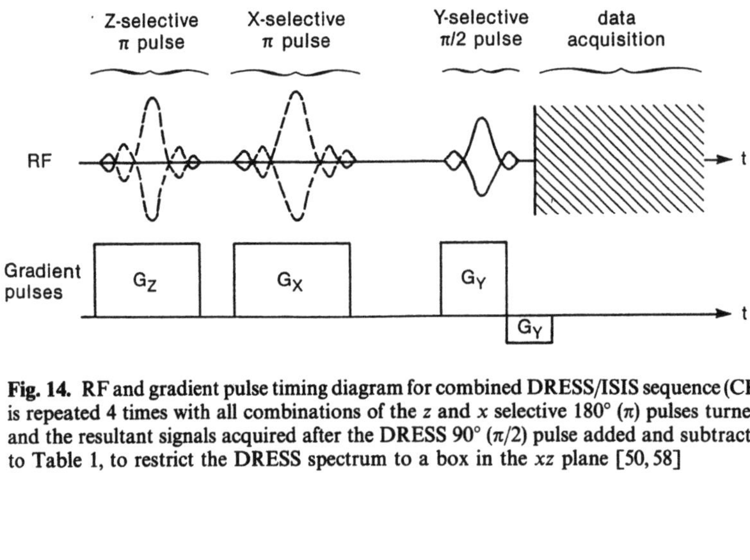

Fig. 14. RF and gradient pulse timing diagram for combined DRESS/ISIS sequence (CRISIS), which is repeated 4 times with all combinations of the *z* and *x* selective 180° (π) pulses turned on and off, and the resultant signals acquired after the DRESS 90° ($\pi/2$) pulse added and subtracted according to Table 1, to restrict the DRESS spectrum to a box in the *xz* plane [50, 58]

Table 1. Combined DRESS/ISIS (CRISIS) cyling sequence for applying selective inversion pulses

Sequence Number	Z-Selective Pulse	X-Selective Pulse	Contribution to total Spectrum
1	off	off	+1
2	on	off	−1
3	off	on	−1
4	on	on	+1

3.2 Advanced NMR Spectroscopy

3.2.1 Relaxation Times

The simplicity of the DRESS sequence renders it an ideal localization platform from which many other types of NMR spectroscopy experiments can be performed. For example, the spin–lattice relaxation times (T_1) of moieties can be measured by the inversion recovery method by preceding the basic DRESS sequence (Fig. 4) with a hard 180° inversion pulse (IR DRESS) at time T_I. T_I is varied for a number of spectral acquisitions so that the moiety has signal $S = S_0(1 - 2\exp[-T_I/T_1] + \exp[-T_R/T_1])$, from which T_1 can be deduced. Alternatively, the saturation recovery method can be employed by repeating the DRESS sequence to acquire spectra at different T_R values so that $S = S_0$ $(1 - \exp[-T_R/T_1])$ [64, 65]. For these measurements, a uniform excitation field should be used and care must be taken to modulate the slice selection pulse so that its profile is substantially rectangular. Otherwise partial saturation effects across the profile will systematically distort the observed T_1 values [64].

Similarly T_2 measurements of un(J-)coupled moieties can be performed by following the basic DRESS sequence (Fig. 4) with a hard 180° NMR pulse at a time $T_E/2$ relative to the selective pulse [65], which is varied for several spectral acquisitions, producing a refocused spin-echo signal $S = S_0 \exp(-T_E/T_2)$ at time T_E (RE DRESS).

3.2.2 Solvent Suppression

Solvent suppression techniques are also easily combined with DRESS, and can be used to acquire in vivo ^1H spectra of some metabolites, mobile membrane constituents and amino acids when the otherwise overwhelming water signal is suppressed [23]. Examples of four such sequences are depicted in Fig. 15 [23].

In Fig. 15a, b, the DRESS sequence is preceded by a long modulated chemical selective pulse tuned to the H_2O resonance with amplitude adjusted to minimize the H_2O signal in the detected ^1H spectrum, which first occurs when the pulse is nominally 90°, if the DRESS pulse is also 90°. Such pulses were first applied in a localization context to ^1H-NMR imaging [66], and have been called CHESS pulses, for chemical selective spectroscopy [67]. A $\sin(t)/t$ or Gaussian RF modulation will suffice with a bandwidth to match the width of the H_2O resonance, applied in the absence of gradients. The DRESS spectrum is then acquired in the usual fashion (Fig. 15a), or as a spin-echo after a further hard 180° NMR pulse (Fig. 15b) which provides the added benefit of attenuating any broad undesirable resonances via T_2 processes [23]. Alternatively, H_2O suppression is accomplished in Fig. 15c with a 180° inversion pulse applied at time T_I prior to the DRESS sequence, where T_I is adjusted to coincide with the null point ($T_I \approx 0.69 T_1$) in the recovery from inversion of H_2O. In the final sequence, Fig. 15d the DRESS spatially selective pulse is replaced altogether

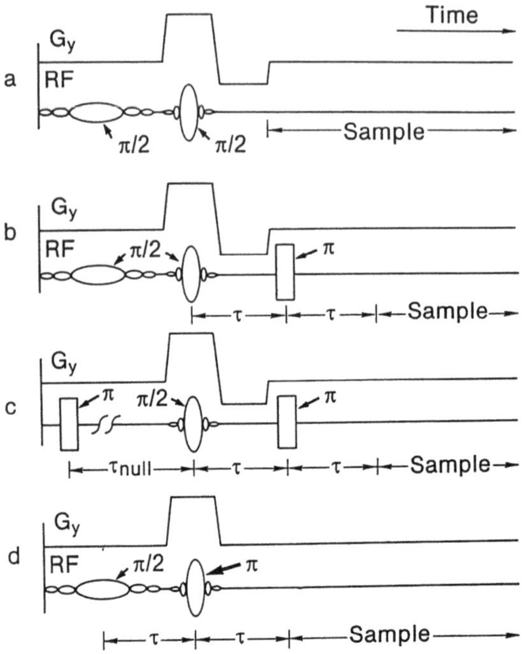

Fig. 15a–d. Sequence timing diagrams for 4 methods of performing solvent suppressed DRESS [23]. Selective RF pulses are depicted as $\sin(t)/t$ modulated envelopes: they are chemical selective when applied in the absence of a gradient pulse, and spatially selective when applied in the presence of one. A negative G_Y lobe is not required in (**d**) if the 180° (π) pulse is symmetric with respect to the positive G_Y pulse. $\tau_{null} \approx 0.69 T_1$ (solvent, or H_2O) is the null point in the inversion recovery of the solvent or H_2O signal.

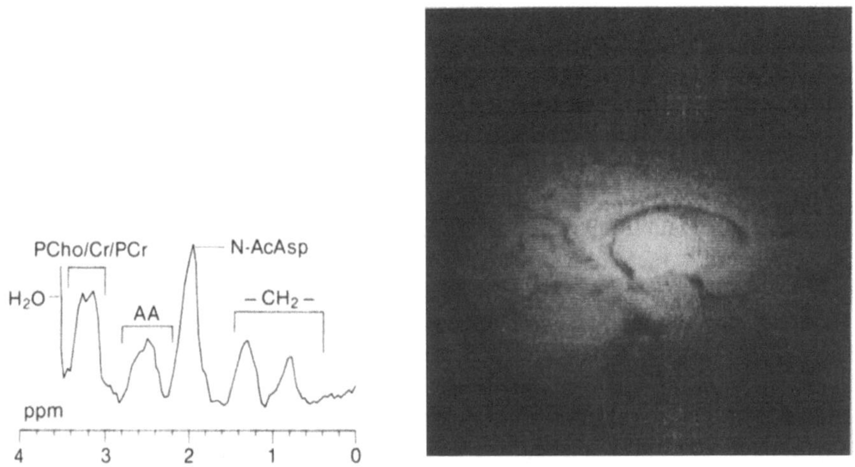

Fig. 16. In vivo H_2O-suppressed ^1H DRESS spectrum (*left*) from a 5-cm deep slice in the normal human brain as imaged at *right*, acquired in just 2 s ($T_R = 1$ s) at 1.5 T using the sequence of Fig. 15B [23]. (PCho/PCr/Cr = phosphorylcholine and total creatine pool; AA = amino acid resonances; *N*-AcAsp = *N*-acetyl aspartate; —CH_2 = lipid methylene resonances)

with a chemical selective pulse tailored to excite only that portion of the spectrum of interest, excluding the H_2O signal. Responsibility for spatial selection of the slice of interest resides with a selective 180° spin-echo refocusing pulse at time $\tau = T_E/2$ later [23]. An early example of a H_2O-suppressed ^1H spectrum from the human brain obtained with the sequence of Fig. 15B is shown in Fig. 16 [23, 50].

3.2.3 *Saturation Transfer*

Saturation transfer is a technically similar experiment designed to measure chemical reaction rates. It is most commonly applied in vivo to the creatine kinase reaction in which phosphocreatine (PCr) is turned over to adenosine triphosphate (ATP) with a first-order rate constant k. If the γ-phosphate peak of ATP is selectively irradiated continuously in a ^{31}P-NMR experiment, its signal saturates and disappears so that ATP contributes no phosphate signal to PCr via the reverse reaction. The PCr signal, S_0, therefore decreases to a new value S as PCr is converted to γ-ATP at a rate k. The rate constant is then calculated from $k = (1 - RS/S_0)/T_{1S}$, where T_{1S} is the T_1 of PCr measured with the γ-ATP peak saturated, and the saturation factor, R, corrects the ratio S/S_0 for partial saturation if the experiment is performed under conditions of $T_R \approx T_1$ or less [68].

The saturation transfer experiment is combined with ^{31}P DRESS to measure spatially localized reaction kinetics, by applying a long duration low-level RF irradiation tuned to γ-ATP (at -2.7 ppm relative to PCr) with amplitude adjusted to the minimum level required to completely saturate it, safety guidelines notwithstanding, and S measured. This irradiation could be applied continuously except during the slice-selection and detection intervals. As a control, S_0 is then measured with the same irradiation applied on the other side of the PCr resonance (at $+2.7$ ppm). Unless relative measures of k will suffice for the study, as might be the case in dynamic experiments, T_{1S} and R must then be measured.

3.2.4 *Decoupling/Overhauser Enhancement*

The benefits of decoupling and nuclear Overhauser enhancement are accessed to DRESS experiments by applying a decoupling irradiation in the usual fashion, treating the DRESS portion of the procedure like a normal excitation pulse. Of course, there will be a glitch in the spin-coupling during the period in which the gradient is applied, but acquisition in the absence of gradients is minimally affected. A ^{1}H-decoupled, Overhauser-enhanced ^{13}C DRESS spectrum from the human heart obtained by applying a continuous low-level broadband noise-modulated decoupling irradiation during the course of a ^{13}C experiment is shown in Fig. 11 [24]. The ^{1}H decoupling RF irradiation was delivered by a surface coil of figure-8 shaped geometry to minimize electrical coupling with a circular ^{13}C surface detection coil. Low pass filters that attenuated the ^{1}H irradiation by more than 100 dB were connected between the ^{13}C coil and the ^{13}C preamplifier [24]. A problem with surface coils for decoupling is their nonuniformity which results in a spatially dependent coupling efficiency that accentuates the sensitivity decrease with depth (Fig. 7A). Also with substantially continuous irradiation, safety guidelines for peak local RF power deposition can be easily exceeded, and must be monitored closely [24].

4 Results

DRESS provided us with some of the earliest glimpses of high-energy phosphate metabolites in the human brain [1], heart, and liver [18], as well as of the low-concentration moieties that are detectable by ^{1}H NMR in the human brain [23]. Here, its scientific contributions to date are summarized.

4.1 *Brain*

We have reported brain ^{31}P-DRESS studies from 9 normal adults, 4 patients with chronic cerebral infarction [16], and 9 patients with senile dementia [17, 69]. These studies were performed in collaboration with BP Drayer and RJ Herfkens at Duke University Medical Center on a 1.5 T GE Medical Systems' (Milwaukee, WI) NMR imaging/spectroscopy research system [6]. Several other patients studied with seizure disorders that showed no significant metabolic abnormalities when studied in the absence of symptoms, or with brain tumors were considered inconclusive and not formally reported. All clinical studies utilized a 6.5-cm diameter surface coil for both excitation and detection, ^{1}H imaging to identify the anatomy of interest during the NMR exam, and a ^{31}P DRESS or SLIT DRESS sequence with 1-cm thick slices, an 8-s T_R to minimize saturation effects, a 3-ms T_D, and an average of about 100 signals per spectrum or SLIT DRESS set. Metabolites were quantified from peak integrals and pH measured from the chemical shift of inorganic phosphate (P_i) [16].

In normal subjects aged 18–50, the PCr/ATP ratio was 1.2 ± 0.3 (SD) and the phosphomonoester (PM) to phosphodiester (PD) ratio was 0.49 ± 0.15 (SD) in predominantly cortical matter above the temple. These values were respectively higher and lower than unlocalized surface coil measurements on infants [16, 70]. Recent work confirms that developmental changes occur in both these ratios: PCr/ATP increases from 0.5–0.9 at birth to about 1.3 at age 6, and PM/PD decreases from 0.6–1.6 to about 0.5 over the same period [71]. There is evidence now that the DRESS values are distorted by T_D artefacts from true adult values of around 1.83 ± 0.07 (SE) for PCr/ATP and 0.32 ± 0.03 (SE) [34, 72] for PM/PD. Nevertheless, all of these PCr/ATP measurements

Table 2. ^{31}P DRESS and SLIT DRESS measurements of phosphate metabolites in neurological disease

			Degenerative disease			
Parameter	Normal	Stroke	Alzheimer's	Vascular	Huntington's	Parkinson's
Individuals	(n = 9)	(n = 4)	n = 1	n = 2	n = 4	n = 2
Spectra	9	10	3	5	6	5
PCr/NTP (SD)	1.2 (0.3)	1.2 (0.3)	0.92 (0.14)	0.92 (0.31)	1.3 (0.4)	1.1 (0.4)
PCr/Pi	7.7 (2.3)	7.1 (2.0)	> 20	> 7.5	> 5	> 8
PD/PCr	1.14 (0.14)	1.04 (0.16)	1.2 (0.1)	1.4–2.0	1.0 (0.4)	1.2 (0.4)
PM/PCr	0.56 (0.15)	0.67 (0.10)	0.63 (0.12)	0.61 (0.14)	0.65 (0.16)	0.50 (0.11)
pH	7.0 (0.1)	7.0 (0.5)	*	7.06 (0.1)	7.0 (.1)	7.0 (0.1)
Total P, %	—	74 (16)	—	—	—	—

SD = standard deviation. NTP measurements are based on the β-phosphate peak of NTP. Total P, % = the ratio of the total integrated P-31 signal in spectra from infarction to that in spectra from the contralateral hemisphere. *Pi too small to measure indicate that the anaerobic cellular energy reserve indexed by PCr is much lower in brain than in muscle where PCr/ATP = 5.6 ± 0.1 (SE) [73]. The developmental changes in PM and PD parallel neurological development and myelination: PM measures some cellular membrane precursors such as phosphorylethanolamine and phosphorylcholine, and PD includes the membrane catabolites, glycerolphosphoroethanolamine, and glycererolphosphorylcholine [16, 26].

In patients aged 18–60 with chronic stable cerebral infarctions of 30–90 cm^3 (as measured on ^1H images) involving the distribution of the right middle cerebral artery at 10-weeks to 7-years post-infarction, no significant alterations in the metabolite ratios PCr/ATP, PCr/Pi, PD/PCr, PM/PCr, or pH were detectable relative to normal. However, the overall levels of NMR-detectable phosphates decreased from 6–40% in infarctions relative to levels measured in DRESS spectra acquired from symmetrically located regions in the contralateral hemisphere, as summarized in Table 2 [16]. The decrease in total phosphates was interpreted as reflecting a loss of metabolically active cells during infarction, and the lack of change in metabolite ratios or pH as an indication that there was no significant area of ischemia persisting in the infarction, unlike ischemic reductions in PCr/Pi seen in animal models of acute stroke [74]. There is now good evidence for a progression of metabolic change in the days following acute through chronic phases of stroke injury [75].

The study group of dementia patients, aged 30–68, comprised 4 with Huntington's, 2 with Parkinson's, 1 with Alzheimer's, and 2 with vascular dementias, all with symptoms including moderate or severe memory loss [17, 69]. The ^{31}P-DRESS results (Table 2) from the cortex above the temple again showed no major changes in metabolite ratios compared with normal, although patient numbers were quite limited. The absence of significant change in high-energy phosphate ratios was interpreted as suggesting that tissue oxygen supply and demand are relatively uncompromised [69]. The possibility that absolute concentrations are altered is presently being investigated on a larger patient group with more recently developed methods of measuring absolute concentrations [34, 72]. Other work showing both no changes of phosphate ratios in dementias [75], and changes sufficient to discriminate Alzheimer's disease from multiple infarct dementia [76] have been reported, a contradiction that emphasizes the need for further work.

Barany et al. [77] reported early findings from H_2O-suppressed ^1H DRESS in 5 tumor patients with meningiomas. Reductions in the total creatine pool, and in *N*-acetyl aspartate, glutamate, and glutamine were noted.

4.2 Heart

The human heart has probably been the most fertile for ^{31}P DRESS applications. After demonstrating feasibility, a succession of experiments were performed to establish (*i*) that regional myocardial ischemia was detectable noninvasively with DRESS using an animal model [19], (*ii*) whether the progression of change in high-energy phosphate metabolism in the first hours and days after coronary occlusion was such that abnormalities in metabolite ratios might be observable in patients studied a few days after onset of ischemia (bearing in mind the results from chronic stroke patients) [20], (*iii*) the normal range of metabolite ratios in DRESS spectra and whether they vary during the cardiac cycle, and (*iv*) whether metabolic abnormalities are detectable in myocardial infarction patients [21]. Again, a 6.5-cm ^{31}P transmit/receive coil, 1-cm thick slices, and a 1.5 T system were employed, but spectra were gated synchronous to the cardiac cycle at $T_R = (\text{heart-rate})^{-1}$. ^1H images provided an anatomical reference, and subjects were oriented prone, rotated left on the surface coil, to orient the anterior left ventricular (LV) wall substantially parallel to the plane of the surface coil.

In animal studies, ^{31}P-DRESS spectra were acquired as a function of depth through the LV in dogs before and after coronary occlusion of the left anterior descending coronary artery (LAD) produced by a surgically-implanted occluder [19, 20]. Spectra were monitored up to 5-days postocclusion. Infarctions were measured by weight in all animals by staining excised hearts. Postocclusion PCr/Pi and PCr/ATP ratios in jeopardized myocardium were significantly reduced from preocclusion values in dogs with infarctions as small as 5g: dogs without infarctions showed no significant metabolic change (e.g. Fig. 8). This was consistent with a loss of oxygen supply and net consumption of high energy phosphates. Acidosis was evident in the first hour postocclusion as tissue pH sunk as low as 5.9. Surprisingly, pH returned to normal 5–15 hr postocclusion, whereas the PCr/Pi ratio determined from peak integrals remained low compared to preocclusion values for the course of the study, although Pi did broaden [20]. The evolution of canine myocardial PCr/Pi and pH is plotted in Figs. 17 and 18. Some fluctuations in metabolite ratios were also noted in the first hour of occlusion. The results suggest that Pi remains substantially localized in the damaged myocardium for days after occlusion whereas either H^+ ions leak out or are somehow neutralized during the first day postocclusion [20].

In normal human heart, we measured a PCr/ATP ratio of 1.6 ± 0.4 (SD) in the anterior LV of 7 subjects aged 30–40 using DRESS [21]. Grist et al. [78] and Sakuma et al. [79] report essentially the same values, 1.6 ± 0.1 (SE) and 1.52 ± 0.15 (SE), respectively, for groups of 5 normal subjects using DRESS. Both we and they found no significant changes in PCr, ATP, and Pi ratios during the cardiac cycle.

We studied four patients (2 male, 2 female) aged 38–70, 5–9 days after anterior myocardial infarction, as evidenced by angina symptoms, elevated serum myocardial enzyme levels, and cardiac catheterization results showing 95–100% stenosis of the proximal or middle segments of the LAD and mild to moderate anterior wall hypokinesia [21]. The ^{31}P DRESS results from epicardially and endocardially displaced slices as summarized in Table 3 [21], show reduced PCr/Pi ratios relative to normal but near neutral pH (7.1–7.2)

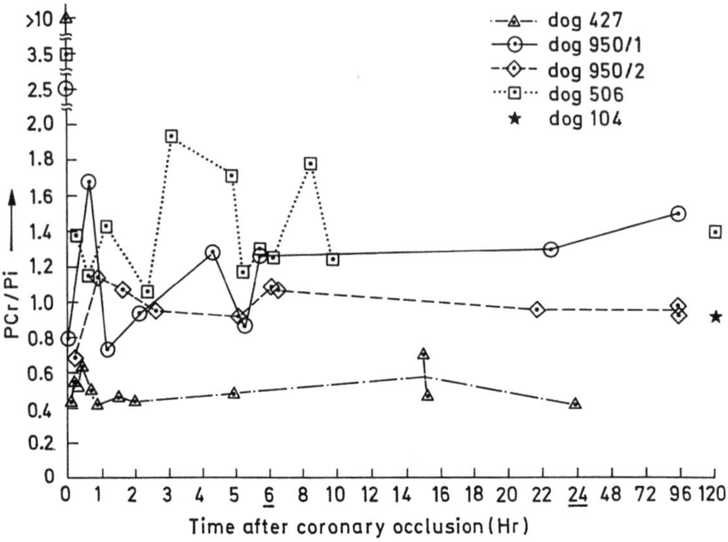

Fig. 17. Evolution of canine myocardial PCr/Pi ratio in the anterior left ventricles up to 5 days after occlusion of the left anterior descending coronary artery, as measured noninvasively by cardiac-gated ^{31}P DRESS with a 6.5-cm surface coil at 1.5 T [20]. Values measured immediately preocclusion are indicated on the vertical axis at 0 hr: note the change in division intervals at 6, and 24 hr postocclusion and for at PCr/Pi > 2. Preocclusion values may underestimate true PCr/Pi values due to contamination by phosphomonoester signals in the vicinity of Pi in the spectra

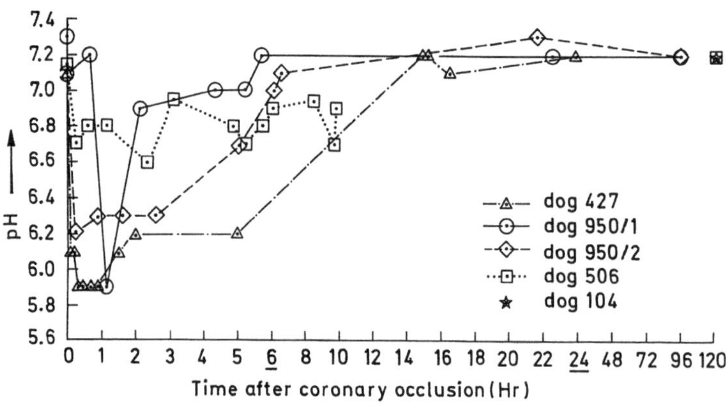

Fig. 18. Evolution of myocardial intracellular pH measured from the chemical shift of Pi in the anterior left ventricular ^{31}P DRESS spectra of the animals in Fig. 17 (error, ± 0.2 pH units), before (at 0 hr; 7.15 ± 0.2 pH units) and up to 5 days after coronary occlusion [20]

Table 3. Myocardial metabolite ratios in ^{31}P NMR DRESS spectra from normal and myocardial infarction patients[a]

Subjects	Time[b]	PCr/β-ATP	PCr/Pi	Pi/β-ATP	PCr/PM	PCr/PD	pH
Normal	—	1.6±0.4	8.8±4.8 >4.0	.16±0.14 (<0.5)	6.5±5 (>1.3)	3.4±2.3 (>1.3)	7.15±0.1
Patient 1	5d						
epi.		1.2±0.2	>20[c]	<.05[c]	4.1±0.4	3.9±0.5	—[c]
endo.		1.5±0.2	1.1±0.1*	1.3±0.3*	3.0±0.6	>20[c]	7.2±0.2
Patient 2	6d						
epi.		1.9±0.1	>20[c]	<.08[c]	2.2±0.2	3.0±0.3	—[c]
endo.		1.5±0.2	1.6±0.2*	1.1±0.3*	2.1±0.3	2.8±0.2	7.2±0.2
Patient 3	5d						
epi.		2.1±0.3	1.0±0.2*	1.5±0.5*	6±1.5	5±2	7.2±0.2
endo.		2.1±0.3	1.1±0.2*	2.0±0.4*	7.6±1	4.2±0.2	7.1±0.2
Patient 4	9d						
epi.		2.2±0.3	1.1±0.1†	2.1±0.3†	—†	3.2±0.3	~7.2
endo.		1.3±0.2	0.86±0.05†	1.3±0.3†	—†	2.3±0.3	

[a] β-ATP, β phosphate of ATP; PM, phosphomonoesters including blood 2,3 diphosphoglycerate; PD, phosphodiesters; epic., from epicardially displaced DRESS spectra; endoc., from DRESS spectra displaced 0.5 cm endocardially. Errors: mean ± SD of 20 spectra from 7 normal hearts, or of 2 to 5 determinations in patients. Inequalities apply to all determinations/spectra (distributions were skew in normal population).
[b] Time between onset of symptoms in patients and commencement of NMR study.
[c] Pi or PD too small to measure.
*Significantly different from normal myocardium ($p < 0.01$, using Student's t test on the patient data considered as a group).
†Includes contributions from PM. PCr/Pi and Pi/β-ATP ratios are significantly different from normal PCr/(Pi + PM) and (Pi + PM)/β-ATP ratios respectively ($p < 0.05$, using Student's t test)

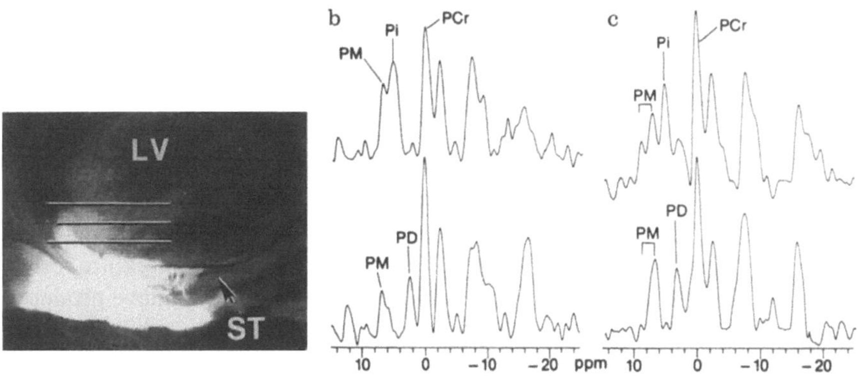

Fig. 19 a. Transaxial surface coil image of the anterior myocardium near the apex of a patient with anterior myocardial infarction 5 days after onset, annotated with the location of endocardial (*top*) and epicardial (*lower*) slices at 0.5-cm intervals studied by cardiac-gated ^{31}P DRESS [21]. (**b**) The DRESS spectra from this patient, acquired in 6 min, show elevated Pi endocardially (*top spectrum*) but not epicardially. (**c**) Endocardial (*top*) and epicardial (*lower*) ^{31}P DRESS spectra acquired in 5–7.5 min from another patient 6 days after myocardial infarction, showing elevated endocardial Pi. T_R values were 0.88 ± 0.13 throughout. A 6.5-cm excite/detect ^{31}P surface coil was used and a normal PCr/ATP ratio consistent with the canine study a day or more postocclusion [20]. Representative spectra are shown in Fig. 19.

Sakuma et al. [79] also studied 13 patients with hypertrophic cardiomyopathy using ^{31}P DRESS, finding significant reductions in myocardial PCr/ATP ratios from 1.52 $\pm$ 0.15 (SE) to 0.97 $\pm$ 0.12 (SE), with greater reductions evident in patients with the lowest ejection fractions. This follows a prior case report by Rajagopalan et al. showing reduced PCr/ATP in a patient with the disease [80].

5 Conclusions

In this chapter I have attempted to give as thorough an account of DRESS as exists: insights into how it came about, how well it localizes, how it is optimized and adjusted, improvements and extensions, and an overview of its findings to date. Of its problems, they would be all too easy to dismiss simply by choosing some alternative spectroscopy localization technique. This would be a mistake. Many of the alternatives utilize surface coils for detection, excitation and for providing localization in the dimensions coplanar with the surface coil [35, 36]. They are therefore subject to the same fuzzy localization problem faced by DRESS. The T_D delay problem occurs in differing degrees in virtually all contemporary clinically-used localization methods save ISIS [61], and chemical shift artefact is a fact of all methods employing selective excitation, including ISIS [1, 59–63]. Moreover, the simplicity of DRESS renders it easily adaptable to many kinds of NMR experiments, and insensitive to motion artefacts since only the region of interest is excited at any given time. The slice it excites is relatively clean and sample-independent compared, for example, to phase-encoding approaches [32]. While we have of late revisited the phase-encoding methods for their advantage of simultaneous multiple volume acquisition [24, 31, 32, 81], which is crucial, for example, for cardiac stress-testing when the precise location of compromised myocardium is unknown [82], such problems might well warrant a review of SLIT DRESS.

Ultimately, the value of all in vivo spectroscopy localization techniques is judged on the strength of their applications. Already a cursory literature review will show that there are many more localization methods than have produced any in vivo data, or will ever produce clinical data. Yet, unlike its sister technology, NMR imaging, the field of in vivo human spectroscopy is still immature in that it has produced no strictly proven clinical applications as of this writing, while its scientific findings are still being hotly pursued. Accordingly, at this early juncture it behooves us not to dismiss any localization method prematurely, and to keep open the capability of choosing a technique based on its suitability to a particular application [83]. After all, without applications, scientific or clinical, localized in vivo spectroscopy could be for nought.

6 References

1. Bottomley PA, Foster TH, Darrow RD (1984) J Magn Reson 59: 338
2. Hoult DI, Busby SJW, Gadian DG, Radda GK, Richards RE, Seeley PJ (1974) Nature 252: 285
3. Jacobus WE, Taylor GJ, Hollis DP, Nunnally RL (1977) Nature 265: 756
4. Damadian R (1971) Science 171: 1151
5. Bottomley PA (1978) PhD Thesis, University of Nottingham, Nottingham UK p 98; Holland GN, Bottomley PA, Hinshaw WS (1977) J Magn Reson 28: 133
6. Bottomley PA, Hart HR, Edelstein WA, Schenck JF, Smith LS, Leue WM, Mueller OM, Redington RW (1984) Radiol 150: 441
7. Bottomley PA (1981) J Phys E: Sci Instrum 14: 1081–1087; Bottomley PA (1982) J Magn Reson 50: 335
8. Ackerman JJH, Grove TH, Wong GG, Gadian DG, Radda GK (1980) Nature 283: 167
9. Nunnally RL, Bottomley PA (1981) Science 211: 177
10. Bottomley PA (1984) US Patent 4,480,228 (filed 1982)
11. Bottomley PA, Edelstein WA (1985) US Patent 4,506,223 (filed 1982)
12. Bottomley PA, Edelstein WA (1984) US Patent 4,484,138 (filed 1982)
13. Bottomley PA, Hart HR, Edelstein WA, Schenck JF, Smith LS, Leue WM, Mueller OM, Redington RW (1983) Lancet ii: 273
14. Brown TR, Kincaid BM, Ugurbil K (1982) Proc Natl Acad Sci USA 79: 3523
15. Bottomley PA, Smith LS, Edelstein WA, Hart HR, Mueller OM, Leue WM, Darrow R, Redington RW (1983) in: Book of Abstracts Society of Magnetic Resonance in Medicine 2nd Annual Meeting August, 16–19, 1983, San Francisco CA, p 53; available from Soc Magn Reson Med, Berkeley, USA
16. Bottomley PA, Drayer BP, Smith LS (1986) Radiol 160: 763
17. Smith LS, Bottomley PA, Drayer BP, Herfkens RJ (1986) in: Book of Abstracts Society of Magnetic Resonance in Medicine 5th Annual Meeting August 19–22, 1986, Montreal Canada, Vol. 4, p 1386; available from Soc Magn Reson Med, Berkeley, USA
18. Bottomley PA (1985) Science 229: 769
19. Bottomley PA, Herfkens RJ, Smith LS, Brazzamano S, Blinder R, Hedlund LW, Swain JL, Redington RW (1985) Proc Natl Acad Sci USA 82: 8747
20. Bottomley PA, Smith LS, Brazzamano S, Hedlund LW, Redington RW, Herfkens RJ (1987) Magn Reson Med 5: 129
21. Bottomley PA, Herfkens RJ, Smith LS, Bashore TM (1987) Radiol 165: 703
22. Bottomley PA, Smith LS, Leue WM, Charles C (1985) J Magn Reson 64: 347
23. Bottomley PA, Edelstein WA, Foster TH, Adams WA (1985) Proc Natl Acad Sci USA 82: 2148
24. Bottomley PA, Hardy CJ, Roemer PB, Mueller OM (1989) Magn Reson Med 12: 348
25. Bottomley PA, Hardy CJ (1987) J Magn Reson 74: 550
26. Bottomley PA (1989) Radiol 170: 1
27. Evelhoch JL, Crowley MG, Ackerman JJH (1984) J Magn Reson 56: 110
28. Haase A, Hanicke W, Frahm J (1984) J Magn Reson 56: 401
29. Schenck JF, Hart HR, Foster TH, Edelstein WA, Hussain MA (1986) in: Kressel HY (ed) Magnetic Resonance Annual. Raven, New York, p 123
30. Sobol WT (1986) Rev Magn Reson Med 1: 181
31. Bottomley PA, Hardy CJ, Roemer PB (1990) Magn Reson Med 14: 425
32. Bottomley PA, Hardy CJ, Roemer PB, Weiss RG (1989) NMR in Biomed 2: 284
33. Garwood M, Schleich T, Bendall MR (1987) J Magn Reson 73: 191
34. Bottomley PA, Hardy CJ (1989) Clin Chem 35: 392
35. Blackledge MJ, Rajagopalan B, Oberhaensli RD, Bolas NM, Styles P, Radda GK (1987) Proc Natl Acad Sci USA 84: 4283
36. Bailes DR, Bryant DJ, Bydder GM, Case HA, Collins AG, Cox IJ, Evans PR, Harman RR, Hall AS, Khenia S, McArthur P, Oliver A, Rose MR, Ross BD, Young IR (1987) J Magn Reson 74: 158
37. Hoult DI (1983) J Magn Reson 43: 69
38. Silver MS, Joseph RI, Hoult DI (1984) J Magn Reson 59: 347
39. Sutherland RJ, Hutchison JMS (1978) J Phys E: Sci Instrum 11: 79
40. Hutchison JMS, Sutherland RJ, Mallard JR (1978) J Phys E: Sci Instrum 11: 217

41. Edelstein WA, Bottomley PA, Hart HR, Smith LS (1983) J Comp Assist Tomogr 7: 391
42. Bottomley PA (1987) US Patent 4,647,858 (filed 1985)
43. Roemer PB, Hickey JS (1988) US Patent 4,737,716 (filed 1986)
44. Fukushima E, Roeder SBW (1981) Expermental Pulse NMR a Nuts and Bolts Approach. Addison-Wesley, Reading (Mass, USA), p 92
45. Allman T, Holland GA, Lenkinski RE, Charles HC (1988) Magn Reson Med 7: 88
46. Barkjuijsen H, De Beer R, Bovee WMMJ, van Ormondt D (1985) J Magn Reson 61: 465
47. Spielman D, Webb P, Macovski A (1988), J Magn Reson 79: 66
48. Laue ED, Skilling J, Staunton J (1985) J Magn Reson 63: 418
49. Nelson SJ, Brown TR (1987) J Magn Reson 75: 229
50. Bottomley PA (1987) Ann NY Acad Sci 508: 333
51. Moonen CTW, von Kienlin M, van Zijl PCM, Cohen J, Gillen J, Daly P, Wolf G (1989) NMR in Biomed 2: 201
52. Bottomley PA, Hardy CJ (1987) J Appl Phys 62: 4284
53. Bottomley PA, Hardy CJ, O'Donnell M, Roemer PB (1987) US Patent 4,812,760 (filed 1987)
54. Pauly J, Nishimura D, Macovski A (1989) J Magn Reson 81: 43
55. Pauly J, Nishimura D, Macovski A (1989) J Magn Reson 82: 571
56. Hardy CJ, Bottomley PA (1991) Magn Reson Med 17: 315
57. Hardy CJ, Bottomley PA, Roemer PB (1988) J Appl Phys 63: 4741
58. Bottomley PA (1988) US Patent 4,733,185 (filed 1987)
59. Muller S, Aue WP, Seelig J (1985) J Magn Reson 65: 332
60. Luyten PR, Marien AJH, den Hollander JA (1986) J Magn Reson 67: 148
61. Ordidge RJ, Connelly A, Lohman JAB (1986) J Magn Reson 66: 283
62. Doddrell DM, Brooks WM, Bulsing JM, Field J, Irving MG, Baddeley (1986) J Magn Reson 68: 367
63. Doddrell DM, Field J, Brereton IM, Galloway GJ, Brooks WM, Irving MG (1987) J Magn Reson 73: 159
64. Edelstein WA, Bottomley PA, Hart HR, Smith LS (1983) J Comp Assist Tomogr 7: 391
65. Bottomley PA (1986) US Patent 4,629,988 (filed 1984)
66. Bottomley PA, Foster TH, Leue WM (1984) i: 1120
67. Matthaei D, Frahm J, Haase A, Schuster R, Bomsdorf H (1985) Lancet i: 370
68. Hsieh PS, Balaban RS (1987) J Magn Reson 74: 574
69. Bottomley PA, Hardy CJ, Smith LS, Drayer BP, Cousins JP (1991) Bull Clin Neurosci 55: 29
70. Hope PL, Costello AM de L, Cady EB, Delpy DT, Tofts PS, Chu A, Hamilton PA, Reynolds EOR, Wilkie DR (1984) Lancet ii: 366
71. Boesch C, Gruetter R, Martin E, Duc G, Wuthrich (1989) Radiol 172: 197
72. Bottomley PA, Hardy CJ, Cousins JP, Armstrong M, Wagle WA (1990) Radiol 176: 407
73. Edwards RHT, Dawson MJ, Wilkie DR, Gordon RE, Shaw D (1982) Lancet i: 725
74. Bottomley PA, Kogure K, Namon R, Alonso OF (1982) Magn Reson Imag 1: 81
75. Welch KMA, Gross B, Licht J, Levine SR, Glasberg M, Smith MB, Helpern JA, Bueri J, Gorell JM (1988) Curr Neurol 8: 295
76. Brown GG, Levine SR, Gorell JM, Pettegrew JW, Gdowski JW, Bueri JA, Helpern JA, Welch KMA (1989) Neurology 39: 1423
77. Barany M, Langer BG, Glick RP, Venkatasubramanian PN, Wilbur AC, Spigos DG (1988) Radiol 167: 839
78. Grist TM, Kneeland B, Rilling WR, Jesmanowicz, Froncisz W, Hyde JS (1989) Radiol 170: 357
79. Sakuma H, Takeda K, Yamakado K, Kinosada Y, Nakagawa T, Okamoto S, Konishi T, Nakano T, Okamoto Y, Nagasawa K (1990) in: Book of Abstracts, Society of Magnetic Resonance in Medicine 9th Annual Meeting August 18–24, 1990, New York, USA, Vol 1, p 248; available from Soc Magn Reson Med, Berkeley, USA
80. Rajagopalan B, Blackledge MJ, McKenna WJ, Bolas N, Radda GK (1987) Annal NY Acad Sci 508: 321
81. Bottomley PA, Charles HC, Roemer PB, Flamig D, Engeseth H, Edelstein WA, Mueller OM (1988) Magn Reson Med 7: 319
82. Weiss RG, Bottomley PA, Hardy CJ, Gerstenblith G (1990) N Engl J Med 323: 1593
83. Bottomley PA, Hardy CJ (1990) Phil Trans R Soc Lond A 333: 531

Image Guided Volume Selective Spectroscopy: A Comparison of Techniques for In-Vivo ^{31}P NMR Spectroscopy of Human Brain

R. J. Ordidge and J. A. Helpern

NIH Center for Cerebrovascular Disease Research, Department of Neurology, Henry Ford Hospital, Detroit, MI, and Department of Physics, Oakland University, Rochester, MI

Table of Contents

This chapter describes methods which have been routinely applied over the last decade to obtain spatially localized ^{31}P spectra of human brain. The Topical Magnetic Resonance (TMR) method found widespread use in some of the early in-vivo studies of human brain, and several examples are presented. The Image Selected In-vivo Spectroscopy (ISIS) method has been developed more recently, and is fully described together with advantages, disadvantages and modifications of this technique.

Finally, the ISIS method is compared with Chemical Shift Imaging (CSI) on the basis of signal efficiency, and contamination with erroneous signal from tissue regions outside the selected volume.

NMR Basic Principles and Progress, Vol. 26

1 Introduction

The brain is one of the most difficult organs of the human body to study. Most investigations of cerebral energy metabolism to date have relied heavily on the use of animal models. NMR has provided the first non-invasive means for studying brain metabolism. The application of in-vivo NMR spectroscopy to the study of human brain is, by far, the most exciting, however, this technology is limited by the specific requirements of magnet bore size, field strength, and a suitable method of signal source localization. It is only within the past few years that a small number of research groups around the world have been able to implement routine human brain NMR spectroscopy [1–17]. To date, most NMR spectroscopic studies have focused on the ^{31}P nucleus and its relevance to various pathologies.

The first spatially localized in-vivo NMR spectra were obtained using surface coils [18]. Surface coils were subsequently used in combination with Topical Magnetic Resonance (TMR) [19, 20] which exploits permanent non-uniformities of the main magnetic field (B_0). Since the development of these earlier methods of spatial localization, the incorporation of switchable linear magnetic field gradients into the design of NMR spectrometers has enabled MR imaging principles to be utilized to achieve more accurate definition of spatially localized regions of interest for in-vivo NMR. The Image Selected In-vivo Spectroscopy (ISIS) technique [21] is one example of many methods which use these imaging principles.

This chapter will first present a brief description of the use of surface coils with TMR and some examples of results obtained from human brain will be presented. The ISIS technique will then be presented and compared with Chemical Shift Imaging (CSI) [22–24] to illustrate the benefits and disadvantages of each approach.

2 Spatial Localization Methods

2.1 *Surface Coils*

Surface coils were originally used by Morse and Singer [25] to measure ^{1}H signals from moving blood in human limbs. Subsequently, Ackerman et al. [18] used surface coils to measure high-resolution ^{31}P NMR spectra from rat leg muscle. The surface coil is basically a circular coil of wire which is placed proximal to the region of interest or the part of the object to be studied, and is positioned in such an orientation that a major component of the B_1 field generated by the coil is orthogonal to the B_0 field. One disadvantage of surface coils is that any part of the sample which lies near the coil will contribute to the

NMR signal. This does not present much of a problem if the tissue of interest is located at the surface of the subject. However, if the region of interest is below the surface of the subject, then it is difficult to asses which part of the NMR signal came from the region of interest and which came from extraneous tissue. For this reason, although surface coils are still used in in-vivo spectroscopy, their applications are usually found in combination with other spatial localization techniques.

2.2 *Topical Magnetic Resonance (TMR) with Surface Coils and ^1H Image Viewfinding*

One of the first techniques developed to tackle the problem of isolation of the signal source was Topical Magnetic Resonance (TMR) [19, 20]. In TMR, non-linear magnetic field gradients are used to generate a spherical region of homogeneous magnetic field (commonly called the *profiled field*) surrounded by a magnetic field which is extremely inhomogeneous. Since high-resolution NMR signals can only be obtained from regions of magnetic fields that are highly homogeneous, the tissue lying within the sphere generates high-resolution NMR signals. The tissue lying outside the sphere (in the inhomogeneous magnetic field) generate excessively broadened NMR signals, which may be removed by subsequent data processing.

The center of the profiled field is fixed at the center of the bore of the magnet. By adjusting the current in room-temperature high order shim coils (mostly Z^4), the size of the profiled field can be adjusted. In the system employed in our laboratory, the profiled field can be adjusted in 1 cm increments from a minimum of a 4 cm diameter spherical volume (dsv) to a 10 cm dsv. The subject

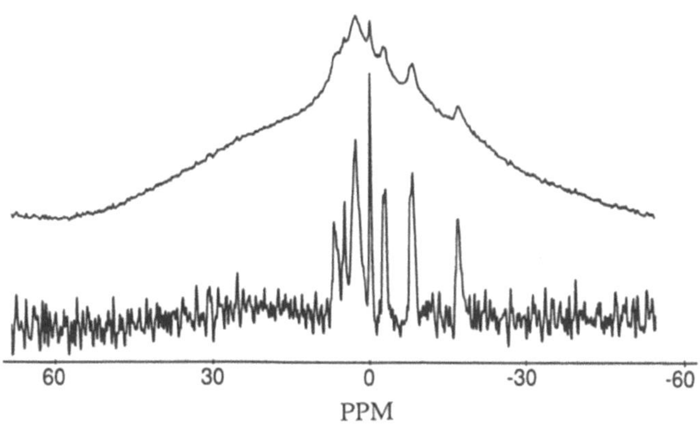

Fig. 1. Human brain ^{31}P NMR spectrum obtained using TMR in a 1.9 T 60 cm bore magnet (256 averages, TR = 1.512 s, 5 cm dsv). The top spectrum is before deconvolution to remove the broad spectral components, and the bottom spectrum is after deconvolution

to be studied is positioned within the bore of the magnet in such a way that the tissue region of spectroscopic interest is concentric with the profiled field. Once located, a surface coil can be used to transmit and receive the RF-signals to and from the region of interest. This technique results in a composite of a high-resolution spectrum superimposed on top of an inhomogeneously broadened spectrum as shown in Fig. 1. The broad "hump", which results mainly from signal originating at the edges of the profiled field, can be removed using deconvolution computer programs. The resonance peaks of the resulting high-resolution spectra can then be analyzed. These peaks, from left to right, are attributed to phosphomonesters (PME), inorganic phosphate (Pi), phosphodiesters (PDE), phosphocreatine (PCr), and the γ-, α- and β-phosphates of adenosine triphosphate (ATP).

It is important to verify the locus of the signal source in in-vivo NMR. For example, contamination of signal from pure brain can come from the overlying scalp muscle which is rich in high-energy phosphates thereby giving a disproportionately intense contribution to the ^{31}P spectrum. Shown in Fig. 2 are two in-vivo ^{31}P spectra obtained from the same individual. The spectrum on the bottom was collected to purposely include extraneous tissue signal (e.g. temporal muscles) and is obviously different from the spectrum on the top which was obtained from pure brain. The pure brain spectrum has lower signal-to-noise as result of the fact that a smaller volume of tissue was selected for this spectrum and because this volume was also located further from the surface coil. Additionally, the concentration of ^{31}P metabolites in the brain is less than that found in muscle.

^{1}H NMR imaging can be applied prior to TMR localized spectroscopy to verify the proper location of the profiled field within the region of interest. The ^{1}H images obtained with a TMR magnet are slightly distorted due to the inherent field inhomogeneities, however, they are quite adequate for localization purposes. Figure 3 shows seven ^{1}H brain images of a normal human volunteer

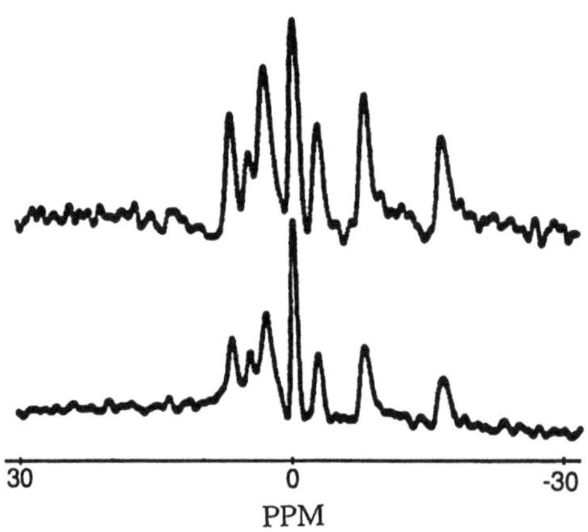

Fig. 2. ^{31}P NMR spectra from the human head obtained using TMR in a 1.9 T 60 cm bore magnet (256 averages, TR = 1.512 s, 5 cm dsv). The top spectrum was obtained from pure brain. The bottom spectrum was also obtained from brain, but intentially included signal from extraneous tissue (e.g. temporal muscles)

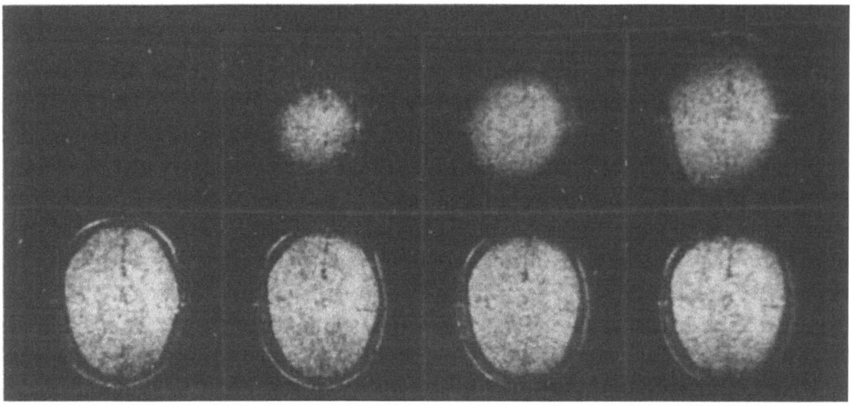

Fig. 3. ^1H human brain images obtained using TMR in a 1.9 T magnet with a progressively increasing dsv (dsv increases in the sequence 4, 5, 6, 7, 8, 9, and 10 cm starting from top left and progressing clockwise to bottom left). Slice thickness was 1 cm, TR = 1.5 s, and image matrix size was 128 × 128 pixels

obtained from progressively decreasing TMR volumes. These images illustrate the approximate extent of the localized volume obtained using TMR.

There are two main disadvantages of the TMR technique. First, the localized TMR volume has a spatial profile which is approximately "bell-shaped" along the diagonal through the center of the localized volume [20]. The edges of the TMR volume are thus poorly defined leading to poor spatial definition of the tissue which contributes to the observed ^{31}P spectrum. The second disadvantage of this method is that for each area of spectroscopic interest, the patient must be realigned within the magnet so that the profiled field of the TMR system is concentric with the desired region of study. Each time the patient is moved, the magnet must be re-shimmed (due to changes in magnetic susceptibility) and the RF coils may need to be re-tuned, thus extending the duration of the experiment.

The complete protocol for TMR with ^1H image viewfinding is as follows. The subjects head is placed inside a ^1H imaging head coil which incorporates a large surface coil used for ^{31}P spectroscopy. This surface coil is placed on the scalp over the region of spectroscopic interest. A thin tube of water (doped with $CuSO_4$) around the outside of the surface coil provides markers in the ^1H image to verify the positioning of the surface coil. The subject is then placed inside the magnet, and the first image is obtained. With accurate knowledge of the strengths of the linear magnetic field gradients used for imaging (i.e. gauss/cm), distances (in centimeters) can be calculated from the image. After this first image is obtained, and the necessary relocation adjustments have been calculated, the patient is repositioned within the bore of the magnet using a movable stretcher. The subject is thus realigned inside the bore of the magnet until the region of spectroscopic interest coincides with the center of the profiled field. Once the subject is appropriately positioned, the surface coil is used to obtain ^{31}P spectra.

To date, over 500 patient studies have been performed at the NMR facility of the Department of Neurology at Henry Ford Hospital using TMR. Typically, the spectra are obtained with a spectral width of 4 kHz, and 4k data points, resulting in a signal acquisition time of 0.512 sec. A recycle delay of 1.0 s is used, resulting in a total interpulse-interval of 1.512 s. The volume of tissue from which the spectra are obtained is approximately 33 cm^3 (with a profiled field of a 4 cm dsv) and this requires the summation of at least 256 transients for adequate signal-to-noise (S/N). Examples of human brain ^{31}P spectra, obtained from stroke patients, are shown in Fig. 4. The diversity of the metabolic profiles shown here demonstrate the variability of human cerebral metabolic response to stroke, and are good examples of the utility of the TMR technique.

One disadvantage of TMR is that only a single localized volume can be investigated in each experiment. Successful application of the TMR technique, or indeed any single volume localization method, implies that a lesion must first be visualized on an ^1H MR image before a volume of interest can be chosen.

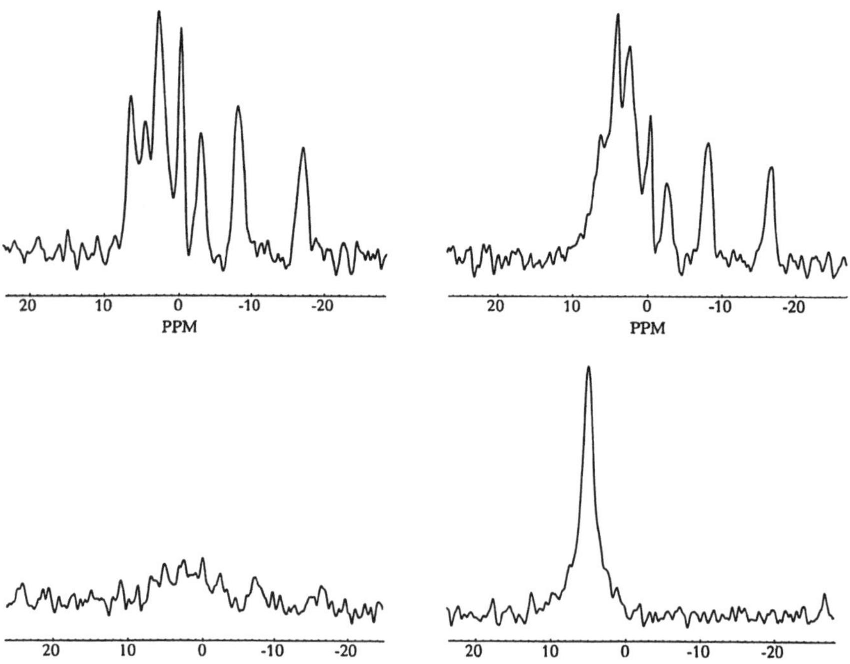

Fig. 4. Four examples of human ^{31}P brain spectra from stroke patients obtained using TMR in a 1.9 T magnet (512 averages, TR = 1.512 s, 4 cm dsv). The spectra range from a normal appearance, (*top left*), to the complete absence of ^{31}P metabolites (*bottom left*). ^{31}P spectra obtained from stroke patients have demonstrated variable levels of inorganic phosphate as exemplified by the *top* and *bottom* spectra on the *right*

Unfortunately, however, in many pathological conditions of the brain, changes in ^{31}P metabolism precede any observable abnormalities in ^{1}H MR images. Another disadvantage of the TMR technique is that it cannot be used for the spatial localization of ^{1}H NMR spectra of metabolites since field profiling prevents removal of the dominant water resonance by use of standard NMR spectral editing techniques.

Despite these disadvantages the TMR technique provides a simple method of producing in-vivo spectra from variable-sized volumes of tissue with the high signal sensitivity associated with the use of surface coils. Since the NMR signal is acquired immediately following a 90° RF pulse, the in-vivo ^{31}P spectrum does not contain the T_2-distortion which is often intoduced during application of several alternative localization procedures.

2.3 Image Selected In-Vivo Spectroscopy (ISIS)

The ISIS technique [21] uses MR imaging principles in combination with the differencing of NMR signals from a sequence of eight experiments to select a cubic volume of tissue for spectroscopic analysis. Upon combination of the free induction decays (FIDs) from the eight ISIS experiments, signal from the desired cube adds, whereas signals from all external regions cancel. The method relies on the principle of selective inversion of the spin population prior to data acquisition. The selective inversion pulses (180° pulses) cause inversion of longitudinal magnetization in selected slices of the sample using amplitude modulated RF excitation pulses in combination with transient field gradients. The selective pulses are applied in sequence along the X, Y and Z axes, and the selected cube is formed at the intersection of the selected planes which are orthogonal to X, Y and Z axes, respectively. The experimental sequence of selective RF pulses for the eight experiments is shown in Table 1, where ON and OFF refer to the presence and absence, respectively, of a selective inversion pulse along the indicated axis. Also shown is the sign (±) used for combination

Table 1. Experimental sequence of selective RF pulses applied during the preparation period

Experiment number	Selective pulse X	Y	Z	Contribution to total spectrum
1	OFF	OFF	OFF	+
2	ON	OFF	OFF	−
3	OFF	ON	OFF	−
4	ON	ON	OFF	+
5	OFF	OFF	ON	−
6	ON	OFF	ON	+
7	OFF	ON	ON	+
8	ON	ON	ON	−

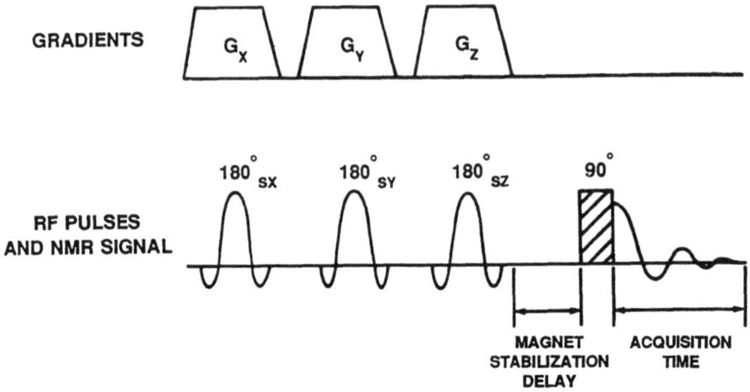

Fig. 5. RF and gradient pulse sequence for experiment 8 of the ISIS sequence (all selective pulses applied)

of the FIDs measured by subsequent 90° RF (non-selective) readout pulses in each of the eight experiments.

The RF and gradient pulse sequence for experiment 8 (all 3 pulses applied) is shown in Fig. 5. The experiment is repeated after a time delay which is sufficient to allow for spin-lattice relaxation of the spin system (typically $5 * T_1$). Signal averaging with a rapid experimental repetition rate may be performed by rapid repetition of each experiment in the sequence in turn. A time delay of $5 * T_1$, however, must be incorporated when the experimental sequence number is changed. This avoids errors in the spatial definition of the localized volume.

The size of the cube can be varied along any dimension by altering the excitation profile width of the respective RF slice selection pulse or the gradient in which this RF pulse is applied. The position of the cube can be altered by adjustment of the individual frequency offset of each of the RF selective pulses.

The usual experimental set-up procedure involves shimming of the main magnetic field followed by the acquisition of standard MR images (coronal, transverse or saggital). The location of the center of the ISIS cube is then chosen from the images, and the X, Y and Z coordinates are translated into frequency offsets which are applied during each respective RF selection pulse. The cube size and position may thus be selected by reference to standard ^1H MR images of the subject. For ^{31}P spectroscopy, the difference in magnetogyric ratios between the ^1H and ^{31}P nuclei must be taken into account when calculating the relevant selective excitation frequency offsets.

The ISIS technique can be combined with surface coils to obtain improved signal sensitivity from regions close to the surface of the subject. Cancellation of signal outside the selected cube is accurate irrespective of the spin nutation angle experienced during each selective preparation pulse. However, selective 180° RF pulses are desirable since they provide the optimum sensitivity from spins within the ISIS cube. Therefore, it is also preferable to use adiabatic selective RF pulses such as the hyperbolic-secant selective inversion pulse proposed by Silver et al. [27]. This ensures that material within the ISIS cube contributes fully in each of the eight experiments, and the optimum signal/noise (S/N) performance is achieved virtually independent of RF field inhomogeneity and RF power level applied during each selective pulse. The experiment can be made completely independent of RF power setting by replacing the 90° readout pulse with a non-selective adiabatic 90° RF pulse. Adiabatic RF pulses are commonly used in the ISIS sequence, even when the RF coil is nominally homogeneous, because they remove the need to accurately calibrate RF nutation angles.

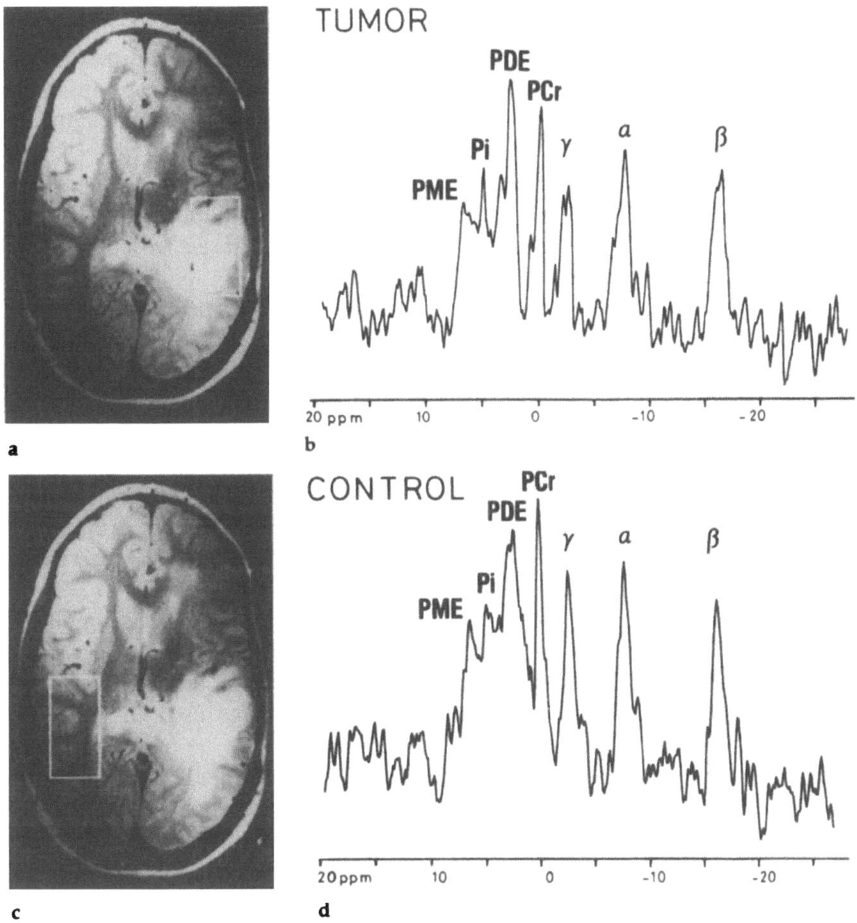

Fig. 6 a–d. MR (TR = 2000 ms, TE = 60 ms) images of a moderately anaplastic astrocytoma (**a**) and the control region (**c**) in the same patient, showing ISIS volume of interest. ^{31}P NMR spectra of tumor (**b**) and control tissue (**d**) show that, except for Pi, the metabolite levels in the tumor are lower (80 ml, 640 scans, TR = 2000 ms). Reproduced with kind permission of Hubesch et al., Ref. 27

The disadvantage of using surface coils is the dramatic decrease in coil sensitivity with increased distance from coil center. This limits the scope for movement of the ISIS cube to new locations, and is a particular disadvantage in human and animal studies where the NMR spectrum of an unaffected region of the organ is also of interest but may be located outside the volume of surface coil sensitivity. An alternative approach is to use a larger homogeneous RF coil which encompasses all areas of interest. For example, a whole-head coil can be used for brain studies so that sequential measurement of equivalent regions from both sides of the brain can easily be made. The concomitant decrease in S/N performance is sometimes a necessary compromise to minimize the experimental duration when studying critically ill patients.

Figure 6 shows MR images of a patient with a moderately anaplastic astrocytoma together with the corresponding ^{31}P ISIS spectra from regions indicated by boxes in the respective images (reproduced with kind permission of Hubesch et al., Ref. 27). The authors report that with the exception of the Pi concentration, ^{31}P metabolite levels are lower in the tumor when compared with control tissue.

The ISIS and TMR techniques are relatively insensitive to T_2-distortion, since a non-selective 90° RF pulse is used for signal excitation prior to data sampling. Both techniques can thus be used for in-vivo ^{31}P spectroscopy, where several metabolites have short T_2 values.

A magnet stabilization delay is included in the ISIS sequence (immediately following the selective inversion pulses) to allow time for eddy currents induced by gradient-switching, to decay. This delay minimizes an associated distortion of the localized spectrum. The use of actively-shielded gradient coils significantly reduces the magnitude of induced eddy currents, however, spectral distortion is still noticable in the absence of a magnet stabilization delay.

The spatial definition of the volume selected using ISIS is superior to that obtained using either TMR or CSI. This will be discussed in more detail in the next section, however, this feature is the main reason for the widespread use of ISIS for in-vivo ^{31}P spectroscopy.

Disadvantages of ISIS are readily apparent but can usually be minimized by appropriate modifications of the technique. Since ISIS is a single volume spatial localization technique, important information may be missed if abnormal tissue cannot be readily identified in the ^{1}H MR image used to guide the localization procedure. However, the ISIS approach may be extended to the acquisition of multiple volumes using the principles of the Hadamard transform [28, 29]. Schemes which have been proposed to achieve this goal have required a dramatic increase in the number of selective spin-inversion pulses per experiment. A more troublesome problem is the imperfect cancellation of signal from undesired sample volumes which receive a variable number of spin inversions and yet are still required to perfectly cancel in the final combination procedure. This condition is particularly difficult to satisfy at the edges of the selected slice profiles which are used to define the multiple cubic elements along each direction.

A second major concern with ISIS is the accurate cancellation of large NMR signals from extensive sample volumes outside the selected cube, particularly when the cube is small compared with the overall dimensions of the sample. Selection to a cube of one tenth the sample size along each direction, with a maximum acceptable contamination in the final spectrum of 10% from the surrounding sample volume, necessitates cancellation of signals outside the cube to a factor of better than 1 in 10000. In practice, the ISIS technique can achieve a cancellation factor of approximately 1 in 1000 with residual errors being caused by spectrometer instability and sample motion. An improved suppression factor may be achieved by presaturation of the unwanted NMR signal from the large volume of surrounding tissue prior to each ISIS experiment. In this manner, the suppression factor may be increased a further 100-fold or better. Two schemes have been specifically proposed for this purpose. Selective noise RF pulses provide an RF power-efficient method of scrambling the longitudinal NMR spin magnetization from unwanted regions of the sample prior to the application of the ISIS experiment [30, 31]. Matson et al. [32] achieve a similar effect using a composite pulse sequence [33] in combination with ISIS.

A disadvantage shared by all localization schemes which employ selective excitation is the presence of a chemical shift related spatial localization error. When nuclei with a broad range of chemical shifts are examined, the associated differences in NMR resonant frequency cause an erroneous and variable displacement of the selected region for each chemical shift value. The spatial displacement error, ΔX is given by the equation:

$$\Delta X = \frac{\sigma B_0}{G_x}$$

where σ is the chemical shift value relative to a suitable reference peak, B_0 is the main magnetic field strength, and G_x is the gradient strength applied during the spatial localization procedure.

This problem may be minimized by using large magentic field gradients for spatial localization. At a field strength of $2\,T$, a localized ^{31}P spectrum (σ range of $\pm$ 15 ppm) obtained with a field gradient of 1 gauss/cm has an associated spatial localization error of $\pm$ 3 mm irrespective of the size of the selected cube.

Incoherent sample motion may cause contamination of the ISIS spectrum with signal from tissue volumes which surround the selected region. The erroneous signal arises through incomplete cancellation of signal from sample which either enters or leaves one of the selected inversion planes during the course of the experiment. This effect does not usually cause problems in human brain examinations.

The ISIS spectrum is slightly T_1-weighted because of spin-lattice relaxation which occurs during the time interval between the selective inversion preparation period and the 90° RF readout pulse. This is only a noticeable effect when the magnet stabilization delay is substantial compared to the T_1 values of the nuclei under investigation.

3 The Relative Sensitivity and Contamination of Spatially Localized Volumes Obtained Using ISIS and CSI Techniques

Contamination is defined as the percentage of signal that appears in the final spectrum which is derived from regions of the sample outside of the spatially selected volume. A contamination of 100% implies equal contributions to the localized spectrum from sample inside and outside of the desired volume. This definition of contamination takes no account of whether the contaminating signal originates from regions immediately outside the selected volume or from regions of the sample which are further afield. Another consideration is that some of the contaminating signals may produce a negative contribution to the final spectrum providing some degree of signal cancellation with positive contributions. Since the sample is, by definition, inhomogeneous (otherwise spatial localization would not be required), the final contamination will vary with the spatial distribution of the NMR nuclei under investigation.

In the following discussion, the absolute contamination will be considered, which is obtained by co-adding the contaminating signal irrespective of the sign of its contribution to the final spectrum. For comparison, the observed contamination in a homogeneous sample will also be presented, which allows signal cancellation of positive and negative contaminating signals. In reality, the true contamination will be between these limits and will depend upon the heterogeneity of the sample.

A large volume of surrounding sample may provide a correspondingly large contamination signal if its contribution to the localized spectrum is not completely removed. Therefore, the relative size of the localized volume compared to the overall observable sample volume must be taken into account when evaluating contamination. An acceptable maximum contamination is defined as 10% of the signal magnitude measured from material within the localized volume.

The efficiency of the spatial localization scheme is defined as the percentage of signal obtained from a localized volume compared with the maximum obtainable signal from that same volume in a fixed experimental acquisition time. When comparing the exact efficiency and contamination of localization schemes for any particular nucleus, it is also necessary to take account of T_1 and T_2 relaxation which have a variable effect on the spectrum obtained by each localization method. Any comparisons should therefore be considered approximate, and the particular properties of the sample may eventually alter the choice of a suitable technique.

In the following comparison, the TMR technique will not be considered because it has been largely replaced by ISIS as a single volume localization technique for ^{31}P spectroscopy. Other localization schemes which rely on selective excitation (e.g. PRESS [34, 35], and STEAM [36]), although similar to ISIS in many respects, are also not considered in this comparison. These

Table 2. Comparison of contamination and efficiency for ISIS and CSI

Technique	Contamination (%) (absolute)	Contamination (%) (observed)	Efficiency (%)
ISIS	8.9	4.2	34.4
CSI[a]	81	43.5	37.1
CSI[b]	8.5	5.2	12.1

[a] 7 Phase encoding steps per dimension
[b] 13 Phase encoding steps per dimension

methods are not generally applicable to the measurement of in-vivo ^{31}P spectra since they cause an unacceptable T_2-weighting of the localized spectra.

Matson et al. [37] have considered the theoretical relative contamination and efficiency of CSI and ISIS techniques for selection of a $4 \times 4 \times 4\,cm^3$ cube within a homogeneous $16 \times 16 \times 16\,cm^3$ sample volume (approximate dimensions of human brain). Table 2 summarizes the results of two comparisons of the ISIS and CSI techniques; one for similar efficiency and one for similar contamination. For similar efficiency, the CSI technique (using 7 phase encoding steps per dimension) results in a ten-fold increase in contamination. For similar contamination, the CSI technique (using 13 phase encoding steps per dimension) results in approximately a 3-fold reduction in efficiency. All CSI experiments utilized 3 dimensional phase encoding and, to minimize contamination, least square weighting (LSW) for combination of signal acquisitions. The ISIS technique utilized a hyperbolic-secant selective inversion pulse.

The CSI technique provides the option of co-adding adjacent volume elements to obtain improved spatial definition from a larger CSI image matrix with several volume elements within the region of interest. However, doubling the CSI image matrix resolution along each axis involves a $\sqrt{2}$ loss of efficiency. For example, combination of $2 \times 2 \times 2$ adjacent cubic volumes along X, Y and Z axes results in a $2\sqrt{2}$ reduction in efficiency compared to a single volume element of equivalent size obtained in a CSI experiment of half the resolution. The combination procedure does result in a substantial reduction of the contamination in the localized signal, although this is still larger than can be achieved using the LSW combination for an equal number of phase encoding steps [37]. The LSW method and the multiple volume summation approach are equally efficient.

4 Conclusion

Techniques for spatial localization of in-vivo ^{31}P spectra have been reviewed. The TMR technique, although initially attractive due to its simplicity, has proved to be too inflexible. The ISIS and CSI techniques are currently the two most popular techniques being used to study in-vivo ^{31}P metabolism.

If the region of abnormal tissue can be visualized by MR imaging, the ISIS technique appears to be superior to CSI since the ISIS cube can be accurately located within the desired tissue volume. If the spatial extent of tissue abnormality cannot be visualized by MR imaging, the CSI technique has the benefit of spatially mapping the ^{31}P spectra with multiple volume elements. In these circumstances it is desirable to produce a ^{31}P metabolite map with reasonably high spatial resolution so that normal and abnormal ^{31}P metabolic levels can be established without partial volume effects.

Spatially localized ^{31}P spectra obtained using CSI are T_2-weighted due to transverse relaxation which occurs during the phase encoding period. The T_2 relaxation times of ^{31}P metabolites in pathological tissues may vary in a similar manner to the well documented changes in T_2 values of the ^{1}H NMR signal. This may cause distortion of the relative changes in metabolite concentration measured by CSI.

Contamination of the ISIS spectrum can be minimized using presaturation RF pulses. The ISIS technique combines high efficiency with low contamination to allow the accurate localization of in-vivo ^{31}P spectra.

5 References

1. Cady EB, Dawson MJ, Hope PL, Tofts PS, Costello AM, Delpy DT, Reynolds EOR, Wilkie DR (1983) The Lancet May: 1059
2. Tofts PS, Cady EB, Delpy DT, Costello AML, Hope PL, Reynolds EOR, Wilkie DR, Gould SJ, Edwards D (1984) The Lancet Jan: 459
3. Younkin DP, Delivoria-Papadopoulos M, Leonard JC, Subramanian VH, Eleff S, Leigh LS, Chance B (1984) Annals Neurol 16: 581
4. Hope PL, Reynolds EOR (1985) Clin Perinatol 12: 261
5. Zimmerman RA, Bottomley PA, Edelstein WA, Hart HR, Redington RW, Bilaniuk LT, Grossman RI, Goldberg HI, Bruno L, Kressel H (1985) Am J Neuroradiol 6: 109
6. Hamilton PA, Hope PL, Cady EB, Delpy DT, Wyatt JS, Reynolds EOR (1986) The Lancet May: 1242
7. Bottomley PA, Drayer BP, Smith LS (1986) Radiology 160: 763
8. Drayer BP, Bottomley PA, Smith LS (1987) Am J Neuroradiol 8: 943
9. Levine SR, Welch KMA, Helpern JA, Chopp M, Bruce R, Selwa J, Smith MB (1988) Ann Neurol 23: 416
10. Black SE, Helpern JA, Kertesz A, Smith MB, Chopp M, Welch KMA (1987) Nuclear magnetic resonance imaging and spectroscopy in stroke. In: Moore WS (ed) Surgery & Cerebrovascular Disease, Churchill-Livingston, New York, NY, p 217
11. Welch KMA, Gross B, Licht J, Levine SR, Glasberg M, Smith MB, Helpern JA, Bueri J, Gorell JM (1988) Magnetic resonance spectroscopy of neurologic diseases. In: Apple SH (ed) Current Neurology, Year Book Medical Publishers, Chicago, IL, vol. 8, chap 9, p 293
12. Brown GG, Levine SR, Gorell JM, Pettegrew JW, Gdowski JW, Bueri JA, Helpern JA, Welch KMA (1989)
13. Cadoux-Hudson TA, Blackledge MJ, Radda GK (1989) FASEB J 3: 2660
14. Twieg DB, Meyerhoff DJ, Hubesch B, Roth K, Sappey-Marinier D, Boska MD, Gober JR, Schaefer S, Weiner MW (1989) Mag Res Med 12: 291
15. Roth K, Hubesch B, Meyerhoff DJ, Naruse S, Gober JR, Lawry TJ, Boska MD, Matson GB, Weiner MW (1989) J Mag Res 81: 299
16. Ramadan NM, Halvorson H, Vande-Linde A, Levine SR, Helpern JA, Welch KMA (1989) Headache 29: 590

17. Martin GB, Paradis NA, Helpern JA, Nowak RM, Welch KMA (1991) Stroke (in press)
18. Ackerman JJH, Grove TH, Wong GG, Gadian DG and Radda GK (1980) Nature 283: 167
19. Gordon RE, Hanley PE, Shaw D, Gadian DG, Radda GK, Stout P, Bore PJ, Chan L (1980) Nature 287: 736
20. Gordon RE, Hanley PE, Shaw D (1982) Progress in NMR Spectroscopy 15: 1
21. Ordidge RJ, Connelly A, Lohman JAB (1986) Mag Res 66: 283
22. Brown TR, Kincaid BM, Ugurbil K (1982) Proc Nat Acad Sci 79: 3522
23. Maudsley AA, Hilal SK, Perman WH, Simon HE (1983) J Mag Res 51: 147
24. Bottomley PA, Charles HC, Roemer PB, Flamig D, Engeseth H, Edelstein WA, Mueller OM (1988) Mag Res Med 7: 319
25. Morse O, Singer JR (1970) Science 170:440
26. Silver MS, Joseph RI, Hoult DI (1984) J Mag Res 59:347
27. Hubesch B, Sappey-Marinier D, Roth K, Meyerhoff DJ, Matson GB, Weiner MW (1990) Radiology 174: 401
28. Bolinger L, Leigh JS (1988) J Mag Res 80:162
29. Ordidge RJ, Bowley R, McHale G (1988) Mag Res Med 8: 323
30. Ordidge RJ (1987) Mag Res Med 5: 93
31. Connelly A, Counsell C, Lohman JAB, Ordidge RJ (1988) J Mag Res 78: 519
32. Matson GB, Twieg DB, Karczmar GS, Lawry TJ, Gober JR, Valenza M, Boska MD, Weiner MW (1988) Radiology 169: 541
33. Levitt MH (1982) J Mag Res 48: 234
34. Bottomley PA (1984) US Patent 4: 480–228
35. Ordidge RJ, Bendall MR, Gordon RE, Connelly A (1985) Volume selection for in-vivo biological spectroscopy. In: Govil, Khetrapal, Saran (eds) Magnetic resonance in biology and medicine, Tata McGraw-Hill, New Delhi, India, p 387
36. Frahm J, Bruhn H, Gyngell ML, Merboldt KD, Hanicke W, Sauter R (1989) Mag Res Med 9: 79
37. Lawry TJ, Twieg DB, Maudsley AA, Weiner MW, Matson GB (1988) Computer Simulation of MRS Localization: Comparison of ISIS and Spectroscopic Imaging (SI) Techniques. Abstract. Proceedings of the Seventh Annual Meeting of Society of Magnetic Resonance in Medicine, San Francisco. p 944. Further details to be published

Localized Spectroscopy Using Static Magnetic Field Gradients: Comparison of Techniques

M. Decorps and D. Bourgeois

Groupe d'Application de la RMN à la Neurobiologie, Unité INSERM U. 318, Université Joseph Fourier, Hôpital Albert Michallon, Pavillon B, BP 217 X, 38043 Grenoble, France

Table of Contents

Techniques for obtaining localized NMR spectra using static magnetic field gradients are described. These techniques are divided into three groups. The first group consists of techniques exploiting inhomogeneities of the main field. The second group consists of techniques using spatially selective pulses and the last group contains phase-encoding techniques. After some general considerations on the signal-to-noise ratio of localized spectra, slice selection and echo techniques, the methods which appear to be routinely useful or promising are discussed in more details.

During the last ten years, several groups have proposed volume selection techniques based upon the use of static magnetic field gradients. The methods allow in vivo spectroscopic information to be obtained from a well defined volume. Aue (1) published in 1986 a review of these techniques and experimental or theoretical comparisons of some of the gradient-encoded techniques have been presented [2–8].

Most of the localization techniques can be divided into three groups.

In the first group, volume selection relies upon the use of a highly homogeneous static field in the region of interest and an inhomogeneous field outside this region. One of the earliest techniques in this group is Topical Magnetic Resonance [9].

In the second group, spatial encoding is achieved by *gradients applied in the presence of frequency selective* rf *pulses*, resulting in spatially selective excitation. Spectroscopic information from one voxel in human subjects or animals is generally obtained after the normal one-dimensional Fourier transform and without further signal processing. The size of the region of interest can be modified by varying the pulsed gradient strength, and its position moved within the object by changing the frequency of the selective pulses. The size and location of the volume of interest (VOI) are defined using a standard NMR image. Following the earlier methods such as VSE (Volume Selective Excitation) [10], DRESS (Depth Resolved Surface-coil Spectroscopy) [11] and ISIS (Image Selected In vivo Spectroscopy) [12], a large number of techniques based on selective excitation in the presence of gradient have been proposed.

In the third group, *gradients are applied during a free precession delay* as proposed by Maudsley et al. [13] or during the acquisition delay [14] resulting in a spatially dependent modulation of the FID. Successive acquisitions are made using different amplitudes of the static field gradients, or using different orientations of the gradient field. After processing of the data array resulting from N spatially encoded experiments, spectroscopic information can be obtained from N voxels simultaneously.

Some of the methods exploiting inhomogeneities of the main field could be useful in suppressing signals near the plane of a surface coil; however, only the methods of the last two groups can ensure flexible localization. Most of the techniques using selective excitation only allow single voxel measurements, but modifications have been proposed to achieve multi-volume spectroscopy. One of them, Hadamard spectroscopy [15, 16], is a spectroscopic imaging (SI) technique since it enables spectra from all the voxels of a region of interest to be obtained simultaneously. Nevertheless, most of the spectroscopic images in clinical or biological applications were acquired using the methods of the third group. In practice, in-vivo spectroscopy frequently uses a combination of methods.

In section 1 some general considerations on signal-to-noise ratio of localized spectra, slice selection and echo techniques will be presented. Section 2 will briefly describe the methods of the first group. Section 3 reviews the methods of the second group, with particular emphasis on those which appear to be routinely useful or promising. Phase-encoding methods are discussed in section 4. An overview of water signal suppression in localized proton spectroscopy is presented in Section 5. Finally, comparison of techniques and some elements of choice are the subject of Section 6.

1 General Considerations

1.1 *Signal-to-Noise Ratio and Spatial Resolution*

Whatever the method, any improvement in spatial resolution results in a large increase in measurement time. The NMR signal obtained from a voxel is proportional to the volume ΔV of this voxel, and the noise varies as $\sqrt{N}$ where N is the number of scans. When the volume of the voxel varies, a scan number proportional to ΔV^{-2} is required to obtain a constant signal-to-noise ratio (S/N), and thus a total experiment time T also proportional to ΔV^{-2}. Therefore, for 3D spatial selection:

$$T \propto \Delta l^{-6} \tag{1}$$

where Δl is the linear resolution ($\Delta V = \Delta l^3$).

If 1 min is the experiment time required to obtain a spectrum from a $1 \times 1 \times 1\,\text{cm}^3$ volume element, then approximately 1 h will be required to obtain a spectrum from a $.5 \times .5 \times .5\,\text{cm}^3$ voxel. The price to be paid for improving the spatial resolution is clearly very high. However, in practice, the loss of S/N may be inferior to the theoretical prediction. The reason for this is that decreasing the voxel size can improve both B_1 and B_0 homogeneities.

1.2 *Signal-to-Noise Ratio and Sample Size*

Most of the spatially resolved experiments were performed on human subjects. For example, a number of ^{31}P spatially resolved experiments have been achieved on human brain. In contrast, most of the ^{31}P spectroscopic studies on rat brain were performed with surface coils without further localization. One reason for this is that if all the linear dimensions (coil diameter, wire diameter, voxel size) are scaled down with decreasing sample size, the S/N achieved from the whole sample *per* unit time, decreases as the sample size decreases.

Consider a sphere of conducting material inside a receiving coil. Let us suppose that all the coil linear dimensions are scaled in relation to sample diameter d. The resistance which models noise due to Brownian motion in the sample is [17]:

$$r_m = \mathrm{k}_1 B_1^2 d^5 \tag{2}$$

where B_1 is the magnetic field produced by the unit current in the coil and k_1 a constant depending on frequency and sample resistivity. The coil resistance r_e is invariant with the scaling of the coil diameter:

$$r_e = k_2 \tag{3}$$

and B_1 given by:

$$B_1 = k_3 d^{-1} \tag{4}$$

The S/N from the whole sample then becomes:

$$\Psi \propto \frac{B_1}{(r_e + r_m)^{1/2}} V \tag{5}$$

where V is the sample volume. Combining Eqs. (2, 3, 4, and 5), the S/N is given by:

$$\Psi = K \frac{d^2}{(1 + \beta d^3)^{1/2}} \tag{6}$$

For small samples $\beta d^3 \ll 1$, and $\Psi \propto d^2$. As the dimensions increase βd^3 increases. In the case of very large samples, the noise arises mainly from the sample and $\Psi \propto d^{1/2}$. Thus, the S/N from the whole sample always decreases as the sample size decreases. Nevertheless, this effect is more significant when magnetic losses can be ignored, i.e. for small samples.

It should be noted that the S/N, Ψ', from a voxel of effective volume ΔV, which would be obtained after N′ scans, can be estimated from the S/N, Ψ, from the whole sample obtained after N scan by using the relation:

$$\Psi' = \Psi \frac{\Delta V}{V} \left(\frac{N'}{N}\right)^{1/2} \tag{7}$$

Since Ψ decreases as the sample size decreases, the fraction of the whole sample which can be analyzed in a localized spectroscopy experiment also decreases.

1.3 *Selective Excitation in the Presence of Gradients*

Selective excitation in the presence of gradients has been extensively studied in the context of NMR imaging [18–23] and is a basic tool in performing spectral localization. Properties of frequency selective pulses have recently been reviewed [24]. A magnetic field gradient spatially encodes the resonance frequencies and the frequency selective pulse excites a spin distribution within the sample. The slice-thickness determines the necessary gradient strength and the pulse band-width. For a given pulse band-width, the higher the gradient strength, the thinner the slice thickness. Slice position is varied by changing on the offset frequency. In the context of localized spectroscopy, a selective pulse can be used (i) to selectively invert a longitudinal magnetization (inversion pulse), (ii) to selectively generate a transverse magnetization (excitation pulse), (iii) to selectively store a transverse magnetization along the x-axis (storage pulse) and (iv) to selectively refocus transverse magnetizations (refocusing pulse). Most of the frequency selective excitation pulses exhibit a frequency dependent phase shift which must be corrected either by reversing the gradient polarity or by adding an appropriate gradient after a refocusing pulse. Storage pulses also have to use gradient reversion, but refocusing and inversion pulses do not. Gradient reversion can be avoided by using 270° pulses [25], or self-refocused pulses [24, 26–29].

1.3.1 *Profile of the Selected Slice*

The profile of the selected slice, using selective excitation in the presence of gradients depends, on the frequency-distribution of the shaped pulse, but also on the on-resonance pulse angle, the relaxation time T_2 and the component of the magnetization upon which it acts. Gaussian-shaped pulses yield relatively poor slice resolution since, within linear approximation, the response is also gaussian. Sinc-shaped pulses offer an improved slice resolution, but due to truncation effects, the frequency distribution of the transverse magnetization is not strictly rectangular and some wobbles can be seen at the edge of the slice profile. Nonlinear effects and T_2 effects further modify the slice profile. Truncation effects can be reduced by multiplying the truncated envelope by a gaussian function or by using Hermitian pulses [30]. Frequency selective inversion can be achieved with a hypersecant pulse [31, 32] which is relatively insensitive to B_1 field strength. It should be noted that the slice profile produced by an excitation pulse greatly differs from that produced by a refocusing pulse [6].

1.3.2 *The Chemical Shift Artifact*

When selective excitation is employed for spectroscopic purposes where several chemical shifts are present, the Larmor frequency depends both on the position and on the chemical shift. Consequently, the position of the selected slice depends on the chemical shift (see for example Refs. [4, 10, 12, 33–35]. The displacement of the slice depends on the gradient strength.:

$$\Delta u = 2\pi \Delta \mathrm{f}/(\gamma \mathrm{G}_u) \qquad (8)$$

where u is the coordinate along the gradient direction. The error can be reduced by using strong field gradients but this will lead to an increase in eddy currents and motion-induced signal losses [36].

1.4 *Echoes*

A number of volume selection methods use Carr–Purcell echoes [37] or stimulated echoes [38].

1.4.1 *Carr–Purcell Echoes*

Since a refocusing pulse can be frequency selective or applied with inhomogeneous B_1, its effect is not a rotation through 180° about an axis of the xy rotating frame plane. To achieve efficient refocusing, without phase or intensity artifacts, when the nominal pulse angle of a refocusing pulse differs from a single and ideal 180°, it is necessary either that a complete dephasing [39] of spins occurs during the spin echo delay or that an Exorcycle [40–42] phase cycling of the refocusing pulse is used.

Consider a refocusing time-symmetrical pulse. Let M_z and $M_{tr} = M_x + iM_y$ be the longitudinal and transverse magnetization before the pulse, and M'_z, M'_{tr} be the longitudinal and transverse magnetization after the pulse. An ideal $180°_x$ refocusing pulse yields:

$$M'_z = -M_z \tag{9}$$

$$M'_{tr} = (M_{tr})^* \tag{10}$$

An imperfect 180° pulse first flips some longitudinal magnetization into the transverse plane, secondly leaves a part of the transverse component M_{tr} in the xy plane without phase reversal, and finally flips back a part of M_{tr} along the longitudinal axis [43]. All these spurious signals can produce artifacts. This is well known in conventional spectroscopy techniques which often use the Exorcycle phase cycling scheme to cancel spurious signals [40]. However, cancellation of spurious signals using phase cycling can be impaired by sample motion or instrumental instabilities, and phase cycling requires the magnetization at the beginning of each transient to be constant. Furthermore, the use of phase cycling increases the minimum duration of the experiment and can prevent shimming on the VOI. Another solution is to achieve a complete dephasing of the magnetization using additional spoiling gradients [44] or "crushers" [45], placed around the refocusing pulse as shown on Fig. 1a. They provide an efficient suppression of artifacts [6] which can be further improved by combining spoiling gradients and phase cycling.

It should be noted that time-symmetrical refocusing pulses which are generally used do not introduce any phase-shift [46, 47].

1.4.2 *Stimulated Echo Sequence*

Stimulated echoes [38] are generated by a three-pulse sequence.

$$\theta^1 - T_E/2 - \theta^2 - T_M - \theta^3 - T_E/2 - \text{Acq} \tag{11}$$

The echo amplitude is proportional to $(\sin\theta^1 \sin\theta^2, \sin\theta^3)/2$. To achieve optimal efficiency the nominal value of each pulse should be 90°. However, whatever the nominal pulse angles, a stimulated echo technique requires a complete dispersion of phases during the first echo delay [38, 48] and destruction of transverse magnetization during T_M. This can be realized by exploiting B_0 inhomogeneities,

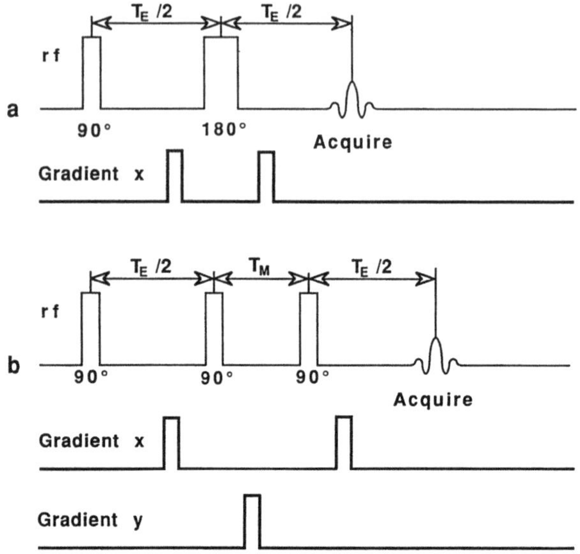

Fig. 1 a, b. Echo sequences and spoiling gradients. **a** Carr and Purcell spin echo sequence **b** Stimulated echo sequence

if the echo time is long enough, through the use of spoiling gradients [6,45,49–51] as shown in Fig. 1b or with the help of an Exorcycle phase cycling of the last two pulses:

$$\theta_x^2, \theta_x^3, \text{add}; \theta_{-x}^2, \theta_{-x}^3, \text{add}; \theta_y^2, \theta_y^3, \text{subtract}; \theta_{-y}^2, \theta_{-y}^3, \text{subtract}, \tag{12}$$

or of the equivalent scheme proposed by Sattin et al [52]. Signals acquired with stimulated echo sequences are T_1 and T_2 weighted. J-modulation is effective during T_E, and zero quantum coherences also evolve during T_M. A major disadvantage of stimulated echo sequences is the loss of a factor of two in S/N as compared to Carr–Purcell echoes.

Spin-echo-techniques are widely used not only for proton spectroscopy where their use is imposed by the requirement of water and lipid resonance suppression, but also for ^{31}P spectroscopy. This is a consequence of the recent introduction of actively shielded gradients [53–55] allowing the use of small spin-echo delays [56]. Spin echo techniques suffer from motion sensitivity [57,6]. Refocusing requires each isochromat to retain the same precession frequency during the spin-echo delay. Motion in a nonhomogeneous magnetic field and modifications in the homogeneity due to coherent or incoherent motion change the precession frequencies, this results in a phase shift which can vary among isochromats leading both to a phase shift to the echo and to an attenuation of its amplitude. The latter can be enhanced by averaging. Motion artifacts can be particularly troublesome in the case of long spin echo delays, and the use of spoiling gradients increases the sensitivity to coherent or incoherent motions.

2 Methods Exploiting Inhomogeneities of the Main Field

Topical magnetic Resonance (TMR) uses nonlinear static field gradients to limit the volume with high homogeneity [9, 58, 59]. The method is known to suffer from poor spectral resolution and difficult control of the voxel volume and position. An alternative to TMR is the sensitive point technique, which uses an alternating magnetic field gradient to eliminate contributions from regions with a high-amplitude time-dependent main field [60, 61]. This method also suffers from problems such as ill-defined boundaries and line broadening. More recently, a different approach was described to eliminate signals from the region near the plane of a surface coil [62–71]. The method involves the application of a surface spoiling B_0 gradient localized, on the surface tissue from which the signal is to be eliminated. Surface spoiling gradients can be electrically [62–68], ferromagnetically [69, 70], or paramagnetically [71] generated. Signals from surface tissue rapidly dephase and go to zero, allowing observation of signals from the deeper-lying region of interest. The use of electrically generated gradients enables the thickness of the rejected surface region to be controlled. This one-dimensional method provides a simple way of eliminating signals from surface tissue and could be useful for a number of applications.

3 Methods Based on Selective Excitation

A large number of techniques based on the application of frequency-selective rf pulses in the presence of static magnetic field gradients, have been suggested to obtain high-resolution spectra from a well-defined volume of a whole sample. These techniques generally acquire information from an image-selected volume element. They can be divided into three categories: (i) difference techniques, (ii) destruction of the magnetization from outside the VOI and, (iii) direct excitation of the spins from inside the VOI. We briefly describe these methods with particular emphasis on three of them which have proven to be routinely useful or promising: ISIS (first group), PRESS, and STEAM (third group).

3.1 *Difference Techniques*

In this group of experiments, volume selection results from an add-subtract scheme. The basic sequence is ISIS [12, 72, 73] which has become one of the most successful technique to achieve spatial localization [32, 74–84]. ISIS enables a spectrum to be obtained from a single voxel. As shown in Fig. 2, the basic experiment begins by a *preparation period* during which the longitudinal magnetization is shaped by means of selective inversion pulses applied in the

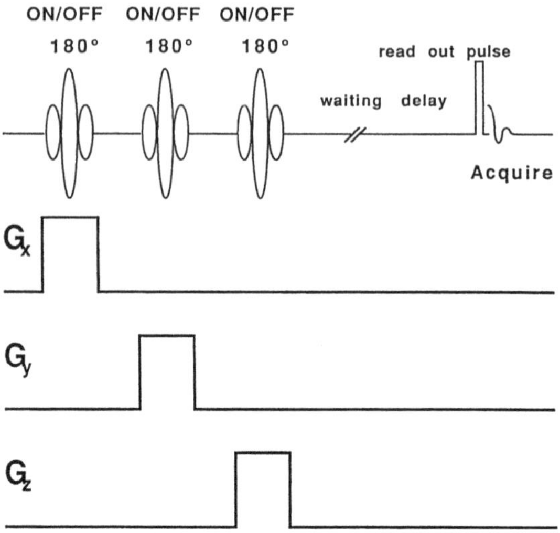

Fig. 2. Basic scheme for Image Selected In-vivo Spectroscopy (ISIS). The three selective pulses are cycled through eight on-off combination and the acquired signal is added or subtracted in order to obtain coherent averaging of the signal from the selected volume

presence of gradient pulses. The preparation period is followed by a *waiting delay* allowing time for eddy currents to decay. Finally, the longitudinal magnetization is sampled by applying an *observe pulse* which can be a pulse sequence designed to perform T_1 or T_2 measurements, water suppression, spectral editing, etc... To achieve a 3D selection, a set of eight acquisitions is performed; each acquisition consists of a different combination of inversion pulses. The receiver is inverted when an odd number of selective pulses are applied. Addition of the eight acquisitions results in coherent averaging of signals from the VOI and the destruction of signals from regions outside the VOI. Since the method is based on an add-subtract process, it is essential to ensure that the magnetization is the same at the beginning of each experiment. In the case of an inhomogeneous radiofrequency field, tip angles vary through the sensitive volume and some longitudinal magnetization remains after the read-out pulse. The value of this component depends on the combination of pulses during the preparation period. The longitudinal magnetization at the end of the recycle delay is therefore dependent on the previous preparation period producing contamination from outside the selected volume. This can be avoided by using long recycle delays (at the price of a reduction of the S/N per unit time), or a saturation recovery period [78, 85].

ISIS is not T_2 weighted. This may be a major advantage for in-vivo ^{31}P spectroscopy where short T_2 values can be found [86]. However, localization is achieved through the use of an add-subtract process, and suppression of magnetization from outside the selected volume can be degraded by limited ADC resolution, instrumental instabilities or sample motion [4]. This limitation can be severe when the region of interest is small compared to the volume of the coil-sensitive region. Techniques such as OSIRIS [87] add to the method a saturation of the outer volume in order to reduce this "subtraction noise" [4]. ISIS can also be employed to restrict the signal to a bar. Spatial selection along the third direction is then achieved using some other technique [35, 88, 89].

3.2 *Destruction of the Magnetization Outside the Volume of Interest*

Three pulses, pulses-sandwiches or pulse-trains, are successively applied in the presence of field gradients to retain longitudinal magnetization only in the region of interest. Magnetization from outside the VOI remains in the transverse plane and is spatially randomized in the gradient field. After the eddy-currents have decreased, a readout pulse is applied. This pulse can be a single $\pi/2$ pulse or a more complex pulse sequence including, for example, water suppression.

The earliest experiment of this kind was VSE [10, 90–93], but a number of modifications of the basic sequence have been described, for example SPARS [94–98], SPACE [3, 99–102], and SPALL [103, 104]. They use pulse sandwiches made of two 90° pulses and one 180° pulse. Two of them are hard, the third is soft. These techniques are limited by the large number of pulses in the pulse sequence (ten for 3D selection if sampling of the z-magnetization is achieved with a single pulse) [5], which give a high sensitivity to errors in pulse angles or inhomogeneous B_1. Errors in pulse angles may result in contamination from the outer volume. Furthermore, all the methods also suffer from potential contamination due to T_1 relaxation during the waiting delay. Extensive phase-cycling can prevent contamination, but its efficiency is impaired by sample motion, instrumental instabilities, and digitization errors. Moreover, the association of soft and hard pulses in a pulse sandwich may introduce dynamic range problems in the transmitter. Other methods use amplitude-modulated pulses in the presence of gradients to saturate the outer volume. Examples are DIGGER [3, 105, 106], FROGS (1D technique [107]), LOCUS [108, 109], ROISTER [110].

Imperfect saturation of the outer volume yields contamination which can be severe in the case of small VOI in a large sample. It is important to realize that the acquired signal results from integration over the whole sample. Consequently, images of the sensitive volume are insufficient to yield an accurate estimation of the degree of contamination. A three-dimensional integration would be preferable [111]. As in the case of difference techniques, the replacement of the nonselective readout pulse with a slice-selective pulse can reduce the contamination artifacts [112]. Inhomogeneous B_1, imperfect saturation with large samples for which the saturation band may be too narrow, may severely impair the efficiency of these techniques. Techniques such as VSE and SPARS were among the first 3D techniques available and they aroused much interest. They have now been superseded by echo techniques. Nevertheless, saturation of the signal from outside the VOI could be associated with a technique such as PRESS or STEAM to improve the final degree of suppression of unwanted signals.

3.3 *Direct Excitation of the Spins Inside the Volume of Interest*

The earliest technique in this group was DRESS [11], a 1D technique which must be associated with a B_1 gradient method. DRESS combines B_1 gradient volume selection and slice selection using selective excitation in the presence of a gradient. Selection of a bar using ISIS or destruction of magnetizations from outside the bar can be combined with a frequency-selective pulse applied in the presence of a gradient along the bar direction. Similarly, FRIVOL [35] uses frequency-selective excitation which eliminates the slice selection chemical shift error.

Two types of techniques are currently used in in-vivo spectroscopy to directly excite the spins inside the VOI and to achieve 3D selection. One of them exploits stimulated echoes [49, 50, 113, 114], the other exploits Carr–Purcell echoes [115–117]. An important advantage of these methods is that a localized spectrum may be obtained in a single scan, which allows shimming of the VOI, and avoids the localization artifacts of subtraction schemes. As the methods described in Sects. 3.1 and 3.2, these techniques use frequency-selective pulses whose offset frequency is adjusted according to the position of the sensitive volume. However, the phase of the signal is affected by random phase discontinuities which might be introduced during the frequency changes. In order to allow coherent averaging, the synthesizer must ensure phase continuity when switched from one frequency to another. Phase-ramped selective pulses can be used to avoid the switching of the carrier frequency and associated phase instabilities [118]. Echo techniques are described below in more detail.

3.3.1 *Stimulated Echoes*

The technique is currently known as STEAM (Stimulated Echo Acquisition Mode) and was independently proposed by several groups [49, 50, 113, 114]. The method employes the pulse sequence of Eq. (11). The three pulses are frequency selective and are applied in the presence of mutually orthogonal field gradients (Fig. 3). As pointed out in Sect. 1.3, phase cycling may be avoided by using gradient pulses in the two $T_E/2$ periods, and a spoiling gradient during the T_M period, in order to disperse the unwanted residual transverse components. The signal is half of that resulting from a Carr–Purcell echo experiment and differs from it in that it is both T_1 and T_2 weighted:

$$S \propto (M_0/2)\exp(-T_E/T_2)\exp(-T_M/T_1) \qquad (13)$$

Multivoxel techniques have been shown to allow two voxels to be obtained without any increase in experimental time [119]. Stimulated echoes can be associated with phase-encoding in one or two spatial directions [120]. Stimulated echoes are now widely used in ^{1}H animal or human studies [45, 51, 120–130].

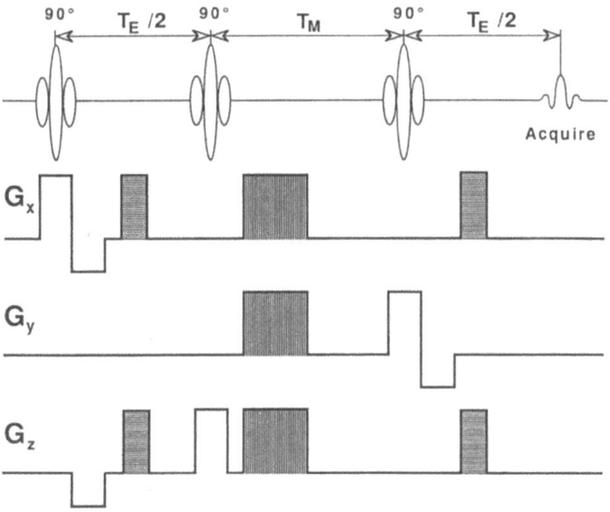

Fig. 3. Pulse sequence and spoiling gradients for stimulated echo localized spectroscopy

3.3.2 *Carr–Purcell Echoes*

The technique is named PRESS (Point Resolved Spectroscopy) or double-spin-echo and was proposed by two groups [116, 117]. Spatial localization is achieved with a three-pulse sequence (Fig. 4):

$$90° - T_E^1/4 - 180° - T_E^1/4 - T_E^2/4 - 180° - T_E^2/4 - \text{Acq} \tag{14}$$

with:

$$T_E^1/4 + T_E^1/4 + T_E^2/4 + T_E^2/4 = T_E \tag{15}$$

The three pulses are frequency selective. The 90° pulse applied in the presence of a gradient (e.g. along the x-axis) selects a slice orthogonal to this axis. The magnetization of the selected slice is then allowed to precess during $T_E^1/4$ and is refocused by a second pulse applied in the presence of a gradient along an orthogonal direction (e.g. the y-axis). At $T_E^1/4$ it refocuses the magnetization in a slice orthogonal to this axis. Finally, after a free precession delay $T_E^2/4$, a second refocusing pulse is applied in the presence of a gradient along the direction orthogonal to the two first directions (z-axis). This last pulse refocuses the transverse magnetization from a slice orthogonal to the first two slices. Therefore, the acquired signal results from the volume element common to the three slices, producing a three-dimensional spatial localization in one single scan. Phase cycling or/and spoiling gradients must be used to cancel phase and intensity errors due to imperfect refocusing pulses. High-resolution localized proton spectra of human or animal tissues have been obtained with Carr–Purcell echo methods [131, 6] which offer the important advantage of a factor of two gain in S/N. Of course, signals are T_2 weighted and J-modulated.

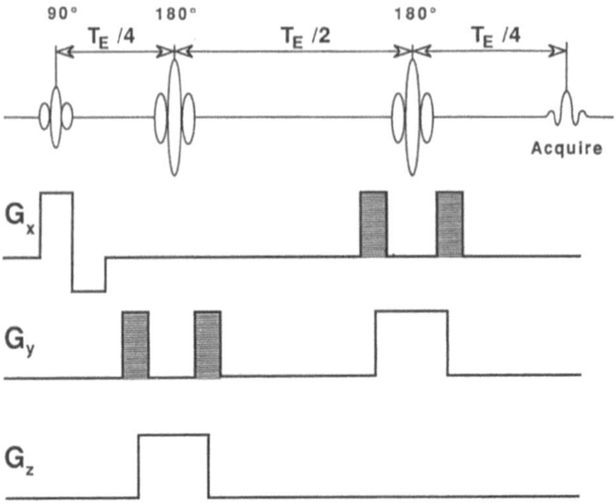

Fig. 4. Pulse sequence and spoiling gradients for double spin-echo localized spectroscopy.

Echo techniques are among the more promising ones and have recently been compared [6]. The observation of short T_2 metabolites with echo methods can be difficult if rapidly switching gradients are not available. However, the feasibility of stimulated echo phosphorus spectroscopy in the human brain has been demonstrated [124] at short echo times (3 ms). Echo techniques are well adapted to proton spectroscopy which can take advantage of long spin-echo-delays to increase water and fat signal suppression. However, proton spectroscopy at short echo times can increase the information content [56]. As pointed out in Sect. 1.4, motion can induce signal losses, but a careful design of gradient waveforms may reduce these losses [36].

3.4 *Multi-Volume Spectroscopy*

Methods such as DRESS, STEAM, and PRESS leave intact the magnetization from outside the three orthogonal slices used for the definition of the VOI. It is then possible to exploit the recovery delay to acquire a spectrum from one or several voxels situated in this region [132]. Tilted gradients can be used to overcome limitations in the choice of the position of the voxels to be acquired in an interlaced mode [131]. The use of frequency-selective pulses with multiple inversion or excitation bands give rise to another class of multi-volume techniques. Multi-volume excitation combined with STEAM or ISIS has been reported [73, 74, 119, 133, 134]. Hadamard spectroscopic imaging (HSI) uses frequency-selective pulses with multiple inversion or excitation bands, and allows N spectra from N volume elements to be obtained [15, 16]. HSI is not affected by truncation artifacts encountered in Fourier techniques. However, there are few studies reported using HSI and it is difficult to assess its performances.

3.5 *Summary*

Excellent localization techniques based on selective excitation in the presence of gradients exist. Difference techniques can be used in spite of gradients with poor switching characteristics. They are not T_2 weighted and can be used to study metabolites with short T_2. However, they suffer from motion sensitivity and/or instrumental instabilities. Localization by saturation of the outer volume can be poor in case of a inhomogeneous B_1, or small VOI in large samples, so direct excitation of the VOI with stimulated echo or double spin echo is often preferred. The development of rapid switching gradients make possible the use of echo techniques for short T_2 species. Techniques like ISIS, STEAM and PRESS are insensitive to B_1 inhomogeneities and can be used with surface coils used as emitter and receiver.

4 Phase Encoding Techniques

A number of methods can be used to simultaneously obtain spectra from N voxels by processing an array of phase- or frequency-encoded FIDs. Spatially resolved spectroscopic information can be extracted from data acquired in the presence of a field gradient [135–140], as first demonstrated by Bendel et al. in 1980 [14]. An important prospect stems from the use of rapid imaging methods using a spin-echo train induced by periodically inverting a field gradient [141–145]. However, these methods have so far not been applied in in-vivo metabolic studies on animals or humans. Most of the techniques currently used employ encoding of the spatial information with a gradient pulse applied during a free precession delay, and acquisition of the NMR signal in an homogeneous field [146–152]. Methods which rely on Fourier transformation of the array of phase-encoded FIDs [153] are most frequently in use. An entirely different way of processing the data set, SLIM, was recently proposed [151, 152] and could be promising. Fourier methods and SLIM use the same pulse sequences to obtain phase-encoded data.

4.1 *Pulse Sequences*

4.1.1 *One-Pulse Techniques*

The simplest pulse sequence is shown in Fig. 5a. It consists of a 90° pulse to generate transverse magnetization, followed by an evolution period during which

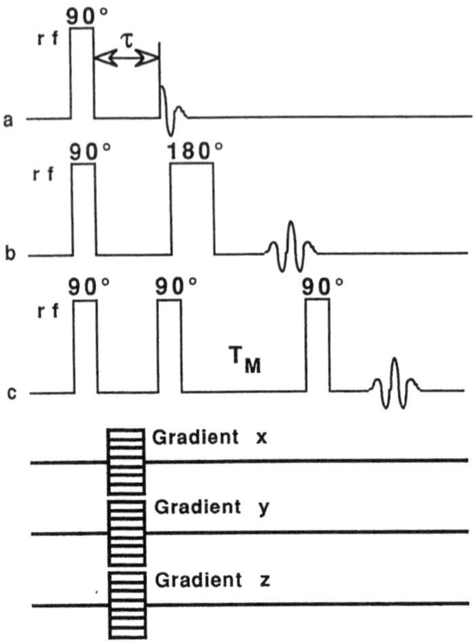

Fig. 5 a–c. Basic schemes for 3D phase-encoded localized spectroscopy: **a** single pulse **b** spin-echo and **c** stimulated echo sequences

one, two, or three gradient pulses along orthogonal directions are applied to phase encode the magnetization. Then, after a homogeneity-recovery delay, the complex signal is acquired. The major drawback of the sequence is the chemical-shift-dependent phase shift due to the delay τ between excitation and acquisition. After Fourier transformation, each line is phase-shifted by $\omega_c\tau$, where ω_c is its *central* frequency in the rotating frame. To obtain the correct spectrum one must correct *the phase of each line* according to its *central* frequency. Such a process can be very time-consuming. What is usually done instead, is to multiply the spectrum by an inverse phase factor, exp $-i\omega\tau$. Each line is therefore corrected by a factor exp $-i\omega\tau$ instead of exp $-i\omega_c\tau$. This introduces some amount of dispersion in the wings, oscillating as a function of frequency. Spectra exhibit base-line distortions which can severely degrade the poorly resolved spectra of in-vivo spectroscopy. Several post-acquisition methods have been proposed to improve this [154, 155].

4.1.2 *Echo Techniques*

The phase shifts can be refocused by adding a 180° pulse to the sequence (Fig. 5b). This is the best method if T_2 is long enough or if the switching time of the gradients is short enough to allow the use of short spin-echo-delays. The use of stimulated echoes (Fig. 5c) is an alternative to the Carr–Purcell spin-echo technique. This pulse sequence could be a remedy in the case of poor switching characteristics of the gradient system when resonances with short natural T_2 are to be observed. Of course, as described in Sects. 1.4 and 3.3, echo techniques require phase cycling or the use of spoiling gradients.

4.2 *Fourier Methods: Spectroscopy Imaging*

4.2.1 *Principle*

Fourier methods are based on the ideas described by Kumar et al. [153]. A Fourier method was used in 1979 by Maudsley et al. [13] to recover the spatial distribution of a resonance frequency in an inhomogeneous static field. Since the first demonstrations of chemical shift distribution imaging by Fourier transform on phantoms [147, 148] and in-vivo [149, 150], a number of methodological works have been published [2, 56, 141, 156–162] and Fourier Spectroscopic Imaging (FSI), is now increasingly used in clinical research [163–169] as well as for animal model studies [44, 170, 171]. Note that phase-encoding of the signal in N spatial dimensions requires a $(N+1)$-dimensional FT. Notations can be somewhat confusing since spatial encoding in N spatial dimensions is frequently called $(N+1)$ SI. In this paper we only retain the number of spatial dimensions. An outline of the method is given below for the simple case of a one-dimensional localization. The generalization to 2D or 3D is straightforward.

FSI uses one of the pulse sequences depicted in Fig. 5. The amplitude of the phase-encoded gradient is generally varied in a systematic fashion from the value-$(M-1)\Delta G/2$ to $(M-1)\Delta G/2$ in equal increments ΔG. A symmetric data set enables pure absorption spectra to be obtained [172–174]. The number of increments, M, determines the digital resolution of the image obtained after two-dimensional Fourier transformation (1 spatial dimension, 1 spectral dimension): M spectra from M voxels. The field of view (FOV) is related to the gradient increment:

$$\mathrm{FOV} = 2\pi/\gamma\Delta G\tau \tag{16}$$

and the voxel size Δl is related to the maximum value of the encoding gradient ($G_{max} = (M-1)\Delta G/2$):

$$\Delta l = \mathrm{FOV}/M = \pi(M-1)/M\gamma G_{max}\tau \approx \pi/\gamma G_{max}\tau \tag{17}$$

The smaller the voxel size, the larger the maximum gradient. The method allows sampling of the k space ($k = \gamma G\tau/2\pi$) in equal increments Δk:

$$\Delta k = \gamma\tau\Delta G/2\pi = 1/\mathrm{FOV} \tag{18}$$

The nth point in the k direction is:

$$k_n = \Delta k[n-(M-1)/2],\ n = 0 \text{ to } M-1 \tag{19}$$

The signal resulting from phase-encoding can be written as:

$$f(t, k_n) \propto \int \rho(t, z) e^{-i2\pi k_n z} dz \tag{20}$$

where $\rho(t, z)$ is the FID from a point with position z. Discrete Fourier transform $F(t, z_m)$ of this phase-encoded data set gives:

$$F(t, z_m) = \frac{1}{M} e^{i\pi m(1-1/M)} \sum_{n=0}^{M-1} f(t, k_n) e^{i2\pi k_n z_m}; m = -\frac{M}{2} \text{ to } \frac{M}{2} - 1 \quad (21)$$

where $z_m = m\Delta l$. When the Nyquist condition is fulfilled and if the signal is not truncated:

$$\sum_{n=0}^{M-1} f(t, k_n) e^{i2\pi k_n z_m} \propto \rho(t, z_m) \quad (22)$$

and

$$F(t, z_m = m\Delta l) \propto e^{i\pi m(1-1/M)} \cdot \rho(t, z_m) \quad (23)$$

Thus, except for the term exp $i\pi m(1 - 1/M)$, reflecting the phase shift due to the shifted acquisition in the k-direction, $F(t, z_m)$ is proportional to the profile of the object. The profile can be obtained either by using the absolute mode, or by using a phase correction. Note that the absolute mode degrades the frequency resolution, except in the case of whole-echo-acquisition in the time domain. The k-space signal is, however, usually severely truncated. Consequently, discrete summation does not yield the exact profile.

4.2.2 *The Spatial Response Function; Truncation Artifacts*

FSI, like Fourier imaging, can be considered as a typical multidimensional spectroscopy method. However, FSI can be differentiated from both 2D or 3D spectroscopic experiments in liquids and NMR imaging by its resolution in the k dimension, which is generally very low. The signal is the k dimension is significantly truncated in the region of high-frequencies (high values of the encoding gradient).

Fourier transform of the untruncated signal (i.e., the profile of the object) is convoluted with the point spread function [8, 175–178]. The physical significance of this is that a point sample at $z = z_0$ is distributed over each voxel. The contamination of a voxel at $z_m = m\Delta l$ by the signal produced by a point sample at $z = z_0$, is given by the *point spread function*:

$$S(z_m, z_0) = \frac{\sin[\pi(z_m - z_0)/\Delta l]}{M \sin[\pi(z_m - z_0)/M\Delta l]} \quad (24)$$

However, the spatial localization properties of Fourier methods are best described by the use of the *spatial response function* which gives the contribution of a point sample at $z = z_0$ to the voxel at $z_m = m\Delta l$. Equation (25) describes both the point spread function (z_m varied, z_0 fixed) and the spatial response function (z_m fixed, z_0 varied). Figure 6 shows the spatial response function in the case $M = 8$ for two voxels of the reconstructed one-dimensional object.

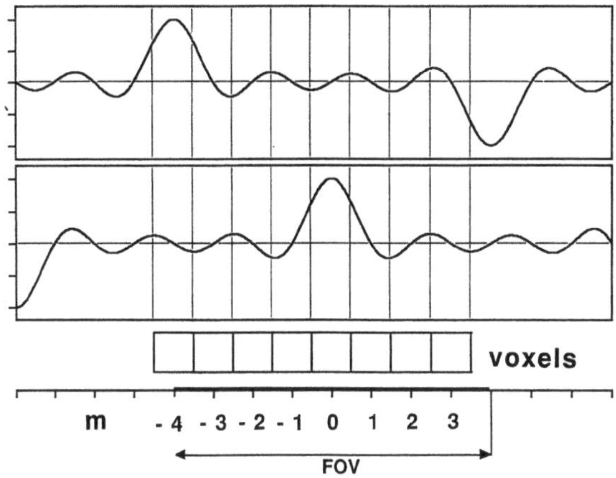

Fig. 6. Spatial response function in the case M = 8, for two voxels of a one-dimensional object

When the profile exhibits discontinuities or sharp edges, the reconstructed profile exhibits overshoots on both sides of the discontinuity. This phenomenon is a source of artifacts in magnetic resonance imaging [179] and is unaccurately [180] known as Gibbs phenomenon. However, due to the limited number of phase-encoding steps, the problem can be more serious in FSI. The following facts are inherent to discrete Fourier transform.

Firstly, as shown in Fig. 6, the voxel at $m = -M/2$ is strongly contaminated by signals from the other end of the sample. This results from the aliasing, inherent to discrete Fourier Transformation. This can be remedied by either shifting the spectroscopic grid or increasing the number of phase-encoding steps at constant spatial resolution. In the later case the S/N per unit time is preserved but leads to an increase of the minimum measurement time.

Secondly, signal "bleed" [9] can arise when the voxel size is larger than the scale of the biological heterogeneities. Due to the shape of the point spread function, the intensity of a point sample at $z_0 = z_m + \delta z$ spreads out over all the other voxels. The intensity of the artifact is strongly dependent on the position of the point sample in the voxel: if the point sample is exactly at the position $z_0 = z_m(\delta z = 0)$ of a sampled point, no contamination is introduced but, if $\delta z \neq 0$, spectral contamination is unavoidable and ghost spectra are introduced in the other voxels. In the case of phosphorus spectroscopy, signal leakages over the spectroscopic image are not a big problem since phosphorylated metabolites generally have an approximately uniform distribution over the volume elements. The situation is different with proton spectroscopy, where the artifact can be particularly troublesome in the case of thin compartments with a high concentration of NMR-visible species, as is the case with subcutaneous fat: the intense lipid proton spectra can spread out over the whole spectroscopic image. Note that partial volume effects also can produce contamination. Contamination due to truncation can be greatly reduced by the use of an apodisation function in the k direction [179]. This point will be discussed in Sect. 4.2.5.

Thirdly, Fourier methods introduce a signal loss compared to ideal selective excitation techniques [181]. This is due to the dispersion of phase shift in a voxel, resulting from the encoding gradient. The evaluation of the signal loss of FSI compared to an ideal selective excitation method can easily be achieved by integrating the spatial response function over the voxel size:

$$\int_{\Delta l/2}^{\Delta l/2} \frac{\sin \pi z/\Delta l}{M \sin (\pi z/\Delta l M)} dz \approx 0.873 \qquad (25)$$

Finally, another consequence of the low spatial resolution is the distortion of the profile. Figure 7 shows the profile of a bar obtained from 8 (Fig. 7a) or 16 (7b) sampling points in the k space. Distortions due to truncation are clearly visible.

4.2.3 *Position of the Spectroscopic Grid*

FSI is generally image-guided and the features of the low spatial resolution spectroscopic image must be compared to those of the high spatial resolution standard-proton-image. It is therefore important to have an accurate knowledge

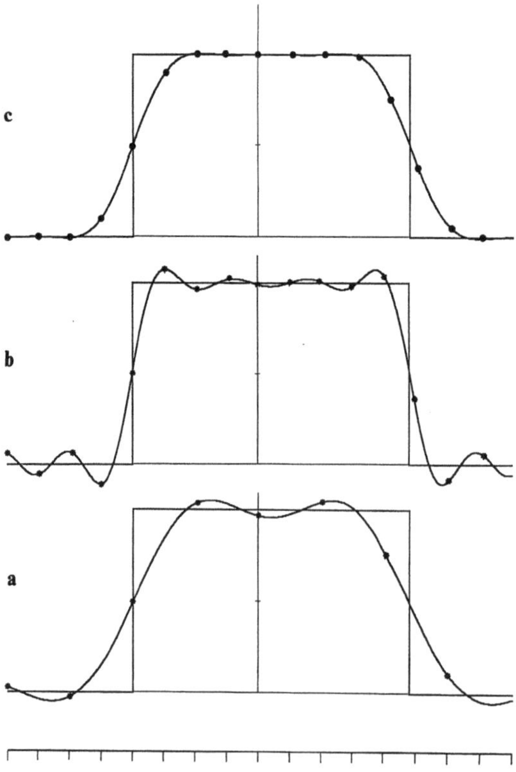

Fig. 7 a–c. Profile of a bar obtained from **a** 8 sampling points in k space, **b** 16 sampling points in k space and **c** 16 sampling points in k space apodized with a Hanning window. The Fourier transformation gives the sampling points indicated by *dots*. The bar extends from 0.25 FOV to 0.30 FOV

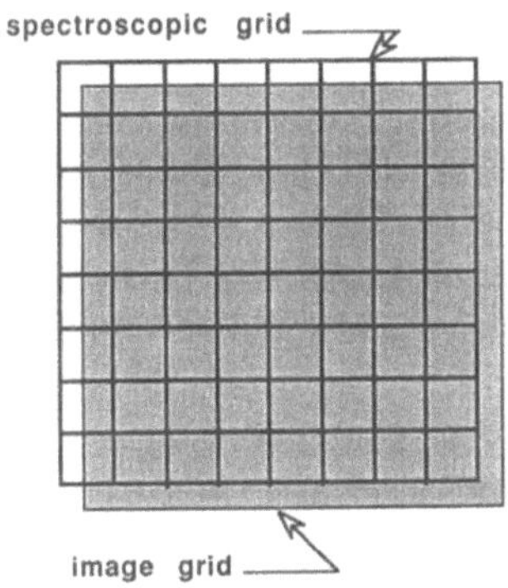

Fig. 8. Relative positions of a low resolution spectroscopic-grid and a high resolution image grid

of the position of the spectroscopic grid with respect to the proton image grid. Consider a spectroscopic image with one spatial dimension. FT in the k direction of the array of phase-encoded FIDs yields spectra from voxels at positions $z_m = m\Delta l = m2\pi/\gamma M\Delta G\tau$, where $m = -M/2, -M/2+1, -M/2+2, \ldots, M/2-1$. Each voxel extends from the value $m\Delta l - \Delta l/2$ to the value $m\Delta l + \Delta l/2$. Consequently, the boundaries of the image are at $(\pm \text{FOV} - \Delta l)/2$ (however, it should be borne in mind that, as pointed out above, the signal from the first voxel $(m = -M/2)$ partly results from an aliasing of the signal in the region comprised between $z = (\text{FOV} - \Delta l)/2$ and $z = \text{FOV}/2$. Since the shift of the field-of-view is determined by the spatial resolution, it follows that the effective field-of-view of the spectroscopic image does not coincide with the field-of-view of the high spatial resolution of the standard proton image. This is illustrated in Fig. 8. This point should be taken into account when the features of the spectroscopic image are to be compared with that of the proton image.

4.2.4 *Shift of the Spectroscopic Grid*

It is well known (shift theorem: see for example Ref. [182, 183]) that the Fourier transform of the function $e^{-i2\pi\Delta zk}f(t,k)$ is equal to $F(t, z+\Delta z)$. It is therefore easy to shift the spectroscopic-grid of Δz by multiplying, before FT, each phase-encoded FID $f(t,k)$ by the quantity $e^{-i2\pi\Delta zk}$ [8, 175, 184, 185]. Such a method can be used to obtain a spectroscopic-grid which coincides with the image-grid. Moreover the use of this technique allows, after acquisition, an adjustment of the voxel position according to anatomical feature (for example, an attenuation of partial volume effects can be obtained if a shift of the grid is practiced in order to maximize the number of voxels entirely situated inside a tumor) [163]. This point is probably a major advantage of FSI over selective excitation. It should be noted that such a post-acquisition shift is fully equivalent to shifting the center of the gradient coil system.

4.2.5 *Shape of Spatial Response and Signal-to-Noise Ratio*

The shape of the spatial response can be modified by using filters in the k space, through a so-called weighting function. k-space filters affects both the shape of the spatial function and the S/N.

Consider a point-sample at the center of the mth voxel. Let $(S/N)_{SV}$ be the S/N per unit time, obtained without gradient encoding. Such a S/N could be obtained with a perfect single voxel technique like ISIS or PRESS. From Eq. (20), the signal from the point sample at the nth phase-encoding step is given by $\exp - i2\pi k_n z_m$. It results from Eq. (21) that an unweighted FSI experiment yields the same S/N as a single voxel experiment:

$$(S/N)_{\text{unweighted}} = (S/N)_{SV} \tag{26}$$

Filtering can be applied during acquisition (acquisition weighting): the number of transients averaged for a given phase-encoding step is proportional to the weighting function [175, 179, 184, 185]. The (S/N) is then given by:

$$(S/N)_{\text{acq weighting}} = \frac{\sum_{k=1}^{M} w(k)}{\sum_{k=1}^{M} |w(k)|} (S/N)_{SV} \tag{27}$$

where $w(k)$ is the value of the weighting function at the kth phase-encoding step, and $|w(k)|$ the corresponding scan number. Clearly, if $w(k) \geqslant 0$, acquisition weighting provides the same S/N as a single-voxel or an unweighted experiment. Acquisition weighting yields an optimum S/N per unit time [179, 185]. The use of negative coefficients (such as those which are needed to achieve a sinc weighting) decreases the S/N.

Filtering can also be applied after acquisition (numerical weighting). In this case, the signal $S(k)$ is multiplied by the weighting function $w(k)$. The S/N is then given by:

$$(S/N)_{\text{num weighting}} = \frac{\sum_{k=1}^{M} w(k)}{\left[M \sum_{k=1}^{M} w(k)^2 \right]} (S/N)_{SV} \tag{28}$$

It can easily be shown that $\sum_{k=1}^{M} w(k) \leqslant \left[M \sum_{k=1}^{M} w(k)^2 \right]^{1/2}$. Equality is obtained when $w(k) = \text{constant}$ (unweighted FSI). Thus numerical weighting degrades the S/N.

The various filtering techniques will now compared in terms of S/N and shape of the spatial function. This can be performed by using a normalized spatial response function, *the sensitivity profile*, defined as the S/N per unit time for a point sample as a function of its distance from the voxel center. Fig. 9 shows the S/N of a one dimensional FSI experiment performed with, (a) 8 phase-encoding

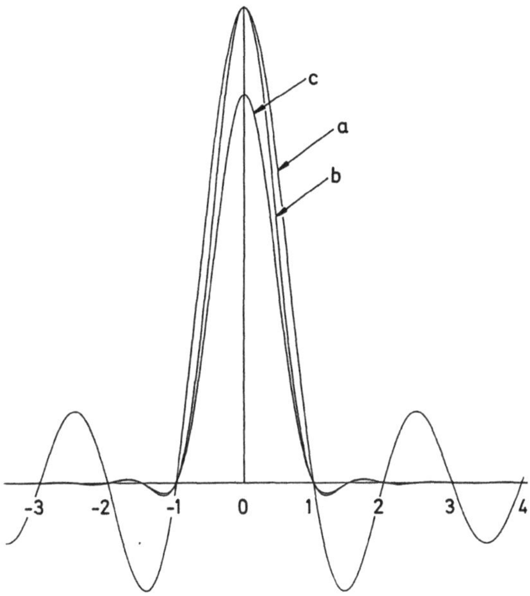

Fig. 9 a–c. Signal-to-noise per unit time for a point sample at position z (sensitivity profile) for a: **a** 8 encoding steps experiment without apodisation, **b** 16 encoding steps experiments using acquisition weighting with a Hanning window and **c** 16 encoding steps experiment using numerical weighting with a Hanning window.

steps without apodisation, (b) 16 phase-encoding steps combined with acquisition weighting and, (c) 16 phase-encoding steps using a numerical weighting post acquisition. Both b, c, are filtered with a Hanning window function. The width of the sensitivity profile is roughly similar for the three experiments but the filtered experiment is not affected by truncation artifacts. The S/N per unit time, at maxima of the sensitivity profile, in the acquisition weighted and in the unfiltered experiments are similar. Figure 9 clearly shows the S/N loss resulting from numerical weighting compared to acquisition weighting. Of course, the S/N per voxel actually results from an integration over the sample volume of the product of the sensitivity profile with the spin distribution.

Weighting eliminates truncation artifacts, but the disadvantage is reduced spatial resolution. For a given width of the spatial response function, a weighted experiment requires a greater number of phase-encoding steps than an unweighted experiment. For example, a 16 encoding steps experiment with Hanning acquisition weighting is free from truncation artifacts as compared to an 8 encoding steps experiment without weighting. The important point is that the width and the height of the sensitivity profile are similar. Figure 7c shows the profile of a bar resulting from 16 sampling points in the k-space and windowing with a Hanning filter. However, acquisition windowing always necessitates signal averaging. In cases where averaging is required to achieve a satisfactory S/N, acquisition weighting should clearly be preferred. This should be the case for most animal models. In cases where S/N is large, numerical weighting should be preferred. However, at a given width of the spatial response function the price to pay when using numerical weighting is an increased number of phase-encoding steps (and thus of the minimum number of acquisitions) and a decreased S/N. An in-depth treatment of weighted phase-encoding can be found in Ref. [179] where the combination of phase-encoding, spatial shifting, and acquisition weighting has been named Selective Fourier Transform.

Finally it should be noted that motion or instrumental instabilities during acquisition of the array of phase-encoded FIDs, yield artifacts [81] similar to t_1 noise in 2D experiments in liquids [186, 187].

4.2.6 *Summary*

FSI seems to be an advantageous localization technique since spectra are simultaneously obtained from a number of volume elements, the position of which can be shifted after acquisition. Spectroscopic images free from truncation artifacts can be obtained by using spatial filtering. A number of high-quality ^1H or ^{31}P spectra have been obtained on humans and animals. The main drawback of the technique is that the minimum number of scans required to obtain the data can be large. For example, 3D FSI may be time-consuming and it can be more efficient to combine interleaved selective excitation of adjacent slices within T_R and 2D FSI of the slice [188]. FSI can also be used in a region of interest selected by a single voxel localization technique [126, 56, 7].

4.3 *Spectral Localization by Imaging (SLIM)*

An elegant method, SLIM [151], was recently proposed to obtain spectroscopic informations from N compartments of arbitrary shape through a set of N or more phase-encoded FIDs. The processing of the data set is based on the use of prior knowledge. It is assumed that the sample is composed of well-defined compartments, each of which is supposed to have uniform distribution of magnetization. The signal resulting from phase-encoding may be written as:

$$f(t, k_n) = \sum_i \rho_i(t) \int_{V_i} \exp - i2\pi k_n r dV \tag{29}$$

where ρ_i represents the signal density in the ith compartment. Given the compartment shapes, the integral over each of the compartment can be calculated. If the number of phase-encoding steps, M, is taken to be equal or greater than the number of compartments N, the problem is reduced to the calculation of the $\rho_i(t)t$ from the system of M linear equations. In-vivo spectra obtained with this method have recently been presented [189, 152]. Inhomogeneous compartments introduce contamination; it was, however, recently demonstrated that localization errors could be minimized by choosing an optimal set of phase-encoding gradients [152]. In conclusion the SLIM technique is promising, but it was proposed too recently have been thoroughly tested until now.

5 Water Signal Suppression

Since early experiment at Yale [190], in-vivo proton spectroscopy has developed rapidly. This increased interest is due to the high sensitivity of this nucleus. However, ^1H spectroscopy requires water signal suppression. A detailed description of water suppression techniques lies outside the scope of this review and can be found in Ref. [47]. The water line saturation technique can be used, which resembles spatial localization techniques employing a saturation of the outer volume (Sect. 3.2). One can also use a selective excitation of the resonances of interest which leaves the water magnetization unchanged; this method resembles spatial localization methods using direct excitation of the spins inside the VOI (Sect. 3.3). Note that the acquisition of a spectrum from a 1-ml VOI in the human brain requires a suppression of the signal from the outer volume by a factor much greater than 10^3. This is quite similar to the requirements for water suppression in spectroscopy.

Water suppression uses frequency-selective pulses. Techniques of the second group (Sect. 3) also employ frequency-selective pulses. The pulses cannot be used for both spatial selection and chemical shift selection, so water suppression must be performed separately.

If the localization results from saturation of all the spins selection outside the VOI (Sect. 3.2), water suppression can be achieved by adding a water suppressing pulse (for example, a binomial refocusing pulse) to the sequence.

When the localization results from selective excitation of the spins inside the VOI, leaving unaffected the magnetizations of a large part of the sample, the use of a water-suppressing refocusing pulse applied after the localization sequence is possible. However, imperfections of the water-suppressing pulse can ruin spatial localization. A better solution is to saturate the water line of the sample. Saturation can be efficiently achieved by using frequency-selective pulses and spoiling gradients [191–195]. Saturation must be applied before the application of the spatial localization sequence. However, further improvement of water suppression can be obtained by saturating during the localization sequence (for example [121]) during the T_M delay of a stimulated echo sequence).

Finally, a spin echo phase-encoding experiment uses rf pulses applied without gradient and water suppression can easily be obtained by replacing the two pulses (or at least one of them) by a frequency-selective pulse such as a binomial pulse. If the experiment uses slice selection, water suppression can be achieved by using a saturation technique.

6 Comparison of Techniques and Elements of Choice

The last ten years have witnessed the design of several techniques to obtain localized spectra from animals or humans. Spatially resolved spectra can now be obtained using one, or a combination of a variety of effective methods. All the methods employing B_0 gradients rely upon the spread of the Larmor frequencies along a spatial axis. With phase-encoding techniques, this spread is achieved during a free precession delay, whereas with selective excitation it is achieved in the rotating frame in the presence of B_1. In the former method, the profile of the selected volume depends on the weighting function, whereas in the latter, it depends on the pulse shape.

Localization techniques are still evolving. However, some techniques have emerged: ISIS which is tolerant to poor switching characteristics of the gradient coils, stimulated echo spectroscopy, double spin echo spectroscopy, and phase-encoding techniques. There is up to date a large interest in phase-encoding techniques which have been compared theoretically [2] and experimentally [7] to single voxel techniques.

The advantages of FSI over single voxel techniques are numerous. The primary advantage is its flexibility. SI produces M spectra from M voxels simultaneously. This is a very important point in the study of diseases or heterogeneous tissues. FSI does not require prior knowledge of the position of the damaged tissue. Shifting the voxel position after acquisition allows the voxel position to be adjusted to a lesion. Furthermore, the distribution of a given metabolite can easily be calculated. The absence of chemical-shift artifact is another important advantage of phase-encoding techniques. Note that HSI is affected by the chemical shift artifact. Finally, the truncation artifacts can be minimized by using a weighting function in the k-space. Apodisation necessarily broadens the spatial response function. To recover the width of the spatial response function, the spatial resolution must be increased; if the weighted function is applied during acquisition, the extension of the sampling process does not result in a loss of signal-to-noise.

The main problems of phase-encoding techniques are the minimum duration of the experiment which can be large, the B_0 homogeneity over the region of interest and the shape of the sensitivity profile. Moreover, the spatial localization achieved with a spectroscopic imaging technique is based on an add process of FIDs. Motion and instrumental instabilities can produce localization errors. Single voxel experiments such as STEAM or PRESS offer the advantage to work with one single scan, allowing shimming on the VOI. Motion or instrumental instabilities may produce some signal loss but the signal from the VOI remains relatively free from contamination. In principal, techniques based on selective excitation in the presence of gradients have the advantage of complete phase coherence and provides a well-defined volumes. However, it has been experimentally found [56] that spectroscopic imaging gives a slightly improved S/N to single-voxel techniques. This result has been attributed to phase variations over the region selected by imperfect spatially selective pulses.

In fact, a combination of methods can be advantageous. For example, spectroscopic imaging can be achieved within a volume selected with a localization scheme using frequency-selective excitation [56, 126]. Similarly, B_1 and B_0 methods can be combined. For example, a slice selection free from chemical shift artifact by exploiting the B_1 inhomogeneity of a surface coil, can be used in combination with spectroscopic imaging within the slice [171]. Finally, spectroscopic information from very small volume elements in a large sample, could be obtained by combining saturation of the outer volume, selective excitation of the region of interest, and spectroscopic imaging within this region.

References

1. Aue WP (1986) Rev Magn Res Med 1:21
2. Granot J (1986) J Magn Reson 66:197
3. Galloway GJ, Brereton IM, Brooks WM, Field J, Irving M, Baddeley H, Doddrell DM (1987) Magn Reson Med 4:393
4. Garwood M, Schleich T, Bendall MR (1987) J Magn Reson 73:191
5. Jackson EF, Narayana PA, Kudrle WA (1988) J Magn Reson 80:23
6. Moonen CT, von Kienlin M, van Zijl PCM, Cohen J, Gillen J, Daly P, Wolf G (1989) NMR in Biomed 2:201
7. Posse S, Aue WP (1989) NMR in Biomed 2:234
8. Bottomley PA, Hardy CJ, Roemer PB, Weiss RG (1989) NMR in Biomed 2:284
9. Gordon RE, Hanley PE, Shaw D, Gadian DG, Radda GK, Styles P, Bore PJ, Chan L (1980) Nature (London) 287:736
10. Aue WP, Müller S, Cross TA, Seelig J (1984) J Magn Reson 56: 350
11. Bottomley PA, Foster TB, Darrow RD (1984) J Magn Reson 59:338
12. Ordidge RJ, Connelly A, Lohman JAB (1986) J Magn Reson 66:283
13. Maudsley AA, Oppelt A, Ganssen A (1979) Siemens Forsch 8:326
14. Bendel P, Lai CM, Lauterbur PC (1980) J Magn Reson 38:343
15. Bolinger L, Leigh JS (1988) J Magn Reson 80:162
16. Goelman G, Subramanian VH, Leigh JS (1990) J Magn Reson 89:437
17. Chen CN, Hoult DI (1989) Biomedical Magnetic Resonance Technology Adam Hilger, Bristol and New York
18. Hoult DI (1979) J Magn Reson 35:69
19. Mansfield P, Maudsley AA, Baines T (1976) J. Phys. E 9:271
20. Hoult DI (1977) J Magn Reson 26:165
21. Mansfield P, Maudsley AA, Morris PG, Pykett IL (1979) J Magn Reson 33:261
22. Frahm J, Hänicke W (1984) J Magn Reson 60:320
23. Morris PG, Dominick JO, McIntyre JO, Rourke DE, Ngo JT (1989) NMR in Biomed 2:257
24. Warren WS, Silver MS (1988) in: Waugh JS (ed) Advances in magnetic resonance. 12:247 Academic Press, New York
25. Emsley L, Bodenhausen G (1989) Magn Reson Med 10:273
26. Ngo JT, Morris PG (1987) Magn Reson Med 5:217
27. McCoy MA, Loiza F, Valentine K, Warren WS (1988) J Magn Reson 80:155
28. Geen H, Wimperis S, Freeman R (1989) J Magn Reson 85:620
29. Roberts TPL, Carpenter TA, Hall LD (1990) J Magn Reson 89: 595
30. Warren WS (1984) J Chem Phys 81, 5437
31. Silver MS, Joseph RI, Hoult DI (1984) J Magn Reson 59:347
32. Segebarth C, Luyten PR, den Hollander JA (1987) J Magn Reson 75:345
33. Volk A, Tiffon B, Mispelter J, Lhoste J-M (1987) J Magn Reson 71:168
34. Hall LD, Tagala SL (1987) J Magn Reson 71:180
35. Volk A, Tiffon B, Mispelter J, Lhoste J-M (1988) J Magn Reson 76:386
36. Gyngell ML, Frahm J, Merboldt KD, Hänicke W, Bruhn H (1988) J Magn Reson 77:596
37. Carr HY, Purcell EM (1954) Phys Rev 94:630
38. Hahn EL (1950) Phys Rev 80:580
39. Rothman DL, Behar KL, den Hollander JA, Shulman RG (1984) J Magn Reson 59:157
40. Bodenhausen G, Freeman R, Turner DL (1977) J Magn Reson 27:511
41. Rance M, Byrd RE (1983) J Magn Reson 52:221

42. Bendall MR, Gordon RE (1983) J Magn Reson 53:365
43. Woessner DE (1961) J Chem Phys 34, 2057
44. de Graff AA, Bovee WMMJ, Deutz NEP, Chamuleau RAFM (1988) Magn Reson Imag 6:255
45. van Zijl PCM, Moonen CT, Alger JR, Cohen JS, Chesnick SA (1989) Magn Reson Med 10:256
46. Blondet P, Décorps M, Confort S, Albrand JP (1987) J Magn Reson 75:434
47. Guéron M, Plateau P, Décorps M (1991) Prog NMR Spectr 23:135
48. Frahm J, Merbolt KD, Hänicke W, Haase A (1985) J Magn Reson 64:81
49. Granot J (1986) J Magn Reson 70:488
50. Kimmich R, Hoepfel D (1987) J Magn Reson 72:379
51. Frahm J, Bruhn H, Gyngell ML, Merboldt KD, Hänicke W Sauter R (1989) Magn Reson Med 9:79
52. Sattin W, Mareci TH, Scott KN (1985) J Magn Reson 64:177
53. Turner R (1986) J Phys D: Appl Phys 19: L147
54. Mansfield P, Chapman B (1986) J Magn Reson 66:573
55. Bowtell R, Manfield P (1990) Meas Sci Technol 1:431
56. Moonen CTW, Sobering G, van Zijl PCM, Gillen J, Daly P in preparation
57. Moonen CT, van Zijl PCM, Le Bihan D, DesPres D (1990) Magn Reson 13:467
58. Gordon RE (1980) Biosciences Rep. 2:701
59. Gordon RE, Hanley PE, Shaw D (1982) Progr NMR Spectrosc 15:1
60. Scott KN, Brooker HR, Fitzsimmons JR, Bennett HF, Mick RC (1982) J Magn Reson 50:339
61. Hinshaw WS (1976) J Appl Phys 47:3709
62. Crowley MG, Ackerman JJH (1985) J Magn Reson 65:522
63. Chen W, Ackerman JJH (1989) NMR in biomed 1:205
64. Chen W, Ackerman JJH (1989) J Magn Reson 82:655
65. Chen W, Ackerman JJH (1989) NMR in biomed 2:267
66. Jehenson P, Bloch G (1991) J Magn Reson 94:59
67. Chen W, Ackerman JJH (1990) NMR in Biomed 3:147
68. Chen W, Ackerman JJH (1990) NMR in Biomed 3:158
69. Hennig J, Boesch C, Gruetter R, Martin E (1987) J Magn Reson 75:179
70. Geoffrion Y, Rydzy M, Butler KW, Smith ICP, Jarrell HC (1988) NMR in biomed 1:107
71. Malloy CR, Lange RA, Klein DL, Jeffrey FM, Peshock RM, Willerson JT, Nunnally RM (1988) J Magn Reson 80:364
72. Ordidge RJ, Mansfield P, Lohman JAB, Prime SB (1987) Ann. NY Acad Sci 508:376
73. Ordidge RJ, Bowley RM, McHale G (1988) Magn Reson Med 8:323
74. Luyten PR, Groen JP, Vermeulen JWAH, den Hollander JA (1989) Magn Reson Med 11:1
75. Hanstock CC, Rothman DL, Prichard JW, Jue T, Shulman RG (1988) Proc Natl Acad Sci USA, 85:1821
76. Hanstock CC, Rothman DL, Jue T, Shulman RG (1988) J Magn Reson 77:383
77. Segebarth C, Grivegnee A, Luyten PR, den Hollander JA (1988) Magn Reson Med Biol 1:7
78. Matson GB, Twieg DB, Karczmar GS, Lawry TJ, Gober JR, Valenza M, Boska MD, Weiner MW (1988) Radiology 169:541
79. Weiner MW, Hetherington H, Hubesch B, Karczmar G, Massie B, Maudsley A, Meyerehoff DJ, Sappey-Marinier D, Schaefer S, Twieg DB, Matson GB (1989) NMR in biomed 2:290
80. den Hollander JA, Luyten P.R., Mariën AJH, Segebarth CM, Balériaux DF, de Beer R, van Ormondt D (1989) Magn Reson Quart 5:152
81. Arnold DL, Shoulbridge EA, Feindel W, Villemure JG (1987) Can. J. Neurol Sc 14:570
82. Robitaille, Merkle H, Lew B, Path G, Hendrich K, Lindstrom P, From AHL, Garwood M, Bache RJ, Ugurbil K (1990) Magn Reson Med 16:91
83. Meyerhoff DJ, Karczmar GS, Matson GB, Boska MD, Weiner MW (1990) NMR in Biomed 3:17
84. Hubesch B, Sappey-Marinier D, Roth K, Meyerhoff DJ, Matson GB, Weiner MW (1990) Radiology 174:401
85. Décorps M, Laval M, Confort S, Chaillout JJ (1985) J Magn Reson 61:418
86. Rémy C, Albrand JP, Benabid AL, Décorps M, Jacrot M, Riondel J, Foray MF (1987) Magn Reson Med 4:144
87. Conelly A, Counsell C, Lohman JAB (1988) J Magn Reson 78:519
88. Jung WI, Lutz O (1988) Z. Naturforsch 43 a: 909

89. Robitaille, Merkle H, Sublett E, Hendrich K, Lew B, Path G, From AHL, Bache RJ, Garwood M, Ugurbil K (1989) Magn Reson Med 10:14
90. Aue WP, Müller S, Seelig J (1985) J Magn Reson 61:392
91. Cross RA, Müller S, Aue WP (1985) J Magn Reson 62:87
92. Müller S, Aue WP, Seelig J (1985) J Magn Reson 63:530
93. Müller S, Aue WP, Seelig J (1985) J Magn Reson 65:332
94. Luyten PR, Marien AJH, Sijtsma B, den Hollander JA (1986) J Magn Reson 67:148
95. Luyten PR, den Hollander JA (1986) Radiology 161:795
96. Luyten PR, den Hollander JA (1986) Magn Reson Imag 4, 237
97. Jensen DJ, Delayre JL, Narayana PA (1986) J Magn Reson 69:552
98. Narayana PA, Jensen DJ, Jackson EF, Brey WW (1988) J Magn Reson 79:11
99. Doddrell DM, Brooks WM, Bulsing JM, Field J, Irving MG, Baddeley H (1986) J Magn Reson 68:367
100. Doddrell DM, Field J, Crozier S, Brereton IM, Galloway GJ, Rose SE (1990) Magn Reson Med 13:518
101. Galloway GJ, Field J, Rose SE, Haseler LJ, Brooks WM, Brereton IM, Bore PJ, Crozier S, Doddrell DM (1989) Magn Reson 9:288
102. Crozier S, Field J, Doddrell DM (1990) Magn Reson Imag 8:277
103. von Kienlin M, Rémy C, Benabid AL, Décorps M (1988) J Magn Reson 79:382
104. Décorps M, Rémy C, Bourgeois D, Jacrot M, von Kienlin M, Benabid AL (1989) Magn Reson Med Biol 2:129
105. Doddrell DM, Bulsing JM, Galloway GJ, Brooks WM, Field J, Irving M, Baddeley H (1986) J Magn Reson 70:319
106. Galloway GJ, Brooks WM, Bulsing JM, Brereton IM, Field J, Irving M, Baddeley H, Doddrell DM (1987) J Magn Reson 73:360
107. Sauter R, Müller S, Weber H (1987) J Magn Reson 75:167
108. Haase A (1986) Magn Reson Med 3:963
109. Granot J, Ben Mair J, Arrive L, Menu Y, Martin N, Nahum H (1990) J Magn Reson 89:139
110. de Crespigny AJS, Carpenter TA, Hall LD (1989) J Magn Reson 85:595
111. Bottomley PA (1989) Radiology 170:1
112. Bottomley PA, Hardy CJ (1987) J Magn Reson 74:550
113. Frahm J, Merbolt KD, Hänicke W (1987) J Magn Reson 72:502
114. McKinnon G (1986) 5th Ann Meeting Soc Magn Reson Med, Works in progress; Book of Abstracts, p 168
115. Bottomley PA (1987) Ann NY Acad Sci 508:333
116. Bottomley PA (1984) U.S. Patent 4 480 228
117. Ordidge RJ, Bendall MR, Gordon RE, Connelly A (1985) in: Govil, Khetrapal and Saran, Tata (eds) Magnetic resonance in biology and medicine McGraw-Hill, New Delhi, p 387
118. Zhang W, van Hecke P (1990) J Magn Reson 89:461
119. Hafner HP, Radü E, Seelig J (1990) Magn Reson Med 15:135
120. Jackson EF, Narayana PA, Flamig DP (1990) Magn Reson 8:153
121. Bomsdorf H, Helzel T, Kunz D, Roëschmann P, Tschendel O, Wieland J (1988) NMR in Biomed. 1:151
122. Frahm J, Bruhn H, Gyngell ML, Merboldt KD, Hänicke W, Sauter R (1989) Magn Reson Med 11:47
123. Bruhn H, Frahm J, Gyngell ML, Merboldt KD, Hänicke W, Sauter R, Hamburger C (1989) Radiology 172:541
124. Merboldt KD, Chien D, Hänicke W, Gyngell ML, Bruhn H, Frahm J (1990) J Magn Reson 89:343
125. Sauter R, Loeffler W, Bruhn H, Frahm J (1990) Radiology 176:221
126. Segebarth CM, Balériaux DF, Luyten PR, den Hollander JA (1990) Magn Reson Med 13:62
127. Desmoulin F, Seelig J (1990) Magn Reson Med 14:100
128. Griffey RH, Flaming DP (1990) J Magn Reson 88:160
129. Frahm J, Michaelis T, Merboldt KD, Hänicke W, Gyngell ML, Chien D, Bruhn H (1988) NMR in Biomed. 2:188
130. Bruhn H, Frahm J, Gyngell ML, Merboldt KD, Hänicke W, Sauter R (1989) Magn Reson Med 9:126
131. Ernst T, Hennig J, Ott D, Friedburg H (1989) NMR in Biomed 2:216
132. Bottomley PA, Smith LS, Leue WM, Charles C (1985) J Magn Reson 64:347
133. Müller S, Sauter R, Webber H, Seelig J (1988) J Magn Reson 76:155

134. Müller S, Hafner HP,, Beckmann N (1989) NMR in Biomed 2:209
135. Manassen Y, Navon G (1985) J Magn Reson 61:363
136. Lauterbur PC, Levin DN, Marr RB (1984) J Magn Reson 59:536
137. Haase A, Matthaei (1987) J Magn Reson 71:550
138. Feig E, Greenleaf F (1985) Appl. Optics 24:3977
139. Park HW, Cho ZH (1986) Magn Reson Med 3:448
140. Hall LD, Sukumar S (1982) J Magn Reson 50:161
141. Doyle M, Mansfield P (1987) Magn Reson 5:255
142. Webb P, Spielman D, Macovski A (1989) Magn Reson 12:306
143. Matsui S, Sekihara K, Kohno H (1986) J Magn Reson 67:476
144. Mansfield P (1984) Magn Reson 1:370
145. Matsui S, Sekihara K, Kohno H (1985) J Magn Reson 64:167
146. Hall LD, Rajanayagam V, Sukumar S (1985) J Magn Reson 61:188
147. Maudsley AA, Hilal SK, Perman WH, Simon HE (1983) J Magn Reson 51:147
148. Brown TR, Kincaid BM, Ugurbil K (1982) Proc. Natl Acad Sci (USA) 79:3523
149. Haselgrove JC, Subramanian VH, Leigh JS, Gyulai L, Chance B (1983) Science 22:1172
150. Pykett IL, Rosen BR (1983) Radiology 149:197
151. Hu X, Levin DN, Lauterbur PC, Spraggins T (1988) Magn Reson Med 8:314
152. von Kienlin M, Mejia R (1991) J Magn Reson 94:268
153. Kumar A, Welti D, Ernst RR (1975) J Magn Reson 18:69
154. Allman T, Holland GA, Lenkinski RE, Charles HC (1988) Magn Reson Med 7:88
155. Derby K, Hawrysko C, Tropp J (1989) Magn Reson Med 12:235
156. Maudsley AA, Twieg DB, Sappey-Marinier D, Hubesch B, Hugg JW, Matson GB, Weiner MW.(1990) Magn Reson Med 14:415
157. Tropp J, Derby KA, Hawrysko S, Sugiura S, Yagamata H (1989) J Magn Reson 85:244
158. Hall LD, Rajanayagham V (1987) J Magn Reson 74:139
159. Hall LD, Tagala SL (1985) J Magn Reson 65:501
160. Martin JF, Wade CG (1985) J Magn Reson 61:153
161. Webb P, Spielman D, Macovski A (1989) Magn Reson Med 12:306
162. de Graaf AA, Bovee WMMJ (1990) Magn Reson Med 15:305
163. Ross B, Narasimhan PT, Tropp J, Derby KS (1989) NMR in Biomed 2:340
164. Bailes DR, Bryant DJ, Bydder GM, Case HA, Collins AG, Cox IJ, Evans PR, Harman RR, Hall AS, Khenia S, McArthur P, Oliver A, Rose MR, Ross BD, Young IR (1987) J Magn Reson 74:158
165. Coutts GA, Bryant DJ, Collins AG, Cox IJ, Sargentoni J, Gadian DG (1989) NMR in Biomed 1:190
166. Glazer GM, Smith SR, Chenevert TL, Martin PA, Stevens AN, Edwards RHT (1989) NMR in Biomed 1:184
167. Bailes DR, Bryant DJ, Case HA, Collins AG, Cox IJ, Hall AS, Harman RR, Khenia S, McArthur P, Ross BD, Young IR (1988) J Magn Reson 77:460
168. Bottomley PA, Hardy CJ, Roemer PB (1990) Magn Reson Med 14:369
169. Twieg DB, Meyerhoff DJ, Hubesch B, Roth K, Sappey-Marinier D, Boska MD, Gober JR, Schaefer S, Weiner MW (1989) J Magn Reson Med 12:291
170. Chatham JC, Blackband SJ (1990) NMR in Biomed 3:190
171. Bourgeois D, Rémy C, le Fur Y, Devoulon P, Benabid AL, Décorps M (1991) Magn Reson Med 21:10
172. Barker PB, Ross BD (1987) J Magn Reson 75:467
173. King KF, Moran PR (1984) Med Phys 11, 1
174. Bax A, Mehlkopf AF, Smidt J (1979) J Magn Reson 35:373
175. Brooker HR, Mareci TH, Mao J (1987) Magn Reson Med 5:417
176. Harris F.J. (1978) Proc. IEEE 66:51
177. den Boef JH, van Uijen CMJ, Holzscherer CD (1984) Phys Med Biol 29:857
178. Haacke EM, Liang ZP, Izen SH (1989) Med Phys 16, 388
179. Parker DL, Gullberg GT, Frederick PR (1987) Med Phys 14:640
180. Whittemore AR (1990) Am J Roentgenol 154:204
181. Maudsley A (1986) J Magn Reson 68:363
182. Papoulis A (1977) Signal Analysis. McGraw-Hill, New York
183. Bracewell R (1965) The Fourier transform and its applications. McGraw-Hill, New York
184. Mareci TH, Brooker HR (1984) J Magn Reson 57:157
185. Mareci TH, Brooker HR (1991) J Magn Reson 92:229

186. Young IR, Cox IJ, Coutts GA, Bydder GM (1989) NMR in Biomed. 2:329
187. Ernst RE, Bodenhausen G, Wokaun A (1987) Principles of nuclear magnetic resonance in one and two dimensions. Clarendon Press, Oxford
188. Bottomley PA, Charles HC, Roemer PB, Flamig D, Engeseth H, Edelstein WA, Mueller OM (1988) Magn Reson Med 7:319
189. Fletcher ES, Gregory CD, Lee H, Greenough WT, Lauterbur PC, Dawson MJ (1990) Abst 9th Meet Soc Magn Reson Med, New York, 1332
190. Behar KL, den Hollander JA, Stromski ME, Ogino T, Shulman RG, Petroff OAC, Prichard JW (1983) Proc Natl Acad Sci (USA) 80:4945
191. Haase A, Frahm J, Hänicke W, Matthaei D (1985) Phys Med Biol 30:341
192. Doddrell DM, Galloway GJ, Brooks WM, Field J, Bulsing JM, Irving MG, Baddeley H (1986) J Magn Reson 70:176
193. Brereton IM, Field J, Moxon LN, Irving MG, Doddrell DM (1989) Magn Reson Med 9:118
194. Brereton IM, Galloway GJ, Field J, Marshman MF, Doddrell DM (1989) J Magn Reson 81:411
195. Moonen CTW, van Zijl PCM (1990) J Magn Reson 88:28

^{1}H NMR Spectroscopy and Spectroscopic Imaging of The Human Brain

Jan A. den Hollander, Peter R. Luyten, Ad J.H. Mariën

From the MR Physics Department, Philips Medical Systems, Best, The Netherlands

Table of Contents

1 Introduction

The first high field ^1H NMR spectroscopy studies of the animal brain were performed in 1983 [1]. These studies showed that ^1H NMR spectroscopy allows in vivo observation of a number of brain compounds, including different aminoacids, creatine, choline residues, and—under certain perturbations—of lactate. Those initial studies have given rise to a large number of studies which apply ^1H NMR spectroscopy to different aspects of brain metabolism, under different kind of perturbations [2]. These studies have shown that a wealth of information can be obtained about cerebral metabolism and physiology by ^1H NMR spectroscopy of the animal brain.

Before ^1H NMR spectroscopy can be applied successfully to the animal brain, one has to face up to a number of complicating factors. Although ^1H NMR spectroscopy is more sensitive than ^{31}P NMR spectroscopy given the same number of spins, acquisition of qualitatively acceptable spectra is considerably more difficult. This has to do with the presence in the ^1H NMR spectrum of living tissues, of two intense signals, namely water and fat. Thus any successful technique in the application of ^1H NMR spectroscopy has, in one way or another, to reduce the intensity of those intense signals, before the low concentration compounds can be observed without interference of the intense signals. Furthermore, spectral resolution is limited because of the small chemical shift range of ^1H NMR spectroscopy compared to other nuclei such as ^{31}P and ^{13}C.

In animal studies these problems are alleviated by the use of high magnetic fields (up to 9.5 tesla) and, sometimes, the use of invasive techniques to remove overlying tissue to reduce the interference of intense lipid signals. However, in order to apply ^1H NMR spectroscopy to the human brain one has to deal with these problems at the relatively low magnetic field of whole body magnets (1.5–2.1 tesla), and one has to rely on non-invasive techniques. Notwithstanding these difficulties the first spectra of the human brain were obtained only a few years after the first animal studies [3, 4, 5]. Although those early studies did demonstrate the feasibility of ^1H NMR spectroscopy of the human brain, they did not lead to a large number of patient studies. The reason was that the techniques used were difficult to optimize, and often led to spectra with unacceptable contamination by lipid and water signals. Over the last few years those problems have been overcome by the optimization of more suitable pulse sequences, and we can now witness a rapid proliferation of human ^1H NMR studies.

The first successful application of ^1H NMR spectroscopic imaging to the animal brain [6] prompted us to explore the potential of applying ^1H NMR spectroscopic imaging to humans. The availability of adequate localization techniques, and water suppression methods was essential before ^1H NMR spectroscopic imaging resulted in patient studies. In the present paper we wish to describe the techniques used in ^1H NMR spectroscopy, and ^1H NMR spectroscopic imaging with a number of applications to volunteers and patients.

2 Localization Techniques

A number of localization schemes have been explored for obtaining ^{1}H NMR spectra of the human brain [7]. Some of those techniques have been based upon suppression of signals from surrounding tissue, while preserving the signals from within the volume of interest [8, 9]. The problem with those techniques is that extreme requirements are posed for the suppression of the signals arising from the outer volume; in particular contaminating residual signals from the intense water and fat signals pose a severe problem. In practice such sequences are limited to use with surface coils or closely coupled head coils. For use with regular head coils the use of a spin echo sequence which selects a localized volume by combining three orthogonally slice selective excitation and refocusing pulses is more favorable under most experimental conditions. By avoiding unwanted excitation of outer volume spins as much as possible by the use of slice selective pulses, the suppression of unwanted signals becomes much less critical. For a given coil geometry these spin echo sequences introduce a fixed ratio between excited spins from the outer volume which have to be suppressed and spins in the region of interest in which one is interested.

The stimulated echo sequence [10] for localized NMR spectroscopy [11, 12, 13, 14] is comprised of three orthogonal slice selective 90 degree pulses. A first pulse excites the spins in a slice. During a time STE/2 the spins dephase in the transverse plane by a magnetic field gradient. At time STE/2 a second slice selective 90 degree pulse is applied, distributing the spins in a plane along the +/ − Z axis and perpendicular to the axis along which the second 90 degree pulse was given. Half of the signal is "stored" along the Z axis, whereas the other half will dephase in the presence of a dephasing gradient during a time Tm. After this time Tm a third slice selective 90 degree pulse will be applied followed by a rephasing gradient, resulting in an stimulated echo signal at a time STE/2 after this last 90 degree pulse. This signal contains half of the original signal at the beginning of the sequence. The volume selection for all three orthogonal directions is performed by the same pulses, resulting in a perfectly symmetrical volume of interest. Moreover, all slice selection pulses are 90 degree pulses which results in relative insensitivity for B1 inhomogeneity and well defined slice profiles. However, the loss half of the signal intensity is a drawback associated with this approach.

The double spin echo technique uses a 90 degrees excitation pulse followed by two refocusing 180 degree pulses (Fig. 1). [15, 16]. By combining these radio-frequency pulses with three orthogonal magnetic field gradient pulses, the first excitation pulse will only affect spins in a particular slice. A 180 degrees pulse applied simultaneously with a magnetic field gradient orthogonal to the first gradient will refocus spins in a second slice, resulting in a spin echo signal from the cross section of the two selective pulses. Finally, a third pulse combined with a magnetic field gradient orthogonal to the other two gradient directions will result in a spin echo signal from the intersection of all three slices. This signal originates

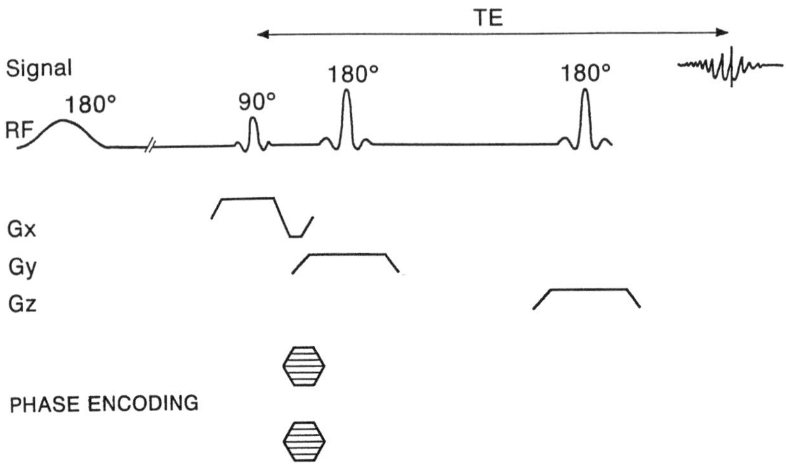

Fig. 1. Schematic representation of an asymmetric double spin echo localization sequence. Before the sequence a single adiabatic inversion pulse is given to suppress the water. After the 90 degrees excitation pulses phase encoding gradients are included for spectroscopic imaging purposes

from a rectangular box, its dimension being determined by the width of the slices.

In principle, the double spin echo sequence preserves all initial magnetization. For uncoupled spins the signal strength is attenuated by T2 and saturation effects only. In order to optimize the amount of signal obtained from the volume of interest, the sequence requires otpimized 180 degree refocusing pulses to guarantee a well-defined slice profile. Numerically optimized 180 degree pulses have been proposed to optimize the slice profiles in imaging [17]. For quantitative purposes the exact volume size definition requires an adequate description of the shape and the width of the selected volume. Often the volume size is defined by the full width at the half height of the excited slice profile. In three-dimensionally localized spectroscopy this may lead to deviations from the desired volume size. Hence, comparing signal to noise or quantitative data obtained by different techniques of volume selection becomes meaningless unless a careful calibration of the exact volume sizes has been made [18].

3 Water Suppression

Many different techniques have been proposed in high resolution NMR spectroscopy to suppress the very intense water resonance in ^{1}H NMR spectra. In principle all those techniques may be applicable for localized in vivo studies. The only important complicating factor is that the use of water suppression pulses may compromise the requirements for a proper localization with respect to outer volume suppression. Different water suppression schemes [19] result in minimum excitation at the water resonance and maximum excitation of the resonances around 1.0–3.0 ppm, including the lipid resonance of subcutaneous fat and bone marrow. These sequences may introduce an interference between the localization and water suppression. For example in proton brain studies this would result in excitation of all the lipid signal within the NMR probe, resulting in a unfavorable ratio between the amount of signal which should be suppressed and the signal from the volume of interest. Recently a method has been suggested using pulses which are both spatially and chemical shift selective [20]. This approach does not compromise outer volume suppression as a result of water suppression.

Another important requirement for water suppression schemes is a uniform excitation over the spectral region of interest. Techniques that meet this requirement and do not interfere with the localization include selective excitations and selective inversions of the water signal. Selective excitations comprise a series of 90 degree pulses at the water resonance followed by (strong) dephasing gradients to dephase the water resonance [21]. Selective inversions of the water resonance can be used to minimize the longitudinal magnetization by inverting the water resonance followed by an inversion delay [22]. Excitation of the volume of interest start at the zero crossing point of the water resonance, resulting in a suppressed water signal in the actual NMR spectrum.

4 Phase Encoding and Spectroscopic Imaging

The most important limitation of localization techniques based on slice selection pulses only is that the spectroscopic information in a single experiment all originate from one or sometimes two particular rectangular volumes. In ^1H NMR applications the size of these boxes may be as small as only a few cubic centimeters. Quite often, pathologies (tumor, stroke, multiple sclerosis) are confined to small lesions. Moreover, even when the lesion is sizable, it is often not very obvious where to locate the box in the brain or other organ of the patient. The affected tissue may be very inhomogeneous or difficult to identify.

Usually, localization of the volume of interest is done on the basis of a magnetic resonance scout image. Particularly for protons, both spectroscopy and imaging can be performed using exactly the same equipment, including the rf coil. Therefore, the patient can be imaged and spectroscopically examined without any repositioning. As the images make use of the same magnetic gradient system as slice selection for spectroscopy, the rectangular volume of interest for spectroscopy can be positioned very accurately. However, slice selection is not the only localization technique. Localized spectral information can also be obtained by using techniques which spatially localize by phase encoding and still preserve chemical shift information by avoiding a read out gradient [23, 24].

These measurements lead to the acquisition of spectra from a large number of voxels in a single experimental session.

Needless to say, the large data sets obtained of two or even three dimensional versions of this technique introduce their own problems. These so called spectroscopic imaging techniques usually result in huge data sets, e.g. a proton spectroscopic imaging experiment using 32 different phase encoding gradient steps in two directions resulted in 1024 spectra in a single slice, each spectrum corresponding to a single volume element (voxel) of 0.5–3.0 cm^3. By measuring each of the 32×32 k values once this results in a total acquisition time of 32 minutes, for a recycling delay time of 2 s. Data processing included gaussian (5 Hz) and exponential (– 4 Hz) multiplication in the chemical shift domain, sine windowing in both the kx and ky domain to reduce overlapping effects from intense residual lipid signals from the skull by the point-spread function characteristics of the truncated Fourier transform [25], and an additional water suppression was achieved by the data-shift accumulation method [26]. Following data windowing, a three dimensional Fourier transform was performed after interpolating the 32×32 into a 128×128 resolution. When the data are acquired as whole echoes, which can be done provided an echo time of 272 ms is used, magnitude spectra can be calculated to give proper absorption mode signals [27] in order to obtain the integrals of the different metabolite signals. The sine windowing in the kx and ky domain introduces an increase of the actual volume size by a factor of 1.5 for each phase encoding direction.

Examining such large data sets is usually impractical. However, the spectral information may be displayed as an image reflecting the spatial density distribution of a particular resonance in the spectrum. In this way metabolite maps may be constructed, which display the biochemical distribution of certain metabolites in the human brain. Examples of such metabolite maps will be shown below.

5 Single Volume ^{1}H NMR Spectroscopy of the Normal Human Brain

Figure 2 gives a single volume ^{1}H NMR spectrum obtained at 1.5 tesla of the brain of a healthy subject. The volume selection method used was the double spin-echo method, with a total echo time of 136 ms, volume size was 60 cm^3. The spectrum shows a number of resonance typically found in the ^{1}H NMR spectrum of the normal human brain, in particular the acetyl group of *N*-acetyl aspartate (NAA) at 2.02 ppm, the methylene resonance of the aspartyl residue of NAA at 2.56 ppm and the methyl groups of creatine and phosphocreatine (Cr) at 3.0 ppm. At 3.2 ppm the methyl group of choline is observed (Cho); this is the sum of different choline containing compounds which remain unresolved at 1.5 tesla. Furthermore, the CH_2 resonance of glutamate/glutamine shows up in the

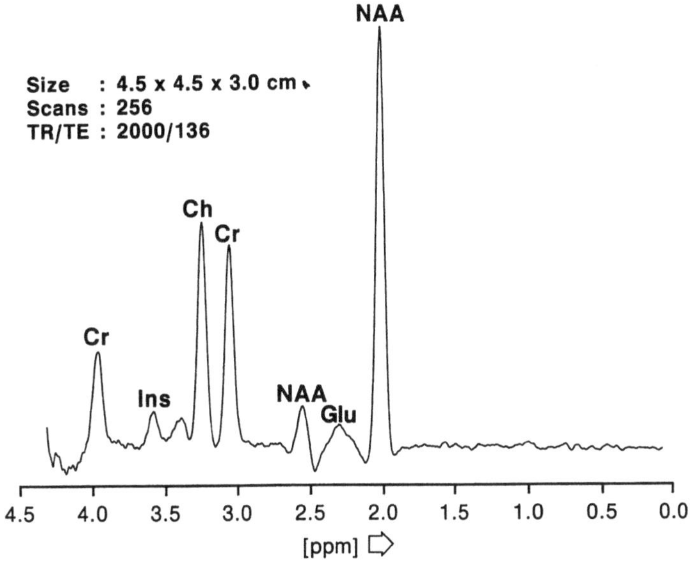

Fig. 2. A single voxel spectrum of a normal volunteer obtained by the double spin echo localization technique. The spectrum is obtained from a 4.5 × 4.5 × 3.0 cm volume, using 256 scans, TE = 136 ms and TR = 2000 ms

spectral region of 2.1–2.3 ppm, the inositol resonance and alpha-CH of amino-acids at 3.6 ppm, and the CH_2 resonance of (phospho)creatine at 3.9 ppm.

The spectrum shown in Fig. 2 is representative of the spectral resolution that can be achieved in vivo ^{1}H NMR spectra of the human brain measured at a field strength of 1.5 tesla; it has to be expected that at higher field strength more resonances can be resolved in the human brain spectrum. In particular the CH_2 region of glutamate can be better resolved at higher field strength [28, 29]. From in vitro high resolution ^{1}H NMR spectroscopy studies of extracts prepared of animal brain a large number of additional resonances have been resolved, in particular of taurine, and aminoacids such as glutamine, aspartate, alanine, and gamma-amino butyric acid (GABA) [30]. In in vivo ^{1}H NMR spectra of the human brain obtained at 1.5 tesla these resonances remain difficult to resolve, except in disorders such as brain tumors, where the concentrations of certain compounds can change profoundly, as will be discussed below in more detail.

For the interpretation of ^{1}H NMR spectra of the human brain it is necessary to understand the biochemical role of the different compounds. It is unfortunate that the biochemical role of some of the main peaks observed in the in vivo ^{1}H NMR spectrum of the human brain is not very well understood. This is particularly true for *N*-acetyl aspartate, which happens to be largest signal in the spectrum. It has been suggested that *N*-acetyl asparate is concentrated in the neurons, but its particular role is not yet very clear [31, 32]. The localization of NAA in the neurons is supported by high resolution ^{1}H NMR studies of brain extracts which showed that the NAA/Cho ratio is higher in grey matter than it is in white matter [30]. Also, NAA is low in extra-axial lesions such as meningiomas and acoustic neuromas [33]. Irrespective of the precise biochemical role of NAA in brain metabolism, its localization in the neurons indicates that this compound may be useful as a marker for neuronal loss or damage in the ^{1}H NMR spectroscopic examination of patients with focal disorders of the brain.

The choline compounds contributing to the choline signal in the ^{1}H NMR spectrum of the human brain (glycerol phosphorylcholine (GPC), phosphocholine (PC)) are intermediates of phospholipid metabolism. It should be noted that the Cho signal in the ^{1}H NMR spectrum of the brain is expected to arise mainly from these small molecules; the ^{1}H NMR signal of choline residues of phosphatidylcholines which are incorporated in membranes are expected to broadenened such that they do not contribute to the in vivo ^{1}H NMR spectrum of the brain. The Cho signal in the ^{1}H NMR spectra of the brain is expected to be related to phospholipid metabolism, and changes in this signal may indicate changes in phospholipid metabolism. The observation that in malignant gliomas the Cho signal is often seen to increase suggests that the ^{1}H NMR visible choline compounds are particularly concentrated in glial cells. In chronic infarcts the choline signal is often seen to persist, while NAA is seen to decrease. This suggests that Cho is also present in gliotic scar tissue, which provided another line of evidence that the Cho signal is mainly concentrated in glial cells.

The role of creatine and phosphocreatine is much better understood. These compounds play a role in high energy metabolism. Creatine is phosphorylated to phosphocreatine through the creatine kinase reaction. However, because we are unable to resolve the creatine and phosphocreatine signals in the in vivo ^{1}H NMR spectrum of the human brain it is not possible to determine the degree of phosphorylation of the total creatine pool by ^{1}H NMR spectroscopy. Thus a decrease in phosphocreatine due to a perturbation of high energy phosphate metabolism is not expected to have an immediate effect on the total creatine signal in the ^{1}H NMR spectrum. Only when the total creatine pool changes as a result of a prolonged perturbation is this expected to show up as a reduction of the ^{1}H NMR signal of the creatine. Both in stroke and in tumors the Cr signal has been found to diminish compared to unaffected tissue.

A much more sensitive indicator of perturbations of high energy phosphate metabolism is provided by the ^{1}H NMR signal of lactate. In the healthy brain the basal level of about 0.5 mM lactate can barely be observed by in vivo ^{1}H NMR spectroscopy. However, lactate is expected to increase in a number of conditions. This can occur either as a result of increased glycolysis or because of a diminished availability of oxygen, or both. Regional increases of lactate have been observed, both in intracranial tumors and in stroke patients. However, also in healthy subjects a small increases in cerebral lactate can be observed by ^{1}H NMR spectroscopy under suitable perturbations. Figure 3 shows a time-series of water-suppressed single volume ^{1}H NMR spectra of the brain of a 29-year-old male subject before, during, and after voluntary hyperventilation [34]. The ^{1}H NMR spectra were obtained from a $8 \times 8 \times 7$ cm selected volume in the left cerebral hemisphere using volume selection by refocussed stimulated echoes, the

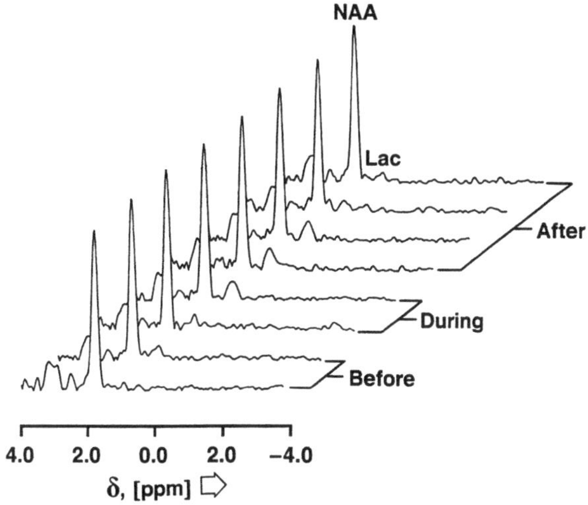

Fig. 3. Sequential ^{1}H NMR spectra of a healthy volunteer before, during, and after voluntary hyperventilation. Each spectrum was acquired in 2 minutes, lactate is seen to increase during hyperventilation (cf. Ref [34])

re-echo time used was 272 ms. Each spectrum is the sum of 64 transients acquired over a period of 2 minutes. Voluntary hyperventilation was maintained by breathing every 2 s. Spectral acquisition was started when paraesthesia and stiffness of the fingers and the perioral region developed as subjective signs of hypocarbia. Two ^{1}H NMR spectra were obtained during continuous hyperventilation, and further spectra were obtained during the subsequent recovery period. In addition to the signals discussed above, the resonance from the doublet of lactate is observed at 1.32 ppm. The lactate signal is seen to increase over the period of hyperventilation, to a constant level, and returns to the baseline level over the subsequent recovery period. The assignment to lactate was confirmed by a lactate editing experiment. On the basis of a coarse calibration using the NAA signal as an internal reference it was determined that lactate rose from a basal level of 0.67 ± 0.05 mmol/liter to 1.41 (± 0.06) mmol/liter during hyperventilation.

The most likely explanation of the increase of lactate during the period of hyperventilation is based upon its connection to an increase of cerebral pH. It is known that arterial pH increases from a normal value of 7.4 up to 7.65. From ^{31}P NMR obtained under similar conditions a small, but significant, increase of cerebral pH was observed, from 7.00 ($\pm$ 0.01) to 7.10 ($\pm$ 0.01). We hypothesize that the increased cerebral pH leads to an increased rate of phosphofructokinase (PFK), leading to an increased glycolytic rate, and increased lactate level. Increased cerebral lactate has also been observed by ^{1}H NMR spectroscopy in the visual cortex during photic stimulation [35]. In these conditions increased cerebral activity is expected to lead to an increased glycolytic rate, with a concomitant increase of lactate.

6 Localized ^{1}H NMR Spectroscopic Imaging of the Human Brain

Single volume ^{1}H NMR spectroscopy measurement can be extended to multivolume measurements. In its most simple form, a bar shaped volume of interest (typically 5 × 2.5 × 10 cm) is selected in the human brain by means of one of the volume selection techniques discussed previously. Then, gradient phase encoding is performed in the direction of the long axis of the volume of interest. After appropriate windowing, both in the time domain and the phase encoding direction, the resulting data set is processed by a 2-dimensional Fourier transform. This results in a series of spectra originating from a series of sections perpendicular to the long axis of the selected bar-shaped region of interest, typically with a nominal thickness of 1 cm). Figure 4 gives the result of such a 1-dimensional ^{1}H NMR spectroscopic imaging measurement of the brain of a healthy subject. The bar-shaped region of interest was selected in the transversal direction through the brain. The stack of ^{1}H NMR spectra shown in Fig. 4 comes from a series of sections of this bar, ranging from the left to the right side of the head. The lower spectrum comes from a section close to the skull; the spectrum

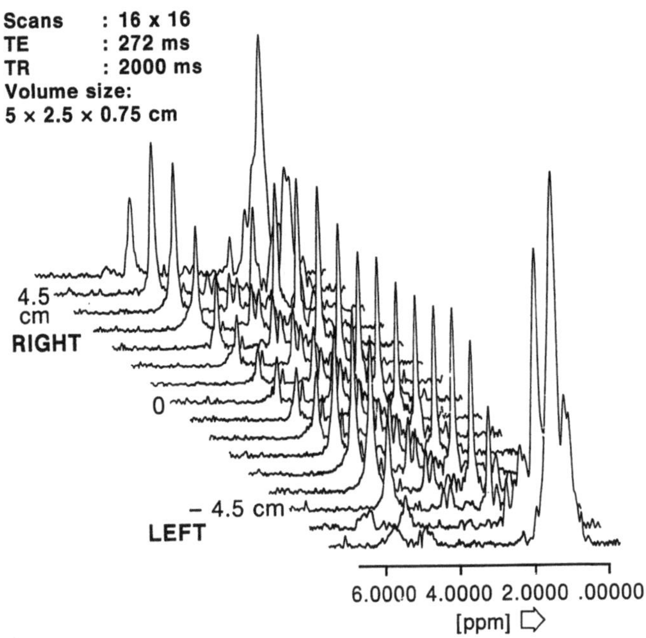

Fig. 4. One dimensional ^{1}H NMR spectroscopic image obtained from a normal volunteer. The stack of spectra pertain to a series of 5 × 2.5 × 0.75 cm volumes ranging from left to right through the head. Note the lipid contributions present in the spectra the voxels near the skull (*lower* and *upper*). The spectra from within the brain are similar to the single volume spectrum of Fig. 2

shows contributions from lipid signals originating from the skull due to imperfect suppression. Following the spectra ranging from left to right through the brain we see the lipid signals disappear, and we see ^{1}H NMR spectra typical for the normal human brain: we observe the signals from NAA, Cr, Cho, and a residual water signal. Finally we arrive at the upper spectrum, which comes from the right side of the head, again showing lipid signals from the skull. The spectra are remarkably uniform for slices going from left to right through the brain. In the center slices a slight decrease is observed in the intensity of the NAA, signal; this decrease is related to partial volume effects caused by the ventricles. Apart from that little spectral variations are observed when following the spectra from left to right through the brain. This lack of variation is most likely related to the fact that the spectra arise from sections with a nominal volume size of 20 cm^3; that volume size is still too large to resolve grey and white matter; thus we still obtain spectra which represent the volume average of grey and white matter.

The next level of complexity is provided by 2D ^{1}H NMR spectroscopic imaging. In this case, we select a slice shaped volume of interest, transverse through the brain. Typical size of the selected section is $9 \times 10 \times 2$ cm. Gradient phase encoding is now performed both in the left–right, and in the anterior–posterior directions, resulting in a 3D data set. This data set is procesed by windowing, both in the time domain, and in the 2D k- domain, followed by 3D Fourier transformation. This results in a 2-dimensional array of spectra; for 16×16 phase encoding steps this results in a total of 256 spectra; for 32×32 phase encoding steps in 1024 spectra. Obviously, this large number of spectra confronts the spectroscopist with the fact that it is impractical to inspect all these spectra on an individual basis. A convenient way to represent these data sets is to reconstruct metabolite images, which display the spatial distribution of the signals observed in the ^{1}H NMR spectra. These metabolite maps are reconstructed by selecting a certain spectral region from the ^{1}H NMR spectra, and integrating that particular spectral region for all spectra of the entire data set. The integrals are then displayed in a grey scale representation to represent the spatial distribution of that particular spectral region. In this way metabolite maps are obtained for NAA, Cho, Cr, lipids, and—in patients—also of lactate. We have also found it advantageous to reconstruct images which represent the ratio of certain metabolites. In particular the NAA/Cho ratio images has been found to be quite useful. In a number of focal pathologies, such as stroke and intracranial tumors, NAA was found to decrease, while Cho remains either unchanged or even increases. In the ratio images these effects are amplified, and the region of abnormal metabolite ratios shows up as an easily recognized region of increased intensity.

Although the metabolite maps, and the ratio images provide a useful way to represent the 2D ^{1}H NMR spectroscopic imaging data sets we feel that it remains absolutely necessary to inspect individual spectra. This is so in order to confirm spectral quality and the assignments of the signals. In particular, lactate can be easily identified in the spectra, on the basis of its chemical shift and the doublet splitting pattern. This information is lost in the integration process employed to

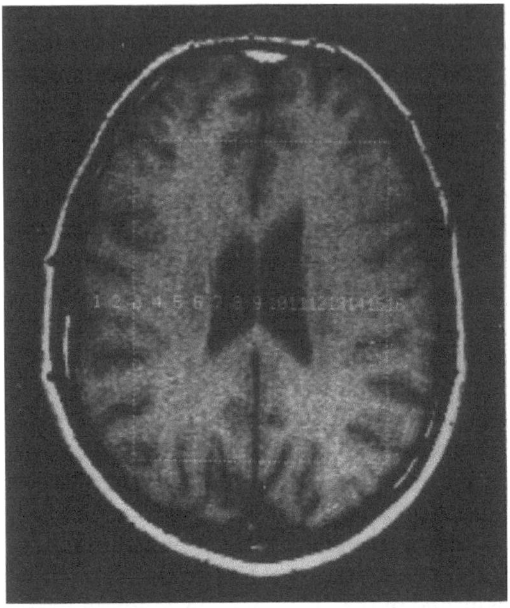

(a)

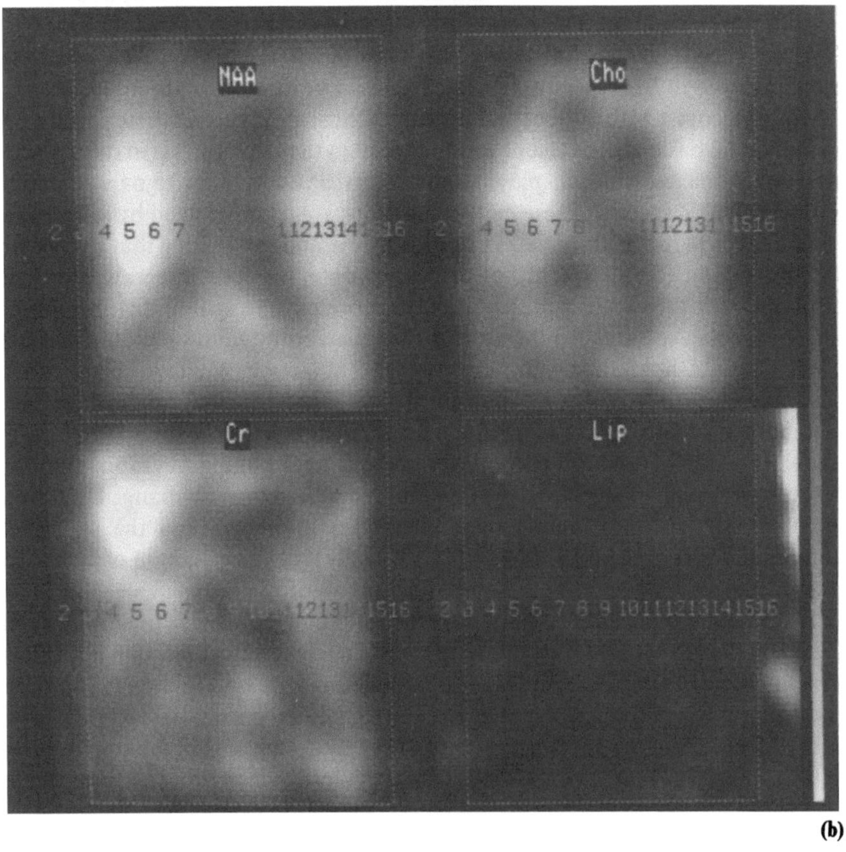

(b)

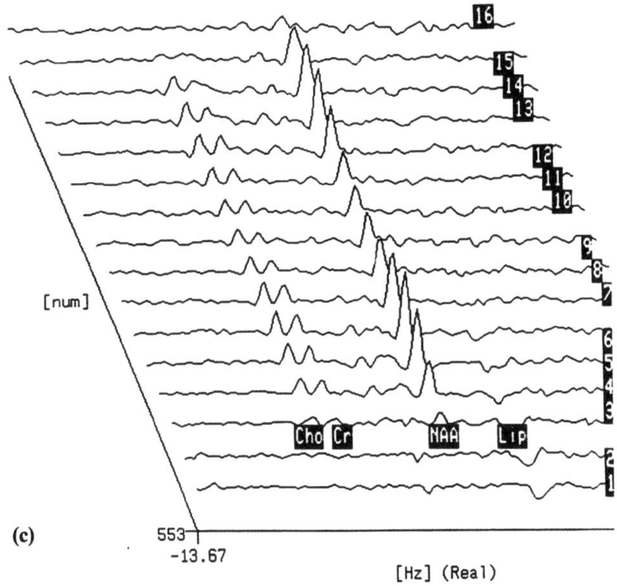

(c)

Fig. 5a. Fast, T1 weighted ¹H NMR scout image obtained of a normal volunteer, the *dashed box* represents the volume of interest selected for the ¹H NMR spectroscopic imaging measurement. **b**) Reconstructed metabolite maps from the ¹H spectroscopic imaging data set obtained of the normal volunteer. Note the butterfly pattern in the NAA image, which corresponds to the location of the ventricles, as shown in the scout image of Fig. 5a. **c**) Series of selected spectra from the 2D spectroscopic imaging data obtained from the volunteer. The location from which the spectra were selected is indicated, both in the scout image 5A, and in the metabolite maps of 5B. Note the decrease in spectral intensity in the spectra which arise from the region of the ventricles (spectra 8–11)

reconstruct the metabolite maps; hence solely on the basis of the metabolite maps it is not possible distinguish lactate from lipid signals which are coresonant. Another reason for inspecting selected spectra is to quantify spectral changes in focal disorders. One may select spectra both from the abnormal region, and from unaffected regions of the brain, and thus obtain quantitative information about metabolite changes in the affected part of the brain.

It is of course impractical to inspect every single spectrum, but a representative selection of spectra can be chosen, and displayed in a stacked curve representation. We have developed a display package which enables one to select spectra from the spectroscopic imaging data set on the basis of the metabolite maps and the scout images. By displaying the appropriate spectrum numbers, both on the scout image and the metabolite maps one has an overview of the spectral changes associated with pathological regions of the brain. Another, very useful way, to visualize correlations between changes in the different metabolite maps on one hand, and the scout image on the other is provided by contour plots. For instance, an intracranial tumor is often associated with increased lactate and/or Cho. The region of increased Cho or lactate can be circumscribed by an appropriate contour line in the corresponding metabolite maps; those contour lines can be projected onto the scout images, to provide a precise correlation of the region of abnormal metabolite concentrations with anatomical changes apparent in the scout images.

An example of a ^{1}H NMR spectroscopic data set obtained from the brain of a healthy volunteer is shown in Fig. 5A–C. Figure 5A shows the fast T1 weighted scout image obtained directly before obtaining the ^{1}H NMR spectroscopic data set. The double echo volume selection method was obtained to select a volume of interest within the boundaries of the skull. The volume of interest is indicated in the scout image by a dashed box; its size was 11×9 cm, with a slice thickness of 2.5 cm. The spectroscopic imaging measurements were performed in 32×32 phase-encoding steps, over a field of view of 22.5 cm; this results in a nominal voxelsize of $0.7 \times 0.7 \times 2.5$ cm; ($\sim 1.2\,cm^3$). Echo time was 136 ms, recycle delay 2 s. The four metabolite maps constructed from this data set are shown in Fig. 5B. Because of the volume selection process, signals only show up with the selected box; the metabolite maps are zoomed to display only the selected region of interest. The NAA, Cho, and Cr images are all rather uniform in this normal volunteer; the most important feature showing up in the images are related to the ventricles. In particular, the NAA image clearly shows the butterfly pattern corresponding to the location of the ventricles. Because the spectroscopic image was obtained over a 2.5 cm slice, rather than the 0.8 cm slice in the scout image the anatomical correspondence is not expected to be quite perfect. The lipid image does not show any signals from within the brain, as expected on the basis of the single volume spectra shown before; however, there is some residual lipid signal along the edges of the lipid image arising from the skull. The stack of selected spectra shown in Fig. 5C come from the locations indicated both in the scout image and in the metabolite maps. Because this measurement was performed with an echo time of 136 ms only half-echoes were collected; hence these spectra are real, phased spectra, rather than the magnitude spectra which are displayed when dealing with whole echo measurements. The selected spectra show the familiar signals from NAA, Cho, and Cr. In the spectra 8–11 there is a clear reduction of spectral intensity; those spectra come from the region of the ventricles, and the reduction in signal intensity is most likely due to partial volume effects. Apart from that, the spectral appearance is rather uniform in going from right to left through the brain. There is, however, a change in the Cho/Cr ratio, albeit rather small. In the spectra from the central region (spectra 7–11) the Cho peak is larger than the Cr peak, while in the spectra from the outer region (spectra 4, 5) the Cr peak is larger than the Cho peak. Those variation are expected as in the spectra from the central region there is a larger contribution from white matter than in the spectra from the outer region. Unfortunately, the cortex was not included in the measurement, otherwise those variations might have been more pronounced because of a larger contribution from the grey matter [30].

7 ^{1}H NMR Spectroscopic Imaging of Patients with Intracranial Tumors

The first example of ^{1}H NMR spectroscopic imaging we wish to show was obtained from a patient who had undergone brain surgery some 7 months before [36]. Subtotal tumor removal was performed, and the patient was treated with radiation therapy. On the basis of pathological examination the tumor was characterized as a grade III astrocytoma. MR imaging performed some time prior to the spectroscopic examination revealed an extensive recurrance of the tumor. The patient was then treated with intravenous chemotherapy. The spectroscopic examination shown in Fig. 6 was performed prior to a first chemotherapeutic dose. The series of spectra come from a 1D ^{1}H NMR spectroscopic examination. A large volume of interest was selected (7.5 cm anterior/posterior direction, 3.5 cm caudal/cranial direction, and 8.0 cm left/right direction). Gradient phase encoding was performed in the left/right direction, 14 cm field of view, 16 phase encoding steps, leading to a series of spectra from adjacent slices 8.7 mm thick. The lower spectra arise from the uninvolved left hemisphere, and are similar to spectra obtained from the brain of a healthy volunteer. Progressing into the tumorous region the NAA and Cr resonances decrease progressively towards very low levels. The Cho line, however, first increases over a few slices, and then decreases toward normal levels at the edge of the volume of interest. In the high field region of the spectrum (1.3 ppm) an intense doublet is observed, which has been assigned to lactate. This assignment to lactate is supported by its J-modulation behavior when the echo- time is reduced to 136 ms (at that echo-time the lactate doublet is observed as being inverted).

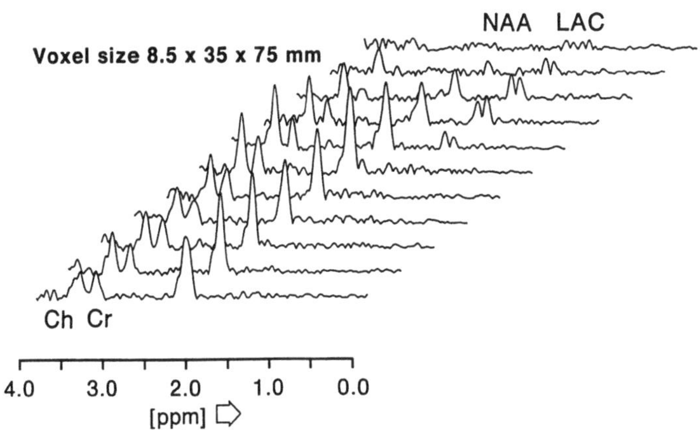

Fig. 6. One dimensional ^{1}H NMR spectroscopic imaging result of a patient with astrocytoma GR-III. From bottom to top the ^{1}H NMR spectra correspond to adjacent sagittal slices (23 ml; 8.5 mm slices) located from the left to right in the selected volume of interest. Resonance line assignments are PCho/Cho 3.22 ppm, PCr/Cr 3.05 ppm, NAA 2.02 ppm, Lac 1.33 ppm. (cf. Ref. [36])

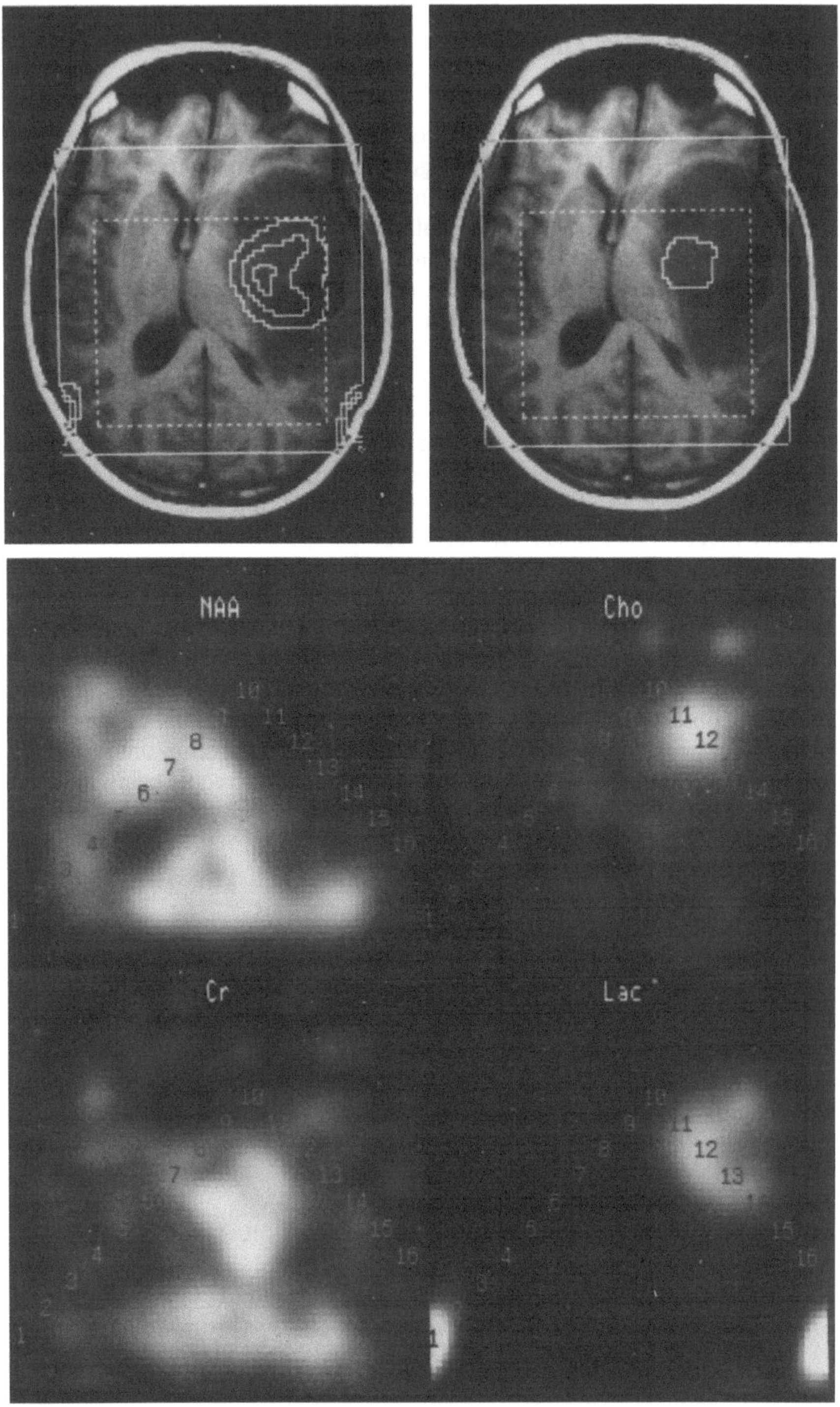

(a) (b) (c)

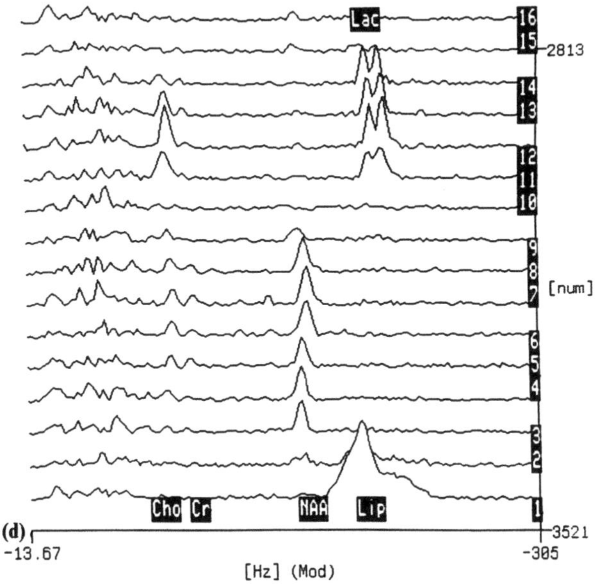

(d)

Fig. 7a. Scout image obtained of a patient who suffered of a grade II astrocytoma. The *outer box* in the scout image corresponds to the field of view of the spectroscopic imaging data set, the *inner dashed box* to the selected volume over which spectroscopic imaging was performed. The contour lines are a projection from the lactate image of Fig. 3C. **b)** Same scout image; however, in this image the contours from the Cho image are indicated. **c)** Metabolite maps reconstructed from the spectroscopic imaging data set obtained from this patient. Note that NAA is reduced in the tumorous region, Cho increases in a relatively small focus within the tumor, while lactate increases, both in the tumor, and in the surgical defect. **d)** Selected spectra from these data. The location from which these spectra were taken are indicated in the metabolite maps of Fig. 7C (cf. Ref. [37])

The explanation of these spectral changes when going from the uninvolved hemisphere to the tumorous region must be connected with variations across the tumorous region. If we assume that lactate is high in the postsurgical defect, while Cho is high in the recurrent tumor it is seen that the spectral variations apparent in the series of spectra of Fig. 6 can be interpreted in terms of varying contributions of uninfiltrated brain tissue, the recurrent tumor, and the postsurgical defect. In this interpretation spectra 6 and 7 from the bottom arise from a region with a large contribution of the recurrent tumor, while spectra 8, 9 and 10 contain large contributions from the surgical defect.

A severe limitation of the 1D ¹H NMR spectroscopic imaging result is related to the fact that the nominal voxel size is still in the order of 23 cm³. Thus the spectra shown in Fig. 6 are most likely the result of partial volume effects, with varying contribution from different regions of the tumor. In order make a more reliable correlation between spectral appearance, and the different regions which can be recognized within the tumor it is necessary to reduce nominal voxel size, to extend the spectroscopic imaging to 2 dimensions, and make correlations between metabolite maps and images.

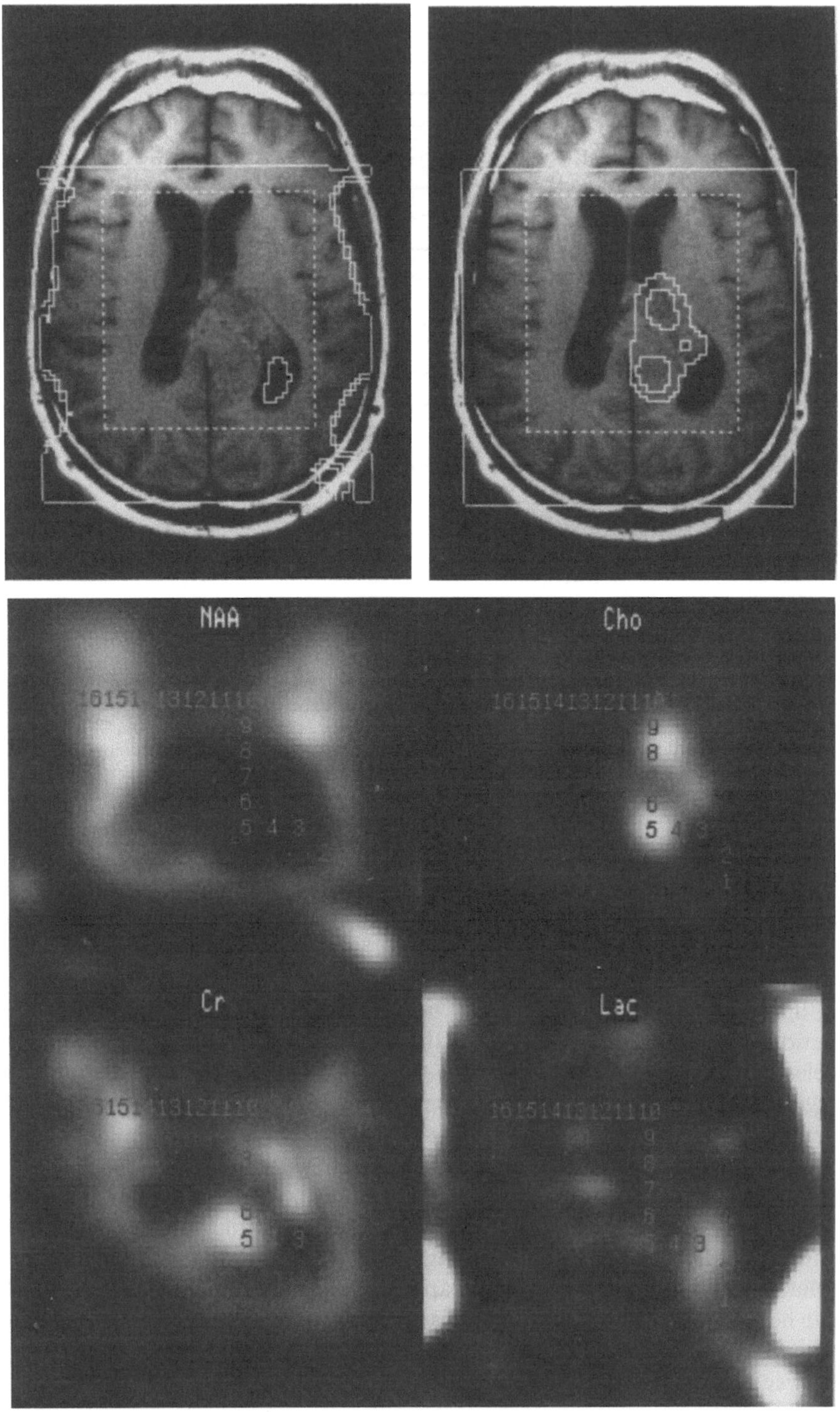

(a) (b)

(c)

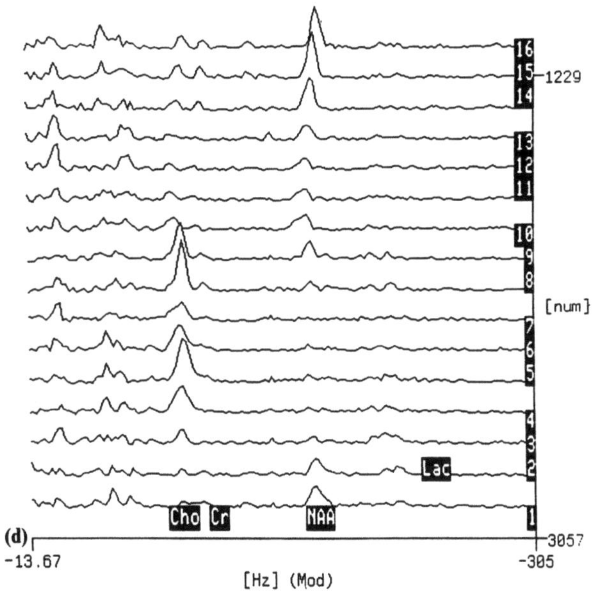

Fig. 8a. Scout image obtained of a patient who suffered from a grade II/III oligodendroglioma. The *outer box* represents the field of view for the spectroscopic imaging measurement, while the *inner box* represents the selected volume. The contour are from the Lac image of Fig. 8C; it demonstrates the location of lactate in the cornu posterius of the left lateral ventricle. **b)** Same scout image, however, here the contours are from the Cho image. The region of increased Choline provides a precise outline of the tumorous region. **c)** Metabolite maps reconstructed from the ^{1}H NMR spectroscopic imaging data set obtained from this patient. Note the decrease of NAA in the tomorous region and the increase in Cho. Lactate is increased in the left lateral ventricle. The high intensity along the outside of the Lactate image comes lipid signals from the skull. **d)** Selected spectra from this spectroscopic imaging data set. The location of the spectra is indicated in the metabolite maps of Fig. 8C. Note the increase Cho signal in the tumorous region (spectra 4–9). No lactate is observed in the spectra from the tumorous region, lactate is confined to the left ventricle (spectra 2, 3). (cf. Ref. [37])

A first example is provided by the 2D ^{1}H NMR spectroscopic imaging result shown in Fig. 7A–D [37]. The data set was obtained of a patient who had undergone resection of a grade II astrocytoma of the left temporal lobe, some 16 months prior to the spectroscopic examination. MR imaging performed prior to the spectroscopic examination helped verify the fluid filled site of the surgery, which was surrounded by a homogeneous space occupying lesion. A volume of interest was selected of 80 × 90 × 25 mm (ap/lr/cc). Spectroscopic imaging was performed over a field of view of 120 × 120 mm, in 16 × 16 phase encoding steps, leading to a nominal voxel size of 7.5 × 7.5 × 25 mm; echo time was 272 ms and whole echoes were collected. The outer box in the scout images 7A and 7B corresponds to the field of view of the spectroscopic imaging measurement, while the dashed inner box represents the selected volume. The metabolite maps of Fig. 7C show that NAA is low in the entire tumor region, as well as in the ventricles. On the other hand, the Cho map shows a relatively small region in

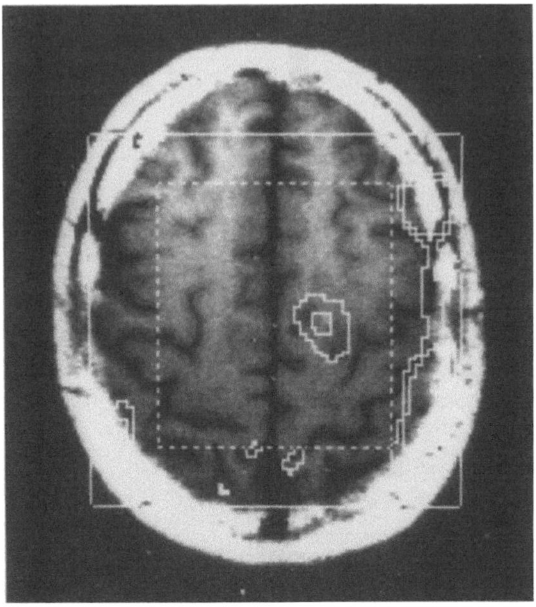

(a)

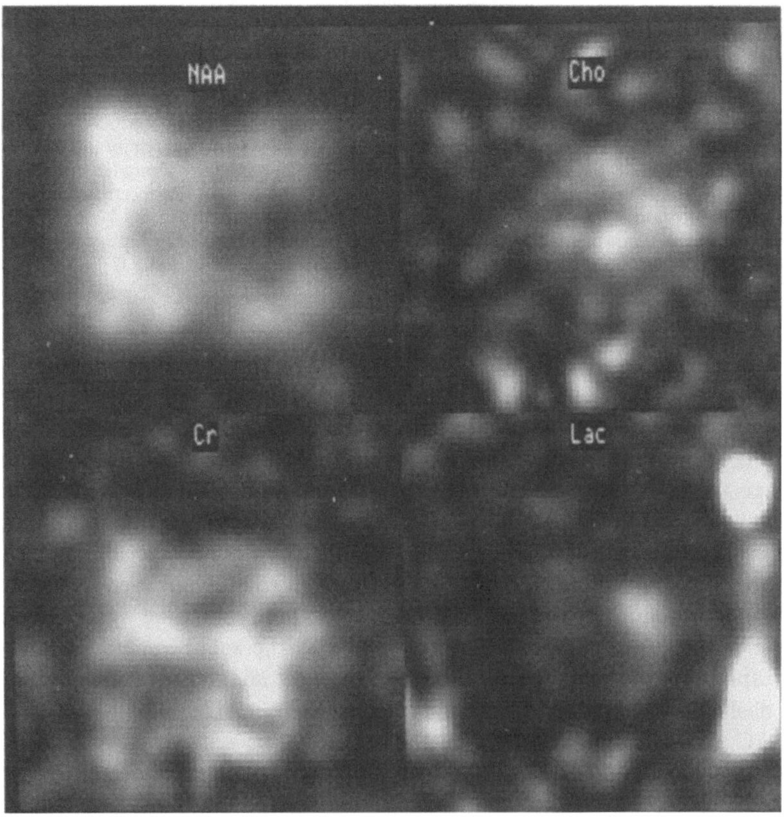

(b)

which the choline level is high, while elsewhere in the tumor region it is lower than in uninfiltrated tissue. The projection of the Cho contours onto the scout image (cf. scout image 7B) shows that Cho is high in the solid part of the tumor, while the lactate contours (scout image 7A) demonstrate that lactate is high in the entire tumorous region, both in the surgical defect and in the tumor tissue. The selected spectra shown in Fig. 7D come from locations which are indicated in the metabolite maps of Fig. 7C. The selected spectra show the doublet signal at 1.3 ppm, characteristic for the lactate signal. Also, the spectra demonstrate the increase in Cho by a factor of 2.6 compared to spectra from uninfiltrated tissue.

A second 2D ^{1}H NMR spectroscopic imaging result is shown in Figure 8A–D [37]. This data set was obtained of a patient who had an inhomogeneous, irregular shaped tumor extending across the midline through the corpus callosum and towards the lateral ventricals. On the basis of CT findings this tumor was characterized as a grade II/III oligodendroglioma. The fast, T1 weighted scout images shown in the figure show the field of view (14 × 14 cm) of the ^{1}H NMR spectroscopic imaging measurement (outer box), while the dashed box indicates the selected volume (10 × 9 × 3 cm). The measurement was performed in 16 × 16 phase encoding steps, echo time was 272 ms. The metabolite maps of Fig. 8C show the decrease of NAA in the tumourous region, and the increase of Cho. The lactate map shows no, or low lactate in the tumorous region. However, somewhat elevated lactate is detected in the cornu posterius of the left lateral ventricle, which is compressed by the tumor. The location of those changes in Cho and in lactate with respect to the scout images is demonstrated by the contour plots in Figs 8A and 8B. The Cho contours of Fig. 8B provide a perfect outline of the tumorous region, while the lactate contours of Fig. 8A demonstrate the presence of lactate in the left lateral ventricle. Figure 8D provides a stack of selected spectra from this spectroscopic imaging data set. The location of those selected spectra is indicated in the metabolite maps of Fig. 8C. Elevated lactate from the region of the left ventricle can be seen in spectra 2 and 3, while the increase in Cho is demonstrated in spectra 4–9. Those spectra arise from the tumorous region; note that in the spectra from the tumorous region lactate is not apparent.

The ^{1}H NMR spectroscopic imaging data set shown in Fig. 9 [37] was obtained from a patient suffering from a grade III oligoastrocytoma in the precentral region of the left hemisphere. Once again, the outer box in the scout image of Fig. 9A depicts the field of view (12 × 12 cm) of the spectroscopic imaging data set, while the dashed box is the selected volume of interest (8.5 × 7.7 × 2.5 cm). Phase encoding was performed in 16 × 16 steps. The metabolite

Fig. 9a. Scout image obtained from a patient who suffered from a grade II oligoastrocytoma. The *outer box* corresponds to the field of view of the spectroscopic imaging measurements, the *inner box* to the selected volume of interest. Also indicated are the *contour lines* taken from the metabolite maps of Fig. 9b. **b)** Metabolite maps reconstructed from the spectroscopic imaging data set obtained from this patient. Note the reduction in NAA, and the increase in Cho and in lactate in the tumorous region. (cf. Ref. [37])

maps of Fig. 9B show the decrease of NAA in the tumorous region, and the increase of lactate. The location of lactate is confirmed by the lactate contour plot upon the scout image of Fig. 9A. The high signal along the edge of the lactate image comes from residual lipid signals from the skull.

In summary, the particular examples of ^{1}H spectroscopic imaging demonstrate a number of aspects of ^{1}H NMR spectroscopy of human intracranial tumors. The first is that in all intracranial tumors the NAA signal is decreased. This reflects the fact that NAA is concentrated in neurons and axons. Therefore, the tumorous region with concomitant neuronal loss, shows up as a region of decreased NAA intensity. On the other hand, choline is seen to increase in the tumorous region. In the examples shown here, the increase in Cho is seen in every single case; however large variations have been observed to occur in the extent of Cho increase [38]. Indeed, in necrotic tumors Cho may be completely absent. Lactate provides a particularly interesting signal. Lactate may increase in tumors due to a variety of rasons: it may be related to a high metabolic rate, or it may be the result of hypoxia or necrosis, and it may also accumulate in defects or cavities. In the few examples shown here, we have already seen a large variation in the lactate signal: In one patient we have seen the presence of lactate in a post surgical defect, but also in the tumorous tissue. In a second patient no lactate is detected in the tumorous tissue, while lactate was detected in the compressed cornu posterius of the left lateral ventricle. In a third patient, lactate was observed in the tumor. These varying results reflect the complexity of tumor metabolism, and once the reasons for those different appearances of lactate are understood it may help in tumor characterization. Studies are in progress which seek to relate PET of ^{18}F fluoro-2-deoxyglucose with ^{1}H NMR of lactate [37, 39]. It is expected that these studies will help elucidate the relation between malignant metabolism and the appearance of lactate.

We have also examined patients with focal cerebral ischemia [40, 41, 42]. The lactate maps obtained from patients after recent stroke demonstrated that lactate is concentrated in the core of the infarction. However, in subacute and chronic stroke, lactate seems to occur in a region next to or surrounding the core of the infarct, while in the core itself lactate is relatively low. This persisting elevated lactate, up to several months after the stroke, shows that at least a certain number of cells within the affected region are still metabolically active; this prolonged anaerobic glycolysis occurs in the periphery rather than in the core of the infarct. Also in stroke, a reduction in NAA has been observed, which once again is indicative of neuronal damage in the affected region of the patient's brain.

^{1}H NMR spectroscopic imaging has also been applied to patients with multiple sclerosis [43]. Here it was observed that in the region of the sclerotic plaques there is a decrease of NAA, once again commensurate with neuronal and axonal damage. Also, small focal regions with increased Cho and Lac were observed, which most likely are related to the active phase of the disease.

8 Conclusion

It was several years after the first ^1H NMR experiments on the animal brain before the method could be extended to studies of human patients. This reflects the considerable technical difficulties which had to be overcome in obtaining high quality ^1H NMR spectra of the human brain, at the relatively low magnetic field of whole body magnets. The progress made is related to improvements in magnet and gradient design, aspects which are crucial in obtaining localized ^1H NMR spectra of the human brain. Those technical improvements have allowed the successful implementation of the spin-echo localization techniques now being used for human ^1H NMR spectroscopy. It should be mentioned that those techniques were among the first spectroscopic localization methods proposed, but were not particularly successful initially. Technical improvements, a better understanding of sequence optimization, and pulse-shape optimization has been essential in the recent success of those techniques.

After the successful implementation of the single volume technique it was possible to proceed towards multivoxel methods. The results shown here demonstrate that spectroscopic imaging can be successfully applied to obtain multiple spectra from a large number of volumes simultaneously. Those techniques are particularly useful for the examination of focal disorders, as demonstrated here for intracranial tumors.

In terms of applications of ^1H NMR spectroscopy and spectroscopic imaging we are currently witnessing a rapid proliferation. Single volume techniques can be most helpful in studying physiological perturbations, such as in hyperventilation [34] or photic stimulation [35]. In addition, single volume techniques are very useful in the study of global disorders, such as in hepatic encephalopathy [44]. On the other hand, ^1H NMR spectroscopic imaging is preferred in the study of focal disorders such as intracranial tumors, focal cerebral ischemia, multiple sclerosis, and epilepsy.

A long term goal of the present studies is to develop clinical applications of ^1H NMR spectroscopy. Although the present work already demonstrates the wealth of information which can be obtained by ^1H NMR spectroscopy and spectroscopic imaging, both in volunteers and in patients, there is still a way to go before we will see routine clinical applications of these techniques. This will require that the techniques contribute to the diagnosis and the clinical treatment of patients. It should also provide information which cannot be obtained by techniques presently in use. One of the first steps in this direction is an implementation of the techniques described here which will allow routine examination of patients, and also ways to routinely display the results. Although important progress has been made over the last few years, further progress is expected, in particular with the recent availability of high field whole body magnets (up to 4.1 tesla), and the development of advanced data processing techniques [46,47]. With those further advancements we expect ^1H NMR spectroscopy to become important, not only as a research tool, but also as a clinical modality.

Acknowledgement: Drs. J. Groen and E.P. van Lohuizen are acknowledged for the design of the optimized rf pulses used in the double spin echo sequence. The hyperventilation study shown in Fig. 3 was done in collaboration with Drs. Tulleken, Van Rijen, Berkelbach van der Sprenkel, Kraaier, and Van Huffelen of the Departments of Neurosurgery and Neurophysiology of the University Hospital in Utrecht, The Netherlands. The patient study showed in Fig. 6 was made in collaboration with Drs. C.M. Segebarth, and D.F. Balériaux from the Erasmus Hospital in Brussels, Belgium. The patient studies represented in Figs. 7, 8 and 9 were carried out in collaboration with Dr. Heindel and Prof. Friedmann from the department of Radiology, University of Cologne, Federal Republic of Germany.

9 References

1. Behar KL, den Hollander JA, Stromski ME, Ogino T, Shulman RG, Petroff OAC, Prichard JW (1983) Proc Natl Acad Sci USA, 80: 4945
2. Behar KL, den Hollander JA, Petroff OAC, Hetherington HP, Prichard JW, Shulman RG (1985) J Neurochem, 44: 1045
3. Bottomley PA, Edelstein WA, Foster TH, Adams WA (1985) Proc Natl Acad Sci USA, 82: 2148
4. Luyten PR, den Hollander JA (1986) Radiology 161: 795
5. den Hollander JA, Luyten PR (1987) Ann. of the New York Academy of Sciences, 508: 386
6. de Graaf AA, Bovée WMMJ, Deutz NEP, Chamuleau RAFM (1988) Magnetic Resonance Imaging 6: 255
7. Luyten PR (1989) Abstr Soc Magn Reson in Medicine 8th Ann Meeting 1989, p 79
8. Luyten PR, Mariën AJH, Sijtsma B, den Hollander JA (1986) J Magn Reson, 67: 148
9. Hanstock CC, Rothman DL, Jue T, Shulman RG (1988) J Magn Reson 77: 583
10. Hahn EL (1950) Phys Rev 80: 580
11. Granot J (1986) J Magn Reson 70: 488
12. Frahm J, Merboldt KD, Hänicke W (1987) J Magn Reson 72: 502
13. van Zijl PCM, Moonen CTW, Alger JR, Cohen JS, Chesnick SA (1989) Magn Reson Med 10: 256
14. Frahm J, Bruhn H, Gyngell ML, Merboldt KD, Hänicke W, Sauter R (1989) Magn Reson in Med 11: 47
15. Ordidge RJ, Bendall MR, Gordon RE, Connelly A (1985) In: Govil, Khetrapal, Saran, (eds) Magnetic resonance in biology and medicine, 387
16. Bottomley PA (1987) Annals of the New York Academy of Sciences, 508: 333
17. Groen JP, Rongen P, in den Kleef J (1986) Abstr Soc Magn Reson in Medicine 5th Ann Meeting p 1442
18. Luyten PR, Groen JP, Vermeulen JWAH, den Hollander JA (1989) Magn Reson in Medicine, 11: 1
19. Hore PJ (1983) J Magn Reson, 55: 283
20. Meyer CH, Pauly JM, Macovski A, Nishimura DG (1990) Magn Reson Med 15: 287
21 Haase A, Frahm J, Hänicke W, Matthaei D, (1985) Phys Med Biol 30: 341
22. Patt SL, Sykes BD (1972) J Chem Phys 56: 3182
23. Brown TR, Kincaid BM, Ugurbil K (1982) Proc Natl Acad Sci U.S.A., 79: 3523

24. Maudsley AA (1986) J Magn Reson 68: 363
25. den Boef JH, van Uijen CMJ, Holzscherer CD (1984) Phys Med Biol 29: 857
26. Roth K, Kimber BJ, Feeney J (1980) J Magn Reson 41: 302
27. Bax A, Mehlkopf AF, Smidt J (1979) J Magn Reson 35: 373
28. Langkowski JH, Wieland J, Bomsdorf H, Leibfritz D, Westphal M, Offermann W and Maas R (1989) Magnetic Resonance Imaging 7: 547
29. Frahm J, Michaelis T, Merboldt K.-D, Höcnicke W, Gyngell ML, Bruhn H (1991) NMR in Biomedicine 4: 201
30. Petroff OAC, Ogino T, Alger JR (1988) J Neurochem 51: 163
31. Koller KJ, Zaczek R, Coyle JT (1984) J Neurochem 43: 1136
32. Birken DL, Oldendorf WH (1989) Neurosci and Biobehav Rev 13: 23
33. Nadler JV, Cooper JR (1972) J Neurochem 19: 313
34. van Rijen PC, Luyten PR, Berkelbach van der Sprenkel JW, Kraaier VJ, van Huffelen AC, Tulleken CAF, den Hollander JA (1989) Magn Reson Med 10: 182
35. Prichard J, Rothman D, Novotny E, Petroff O, Avison M, Howseman A, Hanstock C, Shulman R (1989) Abstracts of the 8th SMRM p 1071
36. Segebarth CM, Balériaux DF, Luyten PR, den Hollander JA (1990) Magn Reson Med 13: 62
37. Luyten PR, Mariën AJH, Heindel W, van Gerwen PHJ, Herholz K, den Hollander JA, Friedmann G, Heiss W.-D (1990) Radiology, 176: 791
38. Bruhn H, Frahm J, Gyngell ML, Merboldt KD, Hänicke W, Sauter R, Hamburger C (1989) Radiology 172: 541
39. Alger JR, Frank JA, Bizzi A, Fulham MJ, DeSouza BX, Huhaney MO, Inscoe SW, Black JL, van Zijl PCM, Moonen CTW, DiChiro G (1990) Radiology 177: 633
40. Berkelbach van der Sprenkel JW, Luyten PR, van Rijen PC, Tulleken CAF and den Hollander JA (1988) Stroke 19: 1556
41. Bruhn H, Frahm J, Gyngell ML, Merboldt KD, Hänicke W, Sauter R (1989) Magn Reson in Medicine 9: 126
42. van Rijen PC, Tulleken CAF, den Hollander JA, Luyten PR (1989) Abstracts 8th Annual Meeting Soc, Magn Reson in Med p 450
43. den Hollander JA, '-Gravenmade EJ, Mariën AJH, Vencken LM, Luyten PR, Oosterwaal LJMP, Minderhoud JM (1990) Ninth SMRM, abstract #1116
44. Luyten PR, den Hollander JA, Bovée WMMJ, Ross BD, Bosman DK, Chamuleau RAFM (1989) Abstr. Eighth Annual Meeting Soc Magn Reson in Medicine, p 375
45. Luyten PR, van Rijen PC, Meiners LC, Mariën AJH, den Hollander JA (1990) Ninth SMRM, abstract #1009
46. de Graaf AA, Bovée WMMJ (1990) Magn Reson Med 15: 305
47. de Beer R, van Ormondt D, Mariën AJH, Luyten PR, den Hollander JA (1990) Ninth SMRM, abstract #210

Spectral Editing

A variety of methods to simplify complex spectra has been developed, exploiting correlations among spin systems such as J-coupling or chemical exchange.

Such methods are especially useful in proton MRS, where the overlap of resonance lines frequently prevents proper identification and quantification. Therefore, homo- and hetero- nuclear proton editing techniques, reviewed by Hetherington, represent important tools for in vivo MR spectroscopy. Exploitation of double quantum coherence allows the selective detection of coupled spin systems (e.g. lactate) with efficient water suppression. This approach is reviewed by Freeman and Hurd.

The resolving power of two-dimensional (2D) NMR techniques makes 2D-NMR an indispensable tool for high-resolution spectroscopy. Obviously, these techniques are of potential utility also for in vivo studies, in spite of some severe methodological problems. The large line widths compared to the coupling constants prevent the resolution of multiplet patterns of coupled nuclei, which in many 2D experiments are detected as antiphase signals. Hence the corresponding contributions will cancel. In addition, the time required to record 2D spectra (hours) may be prohibitive for many in vivo applications. Nevertheless, there are potentially useful applications of 2D methods in in vivo MR. Berkowitz discusses homonuclear techniques, while Navon and coworkers domonstrate the resolving power of 2D heterocorrelated (phosphorus-proton) MRS for the assignment of resonances of phosphorus containing metabolites (phosphomonoesters) in vivo.

Finally, a series of NMR techniques has been developed to study the kinetics of chemical exchange processes. These magnetization transfer methods may be applied to in vivo conditions and offer an unique tool for assessing kinetic parameters such as the turnover rates of endogenous metabolites (e.g. PCr, ATP) in the undisturbed biological medium. Magnetization transfer techniques are discussed by Rudin and Sauter.

Homo- and Heteronuclear Editing in Proton Spectroscopy

H. P. Hetherington

Division of Cardiovascular Disease, Department of Medicine, University of Alabama at Birmingham, USA

Table of Contents

NMR Basic Principles and Progress, Vol. 27

1 Introduction

Due to its high sensitivity, 100% natural abundance and the ubiquity of the ^1H nucleus, ^1H NMR provides a sensitive tool for the study of metabolism in vivo. However, the 100 M water resonance, molar lipid resonances, large number of compounds containing ^1H nuclei, extensive homonuclear coupling and narrow chemical shift range (10 ppm) of biologically active metabolites provides major barriers to uniquely resolving metabolite resonances of interest and therefore the use of NMR in in vivo biochemical studies. Excellent resolution and identification of ^1H resonances of proteins and nucleic acids is commonly achieved by the use of 2-D techniques such as 2 dimensional *J*-resolved spectroscopy and chemical shift correlated spectroscopy. However, use of these additional dimensions of information encoding (chemical shift evolution of coupled resonances or *J*-modulation), severely limits the temporal resolution of the experiment, prohibiting its use to observe rapid metabolic changes. To meet this need spectral editing techniques (homonuclear, heteronuclear editing) and polarization transfer sequences have been developed to retain the temporal resolution of a simple one dimensional experiment yet provide the additional spectral resolution of a 2-dimensional experiment for a single resonance (homonuclear spectral editing), group of resonances (heteronuclear spectral editing) or class of resonances (polarization transfer).

Homonuclear spectral editing sequences achieve selective observation of a single coupled ^1H resonance by exploiting its *J*-coupling interaction with its coupled spin during a spin echo sequence. Specifically, the spins coupled to the resonance of interest are selectively perturbed on alternate acquisitions. The two resultant spectra, perturbed and non-perturbed are then subtracted. Optimally, subtraction of these two spectra results in the cancellation of all resonances, coupled and non-coupled not affected by the selective perturbation. Therefore, only the resonances affected by the selective perturbation will appear in the difference spectrum. This effectively allows low concentration *J*-coupled resonances to be observed against a background of large uncoupled resonances such as the methylene resonance of lipids or other coupled resonances with different coupling constants, multiplicity or resonance positions. Since only two steps or acquisitions are required to achieve resolution, ^1H spectral editing sequences provide a sensitive, rapid and accurate technique for measuring major metabolites with homonuclear *J*-coupled ^1H resonances. ^1H metabolites levels have been used to evaluate metabolic changes in brain [1,2], liver [3], and muscle [4,5] in various physiological states. Specifically, the observation of lactate by spectral editing provides a measure of the extent of mismatch between the glycolytic and respiratory rates. This has proven to be a sensitive indicator of metabolic derangement in brain during hypoxia [6], ischemia [7] and hyperammonemia [8].

Heteronuclear spectral editing sequences are conceptually identical to homonuclear editing sequences, with the exception that the coupled spin is of a different nucleus. These techniques have been most commonly applied to the study of ^1H–^{13}C coupled metabolites [9]. Usually, the ^1H resonance of the metabolite of interest is observed to maximize S/N, while the ^{1}H–^{13}C coupling interaction is altered by perturbation of the ^{13}C spins. If the perturbation is alternately applied in consecutive acquisitions, and the resulting spectra subtracted, a spectrum consisting of only those ^{1}H resonances which are ^{13}C coupled is obtained. Therefore ^{1}H–^{13}C spectral editing has the unique ability to measure metabolic turnover, the incorporation of a labeled substrate into a pool of its products. Specifically, the incorporation of the ^{13}C nucleus of C-1 labeled ^{13}C glucose into the cerebral pools of lactate and glutamate allows the direct measurement of glycolytic and tricarboxylic acid (TCA) cycle fluxes. Measurement of these metabolic rates provides a tool to investigate the primary sources of energy production in biological tissues.

Polarization transfer techniques provide spectral editing by selectively transferring magnetization from a coupled spin to its J-coupled partner. If the spectral region local to the receiving spin is presaturated only the magnetization arising from the transfer will be observed. These sequences have the advantage of providing single scan editing, and therefore being resistant to tissue motion.

Clearly spectral editing and polarization transfer techniques provide powerful methods for the study of in vivo metabolism. A comprehensive description of every homonuclear editing, heteronuclear editing or polarization transfer sequence reported in the literature would be beyond the scope of this review. Additionally such a review would serve only to obscure the theoretical properties and limitations of the fundamental modes in which homonuclear and heteronuclear editing is achieved. It should also be noted that a number of promising techniques achieving spectral editing using multiple quantum techniques have also been described recently [10]. However, these techniques will also not be discussed here such that this review may provide a more critical and focused discussion of spectral editing. Therefore this review will focus on the basic fundamentals of homo- and heteronuclear editing and polarization transfer techniques by using a few but representative sequences as examples. Specifically, this review will provide: 1) a theoretical base for understanding the mechanism and possible limitations of spectral editing, 2) a detailed description and comparison of several typical homonuclear spectral editing schemes and typical applications, 3) a detailed description and comparison of several heteronuclear spectral editing sequences and typical applications, and 4) a brief description of the fundamentals of polarization transfer schemes and two representative but novel applications.

2 Theoretical Fundamentals of Spectral Editing

2.1 *Fundamentals of J-Modulation*

All spectral editing sequences are based on the modulation of J-coupled resonances during spin echo sequences. This modulation results in a sinusoidal dependence of the integrated intensity and phase of the all coupled resonances.

By selectively perturbing this interaction, usually during the spin echo evolution periods, the phase of the desired J-coupled resonance at acquisition can be altered or transferred to another spectral location. Since the desired resonance is commonly overlapped by unwanted resonances, the perturbation to achieve differential J-modulation of the resonance of interest must be applied to its coupled spin, rather than to itself. The exact nature of the perturbation varies depending on the particular sequence, however there are three primary methods for achieving spectral editing, 1) selective suppression of the J-coupling during the echo evolution periods, 2) selective reversal of the coupled spins interaction with the desired resonance and 3) selective transfer of magnetization between coupled spins. To fully understand the principles behind the three basic methods it is useful to begin by reviewing the evolution of a simple weakly coupled homonuclear spin system (IS), to a spin echo sequence,

$$(\pi/2)_x - \tau - (\pi)_x - \tau - AQ \tag{1}$$

where the non-selective excitation $(\pi/2)_x$, and refocusing $(\pi)_x$ pulses are applied with phase x, with an echo evolution period, τ, of $1/2\,J_{IS}$ and J_{IS} is the coupling constant of the IS spin system. Although simple vector diagrams have been used to describe the effects of coupling, spectral editing sequences can be most easily understood and interpreted by the use of product operators [11, 12]. In the following example, the spin of interest, I, is coupled to a nucleus, S with the specific coherence denoted by subscripting the product representations. For example magnetization of the I spin along the Y axis is denoted by the coherence I_y, while the two spin antiphase coherence of the I spin along the x axis is represented by $2I_xS_z$. Initial longitudinal I magnetization, I_z, is transformed by the excitation pulse $(\pi/2)_x$ to the transverse magnetization component $-I_y$,

$$-I_z \xrightarrow{(\pi/2)I_x} -I_y. \tag{2}$$

Ignoring the effects of chemical shift evolution, which will be refocused by the pulse $(\pi)_x$, the transverse magnetization evolves under the influence of the coupling interaction, JI_zS_z, during τ to the 2 spin antiphase coherence $2I_xS_z$

$$-I_y \xrightarrow{J_{IS}I_zS_z} 2I_xS_z. \tag{3}$$

The non selective refocusing pulse, $(\pi)I_x$, refocuses the effects of chemical shift precession and inverts, $(\pi)S_x$, the S_z component of the 2 spin $2I_xS_z$ component

$$2I_xS_z \xrightarrow{(\pi)I_x} 2I_xS_z \xrightarrow{(\pi)S_x} -2I_xS_z. \tag{4}$$

The inversion of the coupled spin effectively reverses the direction of the J-coupling precession during the next echo evolution period. Specifically, the 2 spin coherence $-2I_xS_z$ evolves during the second evolution period τ to yield $-I_y$,

$$-2I_xS_z \xrightarrow{J_{IS}I_zS_z\tau} -I_y. \tag{5}$$

However, a noncoupled spin, I, subjected to this spin echo sequence would be acquired in the I_y state, specifically

$$I_z \xrightarrow{(\pi/2)I_x} -I_y \xrightarrow{\pi I_x} I_y \tag{6}$$

or exactly 180° out of phase with respect to coupled spin. Inspection of the spin echo sequence reveals that phase or J-modulation will only occur when i) the coherence $2I_xS_z$ is created and ii) the pulse $(\pi)_x$ is effective in inverting the S spin. These two requirements form the basis of all spectral editing difference techniques. To achieve spectral editing, on alternate scans, either i) or ii) is inhibited to suppress J-modulation of the desired resonance in alternate scans, and the resulting spectra subtracted from a J-modulated spectrum.

2.2 Homonuclear Editing

In homonuclear editing creation of the 2 spin coherence, $2I_xS_z$, can be suppressed by CW decoupling of the S spin during both echo evolution periods [13]. This effectively sets the coupling constant J_{IS} to 0 resulting in the following transformations

$$I_z \xrightarrow{(\pi/2)I_x} -I_y \xrightarrow{0I_zS_z\tau} -I_y \xrightarrow{\pi I_x} I_y \xrightarrow{\pi S_x} I_y \xrightarrow{0I_zS_z\tau} I_y . \tag{7}$$

Thus, the coherence $2I_xS_z$ is never created, such that the I magnetization undergoes the same transformations as a non coupled spin. Alternately, if J evolution is not suppressed during the echo periods (absence of decoupling), the effectiveness of the $(\pi)_x$ pulse in inverting the S_z component of the $2I_xS_z$ coherence can be altered by use of an I spin selective pulse, for refocusing [14, 15], or inclusion of an additional selective $(\pi)_{-x}$ pulse to S spin so as to make the net inversion angle 0.

$$I_z \xrightarrow{(\pi/2)I_x} -I_y \xrightarrow{J_{IS}I_zS_z\tau} 2I_xS_z \xrightarrow{(\pi)I_x} 2I_xS_z \xrightarrow{0S_x} 2I_xX \xrightarrow{J_{IS}I_zS_z\tau} I_y . \tag{8}$$

Regardless, the subtraction of the spectra acquired from either of these two methods with a spectrum acquired in the which J-modulation occurs, will result in a coherent magnetization of the I spin and no signal from all other resonances.

2.3 Heteronuclear Editing

Heteronuclear editing is conceptually similar to homonuclear editing with the exception that the spin echo refocusing pulse to the I nuclei, usually ^1H nuclei, is not effective in inverting the S spins, usually ^{13}C nuclei. This results in diametrically different approaches for decoupling and selective inversion pulse editing techniques. Decoupling approaches seek to trap the I magnetization in a phase inverted state. If the evolution delay τ is increased to $1/J_{IS}$ the I

magnetization precesses to the inverted state at the end of the first echo evolution state [16]. If *CW* decoupling is applied to the S spin during the subsequent echo evolution period, such that $J_{IS} = 0$, the I magnetization will be trapped in the phase inverted state, $-I_y$ [16].

$$I_z \xrightarrow{(\pi/2)I_x} -I_y \xrightarrow{J_{IS}I_zS_z\tau} I_y \xrightarrow{(\pi)I_x} -I_y \xrightarrow{0 \cdot I_zS_z\tau} -I_y \tag{9}$$

Similarly, selective inversion techniques induce J-modulation rather than inhibiting it. In this case the evolution delay is again chosen as $\tau = 1/2J_{IS}$, but a selective $(\pi)_x$ pulse is applied to the S spins. In this case the spin system undergoes the exact same transformation as described (1–6). Subtraction of this spectrum with a spectrum acquired without the selective $(\pi)_x$ pulse to the S spins (8) results in an edited spectrum.

2.4 *Polarization Transfer*

Polarization transfer sequences are similar to selective inversion based spectral editing in that a spin echo sequence is used and evolution of the antiphase state, $2I_xS_z$, is required for success. However, rather than reversing the direction of precession of the I spin magnetization in the $2I_xS_z$ state, the magnetization is transferred to the S spin, $-2I_zS_x$ by a non selective $\pi/2$ pulse, $(\pi/2)I_y$, $(\pi/2)S_y$. This magnetization is then allowed to precess under J-coupling, refocused with a π pulse and S magnetization is created.

$$I_z \xrightarrow{(\pi/2)I_x} -I_y \xrightarrow{J_{IS}I_zS_z\tau_1} \xrightarrow{(\pi)I_x} \xrightarrow{(\pi)S_x} \xrightarrow{J_{IS}I_zS_z\tau_1} -2I_xS_z$$

$$\xrightarrow{(\pi/2)I_y} 2I_zS_z \xrightarrow{(\pi/2)S_y} 2I_zS_x \xrightarrow{J_{IS}I_zS_z\tau_1} \xrightarrow{(\pi)S_y} \xrightarrow{(\pi)I_y} \xrightarrow{J_{IS}I_zS_z\tau_1} S_y$$

$$\tau_1 = 1/4J_{IS}.$$

As described the technique transfers magnetization from one coupled spin to another. However, for this magnetization to be observed against other resonances in the vicinity, it usually necessary to presaturate the spectral region in to which the magnetization will be transferred.

2.5 *Origin of Imperfections*

Spectral editing techniques appear to present perfect selectivity and yield virtually equivalent results for all techniques. however, as common sense would indicate, not all techniques give equivalent performance, and it is their method in altering the J-modulation that gives rise to differences in performance. Specifically, decoupling and selective inversion techniques require dramatically different durations and amplitudes of RF pulses to the S spins. These pulses in turn have different selection profiles which result in varying degrees of perturbation to other resonances throughout the spectrum. These perturbations subsequently act to alter the phase and intensity of coupled and non-coupled resonances throughout the spectrum. Since these perturbations are applied in only a single set of scans, magnetization from unwanted resonances can appear in the difference spectrum, thereby limiting the reliability of the edited signal. These effects can be grouped into three broad areas 1) *specificity*: the extent to which direct perturbation of the refocusing intensity and phase of coupled and non coupled resonances is altered by pulses applied to the S spin [17], 2) *selectivity*: the size of alterations in the phase and intensity of coupled resonances by perturbation of their coupled spins from pulses applied to the S spins and 3) *water induced anomalies*: alterations in refocusing of coupled and non-coupled spins due to susceptibility changes of the media by alteration of the water magnetization from pulses applied to the S spins [18].

The specificity of an editing sequence can be easily calculated using either rotation matrices, density matrices, or the product operator formalism. For a simple echo sequence (1), the refocusing magnetization in the presence of phase cycling [19, 20] is given by

$$0.5(p_{22} - p_{11}) \tag{10}$$

where

$$P = (-\phi I_z)(\alpha I_y)(\Theta I_x)(-\alpha I_y)(\phi I_z) \tag{11}$$

and P is a rotation matrix describing the effect of the refocusing pulse $(\pi)_x$, p_{ij} are elements of the rotation matrix P, ϕ is the phase of the pulse, α is the off resonance angle and Θ is the effective pulse angle. However, for a selective inversion pulse based homonuclear editing sequence

$$(\pi/2)I_x - \tau_1 - (\pi)S_x - (\pi)I_x - \tau_2 - \tau_1 - AQ \tag{12}$$

where the selective inversion pulse $(\pi)S_x$ is applied on alternate scans, the complete refocusing term is given by

$$P = (-\phi_j I_z)(\alpha_j I_y)(\Theta_j I_x)(-\alpha_j I_y)(\phi_j I_z) \tag{13}$$

where the rotation angles, $\phi_j, \alpha_j, \Theta_j$ reflect the individual perturbations of the selective inversion pulse, πS_x, the refocusing pulse πI_x and the symmetric delay τ_2. It has been shown that the addition of the selective inversion pulse can cause both alterations in the quantity of absorptive refocusing magnetization $0.5(p_{22} - p_{11})$ and also phase shifts due to changes in the dispersive refocusing mangetization $0.5(p_{12} + p_{21})$ [17, 21].

The selectivity of the editing sequence is determined by its ability to resolve the resonance of interest, I, from other nearby resonances whose coupled spins are also near the I coupled S spins. This is easily evaluated by calculating the frequency dependence of the selective inversion pulse $(\pi)S_x$, and evaluating for the degree of inversion, p_{33}, as a function of frequency. The degree of inversion, p_{33}, is combined with the J-modulation properties, (coupling constant, multiplicity, and chemical shift) of the other coupled spin systems in the sample and the expected "error" can be calculated.

The third source of potential "errors", water induced anomalies, arises from any gross differential perturbation of water in the two acquisitions [18]. In this case the selective inversion pulses effect on the longitudinal magnetization of the water induces an alteration in the bulk magnetic susceptibility of the sample during a portion of the spin echo sequence. This in turn results in an alteration of the phase of magnetization throughout the spectrum due to a change in the effective field.

Finally, in assessing any editing sequence it is important to also consider the effects of its use with inhomogeneous coils or inaccurately set pulse amplitudes or durations. Specifically, the pulse angle dependence of the observed edited signal may vary depending on the pulse sequence used. For example, in a homonuclear editing experiment, if the selective inversion pulse applied to the S spins, $(\Theta)S_x$, differs from π, a mixture of zero and double quantum magnetization will be created ($2I_xS_y$), along with the antiphase component I_xS_z. After the final echo evolution period the observable magnetization, I_y is decreased by $\cos(\Theta)$.

$$I_z \xrightarrow{(\pi/2)I_x} -I_y \xrightarrow{J_{IS}I_zS_z\tau} 2I_xS_z \xrightarrow{(\pi)I_x} 2I_xS_z \xrightarrow{\Theta S_x} 2I_xS_z\cos(\Theta)$$

$$-2I_xS_y\sin(\Theta) \xrightarrow{J_{IS}I_zS_z\tau} I_y\cos(\Theta) - 2I_xS_y\sin(\Theta). \quad (14)$$

Thus, the magnetization observed in the edited difference spectrum and its resonance area may represent only a portion of the total quantity of signal arising from that resonance [22]. However, the total effect for inhomogeneous coils can be easily calculated by evaluating the B_1 field profile of the coil used [22, 23].

3 Homonuclear Editing

Homonuclear editing sequences can be grouped into three broad categories: 1) selective decoupling techniques, 2) selective inversion techniques and 3) selective polarization transfer techniques.

3.1 Selective Decoupling Techniques

These techniques were initially described by Campbell and Dobson [13] and first applied in vivo by Rothman et al. [24]. For an IS spin system, the experiment consists of subtracting the signals acquired using non-selective pulses in the presence (A) and the absence (B) of decoupling during a spin echo sequence

[15]. On alternate acquisitions, CW decoupling is applied to the S spins throughout the echo period to prevent J-modulation. The echo periods τ, are chosen to be $1/2J_{IS}$.

$$(\pi/2)_x - \tau - (\pi)_x - \tau - AQ \tag{15}$$

(A) DEC on DEC on

(B) DEC off DEC off

The evolution of magnetization is given by (7) for sequence (A), I_y, and (1–6) for (B), $-I_y$. Additionally, these sequences can be applied to other spin systems, I_nS, I_nS_2, and I_nS_3 where the evolution periods are chosen to be $1/2J_{IS}$, $1/4J_{IS}$, and $1/2J_{IS}$ respectively. For I_nS, and I_nS_3 spin systems, the resulting I magnetization component represents the full intensity of the I spins, whereas for a I_nS_2 spin system, only the outer lines are acquired, representing (1/2) of the total intensity. This sequence is amongst the easiest to apply, with the only major requirements being a second RF source for decoupling the S spins and an accurate value for the chemical shift of the S spins. The selectivity of the sequence is limited by the amplitude of the required decoupling field B_2, $B_2 \gg J$. Specifically, the extent of decoupling is limited by the coupling constant and the number of equivalent S nuclei coupled to the I spins [24]. Although the amplitude of the applied decoupling is usually small, $5-10J_{IS}$, (35–70 Hz for ^1H couplings) the total on resonance rotation of the pulse is large, 1800–3600° due to the lengthy application time $1/J_{IS}$. Since the required amplitude of the decoupling field is independent of the magnetic field strength one may expect that performance of this sequence improves with increasing field strength due to the increased separation in Hz of all resonances. As such, this sequence has been primarily applied to editing widely spaced coupled resonances at high field strength [24]. To its advantage, the pulse angle dependence of these sequence is governed only by the excitation and refocusing pulses (assuming the decoupling is complete over the volume of interest). Finally, since non-selective excitation and refocusing pulses are applied, induction of phase anomalies are governed solely by the frequency profile of the decoupling.

3.2 Selective Inversion Techniques

Shortly after Rothman's report of the in vivo use of CW decoupling techniques, three sequences for spectral editing using selective inversion principles were described [15, 25, 26]. These sequences can be modeled abstractly as:

$$(\pi/2)_x - \tau - (\pi)I_x - \tau - AQ \tag{16}$$

(A) $(0)S_x$

(B) $(\pi)S_x$

where the sequence is composed of two acquisitions, (A) and (B), and $(\pi)I_x$ and $(\pi)S_x$ are selective π pulses applied to the I and S spin respectively.

The sequence reported by Hetherington et al. [15], uses semi-selective pulses for excitation and refocusing, $(\pi)I_x$, and a selective inversion pulse, $(\pi)S_x$, to alter the J-modulation of the resonance of interest, I. Specifically, $(\pi/2)I_x$ and $(\pi)I_x$ were 1–3–3–1 pulses [27], and the selective inversion pulse $(\pi)S_x$ was a DANTE pulse [28]. By using the semi-selective pulse, $(\pi)I_x$, in the absence of the DANTE pulse (A), the lactate C-2 proton resonance (4.11 ppm) is not inverted by the refocusing pulse [14]. Therefore no J-modulation is induced in the lactate resonance, and the I magnetization propagates according to (8). However, the application of the DANTE inverts the coupled spin (C-2 proton resonance) and the lactate C-3 ^1H resonance (1.33 ppm) J-modulates and propagates according to (1–6). Subtraction of the two acquisitions results in the edited lactate signal. Thus in this sequence, the selectivity and specificity of the sequence is governed solely by the frequency profile of the selective inversion pulse. The pulse angle dependence is given by $\sin^3\Theta(1-\cos 2\Theta)$ where the 2Θ dependence reflects the rotation angle experienced by the S spin from the selective inversion pulse.

This sequence has two major advantages over the CW decoupling techniques described earlier. Specifically, it provides improved water suppression by allowing use of semi-selective pulses, and 2) eliminates the high on resonance rotation angle required for decoupling. In this specific application, the use of semi-selective pulses provided spectral editing capability with water suppression factors between 10^4–10^5, thereby minimizing water suppression and the effect of differential perturbation of the water resonance as a possible limitation in spectral editing [15]. Additionally, the use of a selective inversion in place of CW decoupling allows the on resonance angle of the perturbation to the J-coupled spin to be reduced to 180° instead of 1800°–3600°. However, this apparent improvement must be qualified with a word of caution; specifically, the selective inversion pulse should be short compared to $1/2J$, such that the applied B_1 may be larger than that used during decoupling. Thus the specific choice of selective inversion pulse, DANTE, sinc, gaussian or other shape is critical to determining its advantage in specificity and selectivity.

The sequence described by Jue et al. [25] expands on the selection principles of [15]. Specifically, the semi-selective refocusing pulses were replaced with selective DANTE pulses. This affords additional suppression of unwanted resonances, lipid resonances or broad water resonances, which may limit dynamic range. This sequence has proved to be quite effective in resolving lactate in the perfused rat liver at 8.5 T. However, the intrinsic specificity and selectivity of the sequence are still largely governed by the performance of the selective inversion pulse to the coupled spin. Specifically, for lipid resonances, 1.2 ppm, near the lactate resonance 1.33, which are not completely resolved by the selective excitation and refocusing pulses, perturbations in their refocusing phase should be similar in size to that observed in the semi-selective experiment. The selectivity is also governed by the performance of the selective inversion pulse. Therefore this sequence provides only moderate increases in selectivity and specificity for lactate. However, more substantial gains are expected for spectral editing of other resonances where acquisition of distant lipid signals might present severe limitations in dynamic range or spectral quality.

Williams et al. [26], were able to omit the requirement for a separate selective inversion pulse by using a 1–1 semi-selective pulse [27] for refocusing $(\pi)I_x$ in one set of acquisitions (J-modulation not induced) (A) and a non-selective inversion and refocusing pulse (J-modulation induced), $(\pi)I_x$, $(\pi)S_x$ in (B).

$$\text{(A)}\ (\pi/2)_x - \tau - (\pi)I_x - \tau - AQ$$
$$\text{(B)}\ (\pi/2)_x - \tau - (\pi)I_x, (\pi)S_x - \tau - AQ. \tag{17}$$

In this application, the non-selective inversion pulse can be thought of as a pair of pulses containing both the semi-selective refocusing pulse and a selective inversion pulse [16]. In contrast to [15], the two acquisitions must be added to obtain the edited spectrum, due to the phase inversion associated with the evolution delay in the semi-selective refocusing pulse. This technique represents the simplest and most easily implemented of all the homonuclear editing techniques. Specifically, 1) a single frequency source is all that is required, 2) no pulse duration or amplitude calibrations are required for a selective inversion pulse, and 3) due to the relatively broad null of excitation of the 1–1 pulses, exact determination of the lactate C-2 frequency for selective decoupling or inversion is also not required.

However, these advantages also give rise to the limitations of the technique. Specifically, the selectivity and specificity of the sequence are entirely determined by the difference between the frequency profiles of the 1–1 refocusing pulse and the non-selective refocusing pulse. Thus since the 1–1 pulse approximates a $\sin(wt)$ dependence across the spectrum, substantial subtraction "errors" can be expected for frequencies other than at lactate. Clearly in brain, where NMR visible lipid content is minimal, this does not present a problem. However substantial errors may occur in other tissues with greater lipid content. An additional limitation in specificity is encountered due to the non-selective inversion pulse. Specifically, it has been recently noted that the perturbation of the water magnetization results in an alteration of the bulk susceptibility of the metabolites magnetic environment. This in turn results in a phase shift of the acquired magnetization, resulting in subtraction "errors". The selectivity of the sequence is also determined by the difference in frequency profiles of the 1–1 and non-selective inversion pulses. Thus, resolution of lactate and alanine, or compounds with coupled spins in the 1–1 null, is primarily dependent on the chemical shift separation of the resonances to be observed, in contrast to selective inversion or decoupling schemes where both the chemical shift of the observed and coupled resonances to provide selection. However, for tissues with moderate amounts of contaminating signals (moderate specificity demands), at high field with moderate homogeneity, this technique represents a simple and effective means for performing lactate editing.

As described, all editing sequences using differential inversion of the coupled spin result in specificity limitations due to perturbations of the refocusing of unwanted resonances by the selective inversion pulses. Recently two of the mechanisms of specificity limitations and a cycling scheme to theoretically eliminate these errors for a given frequency was reported [17]. Using a matrix formulation and product operators the effect of an arbitrary selective inversion pulse was calculated. The resulting shift in refocusing phase and intensity of at the edited frequency was then identified. By use of pulse symmetry principles [21] a phase cycling scheme was constructed to eliminate these errors at the frequency of interest (chemical shift of the edited resonance). Phase shifts in the refocusing magnetization at that frequency were eliminated by inverting the time order of the selective inversion and refocusing pulse and applying the refocusing and selective inversion pulses with a phase of $(\pi/2 - \phi)$, where ϕ is the phase of the original pulses. Alterations in absorptive refocusing intensity were eliminated by moving the selective inversion pulse symmetrically about the resonance of interest when inversion of the coupled spin was not desired. Finally, for complete theoretical cancellation, the refocusing pulse was required to be applied on resonance to the spin of interest. At 4.7 T, the specificity error at 1.9 ppm, incurred when editing for the C-3 proton resonance of lactate, 1.33 ppm, was reduced by greater than 10 fold. Although relatively complex, this cycling scheme does not increase the number of phase cycles required for spectral editing over that required for standard phase cycling of the refocusing pulse. Therefore no increased demands in system or sample stability are required for the use of this cycling scheme.

3.3 Selective Polarization Transfer

As described, standard spectral editing techniques are based on additions and or subtractions or two or more subspectra to eliminate unwanted resonances. Although this property of producing subspectra is quite useful in allowing comparison of the edited resonance to an internal standard in one subspectrum, the requirement of a subtraction also presents a severe limitation when spectral editing techniques are applied to moving organs such as the heart. Specifically, if tissue motion results in changes in intensity or phase of the acquired resonances in one subspectrum but not in the other, an error will occur in the difference spectrum, thereby compromising the quality of the edited spectrum. To offset this problem, selective ^{1}H polarization transfer techniques utilizing single shot editing, and eliminating the need for subtraction have been proposed. To augment water suppression, and edit for the C-2 ^{1}H resonance of lactate (magnetization observed in the C-3 ^{1}H resonance) von Kienlin et al. [29] used an initial selective excitation pulse $(\pi/2)I_y$ to the C-2 resonance, and incorporated an additional echo evolution period, τ_3 and a selective refocusing pulse, $(\pi)S_y$. Specifically, the sequence was

$$(\pi/2)I_y - \tau_1 - (\pi)I_x,(\pi)S_x - \tau_1 - (\pi/2)I_y,(\pi/2)S_y - \tau_2 - (\pi)I_y(\pi)S_y$$
$$- \tau_2 - \tau_3 - (\pi)S_y - \tau_3 - AQ \qquad (18)$$

where optimal sensitivity due to the IS_3 spin system was obtained using $\tau_1 = 1/10J$, $\tau_2 = 1/4J$. The specificity of this sequence is governed by the extent to which the presaturation pulses can eliminate background signals. The selectivity of this sequence is primarily conferred by the initial selective $(\pi/2)I_y$ pulse and the final selective refocusing pulse $(\pi)S_y$ pulse. Excellent suppression of unwanted resonances was obtained in ischemic rat leg and brain, allowing visualization of the lactate resonance. However, due to the inefficiency of transfer, 38% of total magnetization, and high pulse angle dependence, $\sin^9(\Theta)\cos^2(\Theta)$ make this technique is inherently insensitive when used with an inhomogeneous coil.

3.4 Applications to Animals

The use of homonuclear editing techniques to resolve specific biochemical question is becoming common in vitro systems. Crockard et al. [30] has used the editing sequence described by Williams et al. [26] to resolve lactate from gerbil brain during acute ischemia and reflow. The appearance and clearance of lactate was correlated with ^{31}P NMR measurements of energy metabolism and blood flow. Hanstock et al. [31] has demonstrated that spectral editing can be combined with DEPTH pulse localization [20] to monitor regional lactate appearance after cold lesion in cat brain. To date, most ^{1}H homonuclear editing experiments have been applied to resolve lactate and observe the effects of anaerobic metabolism, Crawford and Sanders [32] have demonstrated the broader utility of spectral editing techniques in vivo metabolism studies. Specifically, they have used a decoupling based editing sequence to resolve glutathione in packed red blood cells, enabling a measure of the redox potential.

3.5 Applications to Humans

Due to the combined limitations of available field strength and adequate bore size for human studies, there have been relatively few applications of spectral editing sequences to human studies. At 4.7 T, Pan et al. [5, 34] used semi-selective pulses with selective inversion pulses and phase cycling along with surface coil localization techniques [33] to measure buffering capacity, by correlating lactate and pH clearance in the human forearm after exercise. Hanstock et al. [35] has demonstrated the ability to edit lactate from healthy human brain at 2.1 T. This study indicated the feasibility of performing spectral editing at moderately low field strengths. Although lactate has been observed from human brain by several volume localization methods [36–38] without spectral editing techniques, the possibility of alanine or small lipid resonances overlapping the lactate resonance may limit the accuracy of non-spectral editing methods of lactate quantitation.

3.6 Quantitation

The primary goal of most spectral editing methods is to provide an error free method for evaluating changes in metabolite resonances which have significant overlap with other resonances. As such, the assignment of millimolar quantities to these edited resonances is a natural extension. However, the conversion of these well resolved signals to millimolar quantities requires knowledge of: 1) an internal or external standard, 2) the T_1 and T_2 of the metabolite and reference, and 3) the pulse angle dependence of the metabolite and reference signals. In animal studies, where chemical extraction is possible at the conclusion of the experiment, typical values for lactate were in good agreement with that expected when the above factors were considered [39]. Finally, Pan et al. [5] using the total creatine pool as an internal standard in human muscle, and correcting for differential T_1, T_2 and pulse angle dependence responses, obtained post exercise lactate concentrations which were in agreement with literature values.

4 Heteronuclear Editing

Due to the low sensitivity of the ^{13}C nucleus, yet its importance in facilitating measurements of metabolic turnover, a variety of heteronuclear editing methods have been pursued. These techniques are designed to measure the quantity of ^{13}C incorporation into a specific metabolite, and use the superior sensitivity of the ^{1}H nucleus to achieve improved time resolution. Similar to homonuclear editing, heteronuclear editing can be grouped into three broad categories, 1) CW decoupling based schemes, 2) inversion of coupled spins, and 3) polarization transfer. The majority of these sequences find their basis in earlier work in solution chemistry [40–42].

4.1 Decoupling Techniques

Unlike homonuclear editing sequences, the large difference in the frequency of the decoupled resonance ^{13}C, to that of the observed resonance ^{1}H, provides complete specificity for virtually any perturbation to the coupled spin. This complete isolation was initially used by Silerud et al. [43] to resolve the ^{13}C–^{1}H coupled ^{1}H resonances. Specifically, ^{1}H pulse acquire measurements were made in the presence and absence of ^{13}C decoupling during acquisition. Subtraction of the resulting spectra yielded a difference spectrum containing signals arising from the ^{13}C satellites in the non decoupled spectrum and a central inverted resonance derived from the decoupled ^{13}C satellites. Thus the edited ^{13}C signals are spread over approximately 120 Hz, with the maximum intensity (the inverted resonance) reflecting 1/2 of the total $^1H-^{13}C$ signal available. To improve spectral interpretation and S/N, Rothman et al. [44] developed a heteronuclear spin echo version of the homonuclear CW decoupling experiment [15]. Since the ^1H refocusing pulse does not affect the ^{13}C nuclei, no J-modulation can be induced. Therefore, to achieve a phase inversion of the ^{13}C coupled ^1H resonances, their magnetization must be trapped in the J-evolved, phase inverted state, I_y. In this sequence trapping is achieved by increasing the echo evolution delays, τ, to $1/J_{IS}$ and applying decoupling, ($J_{IS} = 0$) during the second echo evolution period

$$I_z \xrightarrow{(\pi/2)I_x} -I_y \xrightarrow{(J_{IS}\tau)I_zS_z} I_y \xrightarrow{(\pi)I_x} -I_y \xrightarrow{(0.\tau)I_zS_z} -I_y. \tag{19}$$

With the phase difference encoded in the magnetization state prior to acquisition, the ^1H spectrum can then be ^{13}C decoupled. Thus only a single ^{13}C line is observed in the $^1H-^{13}C$ edited difference spectrum thereby achieving full sensitivity and improved spectral resolution. The difference spectrum is generated by subtracting the spectrum acquired with decoupling of the S spins during the second evolution period, (A) from that acquired without any decoupling during either echo evolution period (B).

$$(\pi/2)I_x - \tau - (\pi)I_x - \tau - AQ \tag{20}$$

(A) DEC off DEC on DEC on

(B) DEC off DEC off DEC on.

Unlike $^1H-^1H$ coupling constants (5–10 Hz) many ^{13}C coupling constants are on the order of 125 Hz, such that substantially higher decoupling powers are required for complete decoupling, 625–1250 Hz. This effectively limits the selectivity of the technique in the ^{13}C spectrum. Specifically, any $^1H-^{13}C$ coupled ^{13}C resonance within a kHz, (11 ppm at 8.5 T, 45 ppm at 2 T) of the target ^{13}C coupled resonance may also be observed in the difference spectrum. Thus although the specificity is complete, and no water-induced anomaly occurs, the selectivity of this technique is quite poor.

4.2 Inversion Techniques

Due to the inherent lack of selectivity of the spin echo decoupling technique, a broad band pulse technique for editing offers the possible advantages of decreased power deposition and or increasing the number of ^{13}C metabolites that can be observed in a given measurement through broader band of ^{13}C perturbation [9]. Inversion based sequences are obtained by a simple and straight forward extension of the same principles utilized in the selective inversion homonuclear editing experiment. Specifically, a non-selective ^{13}C pulse is applied during alternate acquisitions of a ^1H spin echo sequence. An evolution delay, τ_1, of $1/2J$ ($^1H-^{13}C$) is used.

$$(\pi/2)I_x - \tau_1 - \tau_2 - (\pi)I_x - \tau_2 - \tau_1 - AQ \tag{21}$$

(A)	$(\pi)S_x$	DEC on
(B)	$(0)S_x$	DEC on.

Unlike the heteronuclear CW decoupling technique any ^{1}H spin echo evolution period may be used, $(\tau_1 + \tau_2)$. However, T_2 or ^{1}H–^{1}H J-modulation losses can be reduced if the total echo time, $\tau_1 + \tau_2$ is kept near $1/2J$ (^{1}H–^{13}C).

Similar to the homonuclear selective inversion technique, the selectivity of this sequence is completely governed by the frequency characteristics of the inversion pulse $(\pi)S_x$. Due to the short ^{13}C–^{1}H evolution times, ≈ 4 ms, resolution in the ^{13}C spectrum via editing is not feasible. Thus if considerable overlap of ^{13}C edited signals occurs in the ^{1}H spectrum, additional methods of spectral resolution may be required.

To address this problem Jue [45] has described an alternative sequence combining ^{1}H–^{13}C and ^{1}H–^{1}H editing sequences for simultaneous selection of only the ^{13}C coupled ^{1}H resonance arising from only the desired ^{1}H resonance. Specifically the sequence

$$^1\mathrm{H}(\pi/2)I_x - \tau_1 - \tau_2 - \tau_3 - (\pi)I_x - \tau_1 - \tau_2 - \quad \tau_3 - AQ$$

(A)	$(\pi)S_{1x}$	$(0)S_{2x}$
(B)	$(0)S_{1x}$	$(0)S_{2x}$
(C)	$(\pi)S_{1x}$	$(\pi)S_{2x}$
(D)	$(0)S_{1x}$	$(\pi)S_{2x}$

was used, where I refers to the ^{1}H resonance to be observed, S_1 its homonuclear J-coupled partner, and S_2 its heteronuclear ^{13}C J-coupled partner. The evolution delays, were $\tau_1 = 1/2J_{1S}$ (^{1}H–^{1}H), $\tau_2 + \tau_3 = 1/2J_{1S}$ (^{1}H–^{1}H), and $\tau_3 = 1/2J_{IS}$ (^{1}H–^{13}C). As described previously, homonuclear editing is achieved by subtracting the pairs $[(A)-(B)]$ and $[(C)-(D)]$. ^{13}C editing is achieved by subtracting pairs of acquisitions acquired in the absence and presence of the ^{13}C inversion pulse. $[(A)-(C)]$ and $[(B)-(D)]$. Thus the complete, double edited spectrum is given by $[(A)-(B)]-[(C)-(D)]$. This technique provides a powerful tool for heteronuclear editing when the desired ^{13}C resonance can not be resolved by simple ^{1}H–^{13}C editing.

In this example the selectivity of the sequence is largely governed by the selectivity of the ^{1}H editing. However, this increase in selection comes at the cost of 1) increased T_2 losses through longer echo evolution periods, 2) possible J-modulation losses for non AX ^{1}H–^{1}H spin systems, and 3) losses in editing efficiency if the applied B_1 is spatially inhomogeneous due to the $\sin^7(\Theta)$ dependence. Therefore, in the instance that suitable ^{13}C editing can be achieved without the additional homonuclear editing, the simpler heteronuclear experiment provides a simpler alternative that may yield better edited signal intensity.

4.3 Polarization Transfer

Heteronuclear polarization transfer techniques have also been used to perform heteronuclear editing. Bendall et al. [46] has used a polarization transfer sequence to selectively observe ^{13}C labeled metabolites of different ^{1}H multiplicity from a ^{13}C labeled ex vivo liver. The sequence

$$(\pi/2)I_x - \tau - (\pi)I_x, (\pi/2)S_x - \tau - (\pi)S_x - \tau - AQ(^{13}C)DEC(^{1}H) \quad (23)$$

$$\begin{array}{ll} (A) & (\pi/4)I_y \\ (B) & (\pi/2)I_y \\ (C) & (3\pi/4)I_y \end{array}$$

was used, where $\tau = 1/2J$ ($^{1}H-^{13}C$), the transferred ^{13}C magnetization, S, was acquired and ^{1}H decoupling was applied during the acquisition. Subspectra containing only ^{13}C labeled CH, CH_2 and CH_3 resonances was obtained by constructing linear combinations of spectra acquired with different ^{1}H pulse angles Θ, ($\Theta = \pi/4$, $\pi/2$ and $3\pi/4$).

$$\begin{array}{ll} CH: & \pi/2 \\ CH_2: & \pi/4 - 3\pi/4 \\ CH_3: & \pi/4 + 3\pi/4 - .707\pi/2 \end{array} \quad (24)$$

The accuracy of the filtering and suppression of unwanted spin systems is strongly dependent on both the applied pulse angle and the coupling constant. Specifically, contamination of subspectra will arise due to the variation in $^{1}H-^{13}C$ coupling constants.

4.4 Applications in Animals

Measurements of ^{13}C metabolites in cells [47], tissues [48], and intact animals [49] by using simple ^{13}C pulse and acquire techniques have been extensively employed throughout the in vivo literature. Although these measurements provide detailed information regarding relative metabolic fluxes of different pathways, the required acquisition times can be long (due to the low sensitivity of the ^{13}C nucleus) prohibiting multiple measurements prior to attaining steady state ^{13}C enrichment. Additionally, since only the ^{13}C signal is observed, the actual level of fractional enrichment can not be determined. Thus, $^{1}H-^{13}C$ editing techniques have been applied to 1) to enable rapid measurements of ^{13}C incorporation to enable absolute rate measurements of metabolic pathways [9] and 2) obtain fractional enrichment information to assess the activity of various pathways [44].

Fitzpatrick et al. [9] measured the cerebral TCA cycle flux by observing the incorporation C-1 labeled ^{13}C glucose into the C-4 ^{1}H resonance of glutamate. Time resolution of 90 s was achieved at 8.4 T on the brain of an intact rat in vivo. The measured flux was determined from the exponential rate constant of the observed data along with literature values for the total glutamate pool. The calculated flux was in excellent agreement with other standard biochemical techniques for determining cerebral TCA cycle rates. Rothman et al. [44] used the temporal variation of the ^{13}C fractional enrichment of lactate during anoxia to evaluate the breakdown of unlabeled glycogen in rat brain. In this example, lactate produced during the initial portion of the anoxic period was labeled according to the glucose fractional enrichment. During the later stages of the anoxic period, unlabeled brain glycogen breakdown and subsequent metabolism to lactate, resulted in a decrease in the ^{13}C fractional enrichment of lactate.

4.5 Applications in Humans

Recently Rothman et al. [50] demonstrated that measurement of TCA cycle rates in the human brain by ^{1}H–^{13}C heteronuclear editing is also feasible. Rothman and colleagues observed the effects of TCA cycle flux by observing the ^{1}H–^{13}C edited C-4 ^{1}H resonance of glutamate using a volume localization technique with a heteronuclear inversion pulse editing sequence. Although in these initial applications time resolution was only 6 min at 2.1 T, thereby limiting the accuracy of the calculated TCA cycle rate, the possibility of using larger volumes, improvements in coil design and availability of higher field human systems suggest the possibility of highly accurate measurements of TCA cycles rates in humans in various disease states. Finally, as recently suggested by Rothman et al. [51], the ^{13}C fractional enrichment of the lactate C-3 proton resonance after administration of ^{13}C labeled glucose may provide a sensitive indicator of ongoing anaerobic activity or slow lactate washout in stroke patients. Specifically, in patients with ongoing anaerobic activity due to vascular compromise, lactate will become ^{13}C labeled. In instances of infarction and slow washout, the elevated lactate pool will not be labeled. This application of spectral editing techniques may provide a clinically diagnostic measure of the metabolic status of the stroke.

5 Summary and Conclusions

Although a wide variety of both homonuclear ^{1}H and heteronuclear ^{1}H–^{13}C editing and polarization transfer sequences have been described in the literature with various advantages and disadvantages, the vast majority of these sequences represent simple variations on a theme. Specifically, most sequences can be thought of as spin echo sequences where either *J*-modulation is selectively inhibited by decoupling of the coupled spin, or its effects reversed through inversion of the coupled spin while it is in an anti-phase state (I_xS_z or I_yS_z).

Coherence transfer methods rely on the creation of an antiphase magnetization state followed by transfer of the magnetization from the coupled to the desired spin. Standard spectral editing sequences rely on subtraction of alternate scans to yield an edited spectrum which is therefore sensitive to random fluctuation such as field instability or sample motion. Although homonuclear coherence transfer techniques have been utilized to offset this limitation, their high pulse angle dependence, and necessity to completely eliminate resonances in the region of the desired resonance can result in poor sensitivity when used with inhomogeneous coils.

As discussed, the performance of various editing and coherence transfer sequences can be conceptually and numerically evaluated by assessing the specificity and selectivity of the sequence. This allows the experimentalist to choose the most appropriate sequence for his own unique application and experimental needs. These technical concerns include: 1) the difficulty of implementation, 2) sensitivity to random motion or field instability, 3) sample characteristics including: spectral positions, pattern of coupling, coupling constants and concentrations of observed metabolites and 4) the type of coil used.

Acknowledgement: The author would like to thank Dr. J.W. Pan for helpful comments and suggestions in preparing the manuscript.

6 References

1. Behar KL, Rothman DL, Shulman RG, Petroff OAC, Petroff JW (1984) Proc Natl Acad Sci 81: 2517
2. Rothman DL, Behar KL, Hetherington HP, Shulman RG (1984) Proc Natl Acad Sci 81: 6330
3. Jue T, Arias-Mendoza F, Gonnella NC, Shulman GI, Shulman RG (1985) Proc Natl Acad Sci 82: 5246
4. Williams SR, Gadian DG, Proctor E, Sprague DB, Talbot DF, Young IR, Brown FF (1985) J Magn Reson 63: 406
5. Pan JW, Hetherington HP, Hamm JR, Shulman RG (1991) Magn Reson Med Vol 20, No 1, p 48
6. Behar KL, den Hollander JA, Stromski ME, Ogino T, Shulman RG, Petroff OAC, Prichard JW (1983) Proc Natl Acad Sci 80: 4945
7. Gadian DG, Frackowiack RSJ, Crockard HA, Proctor E, Allen K, Williams SR, Russel RWR (1987) J Cereb Blood Flow and Metab 7: 199
8. Fitzpatrick SM, Hetherington HP, Behar KL, Shulman RG (1989) J Neurochem 52: 741
9. Fitzpatrick SM, Hetherington HP, Behar KL, Shulman RG (1990) J Cereb Blood Flow and Metab 10: 170
10. Sotak CH, Freeman D (1988) J Magn Reson 77: 382
11. Van de Ven FJM, Hilbers CW (1983) J Magn Reson 54: 512
12. Ow Sorensen, Eich GW, Levitt MH, Bodenhausen G, Ernst RR (1983) Prog Nucl Magn Reson Spectrosc 16: 163
13. Campbell ID, Dobson CM (1979) Meth Biochem Anal 25: 1
14. Brindle KM, Porteus R, Campbell ID (1984) J Magn Reson 56: 543
15. Hetherington HP, Avison MJ, Shulman RG (1985) Proc Natl Acad Sci 82: 3115
16. Rothman DL, Behar KL, Hetherington HP, den Hollander JA, Bendall MR, Petroff OAC, Shulman RG (1985) Proc Natl Acad Sci 82: 1633
17. Hetherington HP, Hamm JR, Pan JW, Rothman DL, Shulman RG (1989) J Magn Reson 82: 86

18. Edzes HT, Zuurbier CJ, Huijing PA, Bulthuis J (1990) J Magn Reson 87: 395
19. Bodenhausen G, Freeman R, Turner DL (1977) J Magn Reson 27: 511
20. Bendall MR, Gordon RE (1983) J Magn Reson 53: 365
21. Hetherington HP, Rothman DL (1985) J Magn Reson 65: 348
22. Hetherington HP, Wishart D, Fitzpatrick SM, Cole P, Shulman RG (1986) J Magn Reson 66: 313
23. Evelhoch JL, Crowley MG, Ackerman JHH (1984) J Magn Reson 56: 110
24. Rothman DL, Arias-Mendoza F, Shulman GI, Shulman RG (1984) J Magn Reson 60: 430
25. Jue T, Arias-Mendoza F, Gonnella NC, Shulman GI, Shulman RG (1985) Proc Natl Acad Sci 82: 5246
26. Williams SR, Gadian DG, Proctor E (1986) J Magn Reson 66: 562
27. Hore PJ (1983) J Magn Reson 55: 283
28. Morris GA, Freeman R (1978) J Magn Reson 29: 433
29. Von Kienlin M, Albrand JP, Authier B, Blondet P, Lotito S, Decorps M (1987) J Magn Reson 75: 371
30. Crockard HA, Gadian DG, Frackowiak RS, Proctor E, Williams SR (1987) J Cereb Blood Flow Metab 7: 394
31. Hanstock CC, Bendall MR, Hetherington HP, Bosviert DP, Allen PS (1987) J Magn Reson 71: 349
32. Crawford A, Sanders JKM (1988) J Magn Reson 80: 143
33. Pan JW, Hetherington HP, Hamm JR, Rothman DL, Shulman RG (1989) J Magn Reson 81: 608
34. Pan JW, Hamm JR, Hetherington HP, Rothman DL, Shulman RG (1991) Magn Reson Med Vol 20, No 1, p 57
35. Hanstock CC, Rothman DL, Prichard JW, Jue T, Shulman RG (1988) Proc Natl Acad Sci 85: 1821
36. Frahm J, Bruhn H, Gyngell ML, Merboldt KD, Hanicke W, Sauter R (1989) Magn Reson Med 9: 79
37. Luyten PR, Segebarth C, Baleriaux D, den Hollander JA (1987) Radiology P 165: 66
38. Sappey-Marinier D, Calabrese G, Hugg J, Deicken R, Fein G, Weiner MW (1990) In: Proceedings of the Society of Magnetic Resonance in Medicine New York, p 106
39. Williams SR, Proctor E, Allen K, Gadian DG, Crockard HA (1988) Magn Reson Med 7: 425
40. Freeman R, Hill HDW (1971) J Chem Phys 54: 301
41. Bendall MR, Pegg DT, Dodrell DM, Fields J (1981) J Am Chem Soc 103: 934
42. Freeman R, Mareci TH, Morris GA (1981) J Magn Reson 42: 341
43. Sillerud LO, Alger JR, Shulman RG (1981) J Magn Reson 45: 142
44. Rothman DL, Behar KL, Hetherington HP, den Hollander JA, Bendall MR, Petroff OAC, Shulman RG (1985) Proc Natl Acad Sci 82: 1633
45. Jue T (1988) J Magn Reson 78: 161
46. Bendall MR, den Hollander JA, Arias-Mendoza F, Rothman DL, Behar KL, Shulman RG (1985) Magn Reson Med 2: 56
47. Cohen SM, Ogawa S, Shulman RG (1979) Proc Natl Acad Sci 76: 1603
48. Cohen SM, Shulman RG (1982) In: Cohen J (ed) Noninvasive probes of tissue metabolism, 00, 00 (Publisher, Location) p 119
49. Ganioni P, Alger JR, Shulman RG (1983) Biochem 22: 4974
50. Rothman DL, Novotony EJ, Shulman GI, Howseman AM, Mason GF, Petroff OAC, Nixon T, Hanstock CC, Prichard JW, Shulman RG (1990) In: Proceedings of the Society of Magnetic Resonance in Medicine, New York, p 986
51. Rothman DL, Howseman AM, Graham GD, Lantos G, Petroff OAC, Brass LM, Fayad PB, Shulman GI, Shulman RG, Prichard JW (1990) In: Proceedings of the Society of Magnetic Resonance in Medicine, New York, p 107

Metabolite Specific Methods Using Double Quantum Coherence Transfer Spectroscopy

D. Freeman and R. Hurd

General Electric NMR Instruments, 255 Fourier Ave., Fremont, CA 94539, USA

Table of Contents

One of the great challenges in proton in vivo nuclear magnetic resonance (NMR) is editing for metabolites which occur during tumor metabolism, or as the end products of hypoxic or ischemic insult to normal tissue. These compounds may include lactate, alanine, glutamine and taurine. Lactate is of particular interest since it is present in levels of less than one millimolar in most normal tissue, but can rise to thirty millimolar in abnormal tissue. In order to observe lactate in tissue other than brain the overlapping lipid signal must be removed, requiring some sort of editing technique, in addition to efficient water suppression. The above compounds have the property that they are 'J' coupled and are hence amenable to J-modulation, polarization transfer, and multiple quantum editing schemes.

There have been a variety of methods suggested for lactate editing, each with advantages and disadvantages. This article will review the significant contributions to the development of these methods. The article will focus on the most successful of these methods and the theory behind them. In general, the most efficient methods for editing lactate have involved the use of gradient selection of multiple quantum coherence. The efficiency of these techniques for the suppression of water, and editing of lipid resonances will be examined.

Double quantum coherence transfer techniques have been used to obtain lactate edited spectra and images *in vivo*. They are currently being tested as non-invasive methods for monitoring the response to drug treatment in tumors in the clinical setting.

This type of gradient enhanced spectroscopy technique has been extended to other important applications such as proton detected carbon spectroscopy and imaging, and to high resolution NMR experiments. Common advantages include water suppression efficiently and reduced sensitivity to motion.

NMR Basic Principles and Progress, Vol. 27

1 Introduction

Multinuclear spectroscopic analysis combined with imaging of normal and diseased tissue *in vivo* is beginning to provide useful information for the clinician on the extent, severity, and time course of treatment in animal models [1,2] and human patients [3–6]. In most instances the nucleus of choice has been phosphorus, fluorine, carbon or sodium, any of which has advantages and disadvantages depending on the study. Proton NMR, however, should provide the ideal nucleus for spectroscopic examination (providing the overriding signal from water, and interfering lipids can be removed) due to its abundance, gyromagnetic ratio, and the ease of combination with imaging protocols. Compounds of clinical or biochemical interest found at tissue levels observable by proton NMR include lactate, taurine, *N*-acetylaspartate and glutamine. Though lactate is an end product of glycolysis, it is of particular interest because of its biochemical and pathological importance. One might expect an increased level of lactic acid during anaerobic glycolysis, as is found in certain solid cell tumors; in conditions of reduced blood flow, in inborn errors of metabolism; or even in drug related incidences.

Observation of lactate *in vivo* presents a real challenge compared to other compounds due to the solvent peak (tissue water, and the large peak from lipid observed in most tissue (except normal brain).) The requirements for an efficient *in vivo* spectroscopic and imaging technique for lactate editing are thus stringent, demanding that the method (a) be efficient in terms of selectivity for lactate (good lipid suppression) (b) be efficient in sensitivity (c) be easy to set up and fast to run bearing in mind patient constraints (d) require minimal power deposition (i.e. avoid decoupling techniques if possible) and (e) allow for localization of the signal to the region of interest possibly deep within the body, or provide an image of the metabolite distribution.

There have been many attempts to produce an efficient method for lactate edited spectroscopy and imaging, some more successful than others, some merely limited by the technology at hand. We will review the most significant of these efforts, and list their advantages and limitations. The most successful methods have employed gradient selection of multiple quantum coherence. The extension of gradient selection of coherence into other applications in NMR will be presented.

A method has recently been demonstrated which achieves a further increase in lactate signal via gradient selection of both zero and double quantum signal [24]. Because of gradient symmetry, however, this method appears to require relatively good B_1 homogeneity and good frequency selectivity to keep unwanted single quantum water and lipid signal dephased. The *in vitro* results are encouraging, but the advantages have not yet been realized *in vivo* (Table 1).

Table 1. History of lactate editing methods

First Author	Method	Suppression	
		Water	Lipid
Rothman et al. (1984) J Magn Reson 60: 430	Homonuclear J modulation difference technique	Presat.	Subtraction
Williams et al. (1985) J Magn Reson 63: 406	Homonuclear J modulation difference technique	T_2 weighted	Subtraction
Von Kienlin et al. (1987) J Magn Reson 75: 371	Homonuclear polarization transfer	T_2 weighted	Non selective $\pi/2$ on CH for polarization transfer. Presat lipid with DANTE pulse
Hardy and Dumoulin (1987) Magn Reson Med 5: 58	Selective homonuclear polarization transfer	Binomial	Selective pulse on CH or CH_3 for polarization transfer
Sotak and Freeman (1988) J Magn Reson 77: 382	Zero quantum coherence difference technique	T_2 weighted & zero quantum coherence	Zero quantum
Sotak et al. (1988) J Magn Reson 78: 355	2 dimensional double quantum coherence	DQ filter	DQ & 2D for residual lipid
Webb et al. (1988) Soc Magn Reson Med 2: 758	Volume localized double quantum filter	DQ filter	DQ filter, long DQ evolution
Keller et al. (1988) Soc Magn Reson Med 2: 757	Double quantum filter	DQ filter	Short DQ prep time to reduce strongly coupled lipids
Eland et al. (1988) Soc Magn Reson Med 2: 818	3D chemical shift selective	1331	NO
Doddrell et al. (1989) Magn Reson Med 9: 132	One dimensional zero quantum coherence transfer	Zero quantum coherence	Gaussian selection of CH
McKinnon et al. (1989) Soc Magn Reson Med 1: 222	Double quantum coherence transfer	DQ filter	Coherence transfer selective 1331 excitation
Knuttel and Kimmich (1989) J Magn Reson 83: 335	Cycl. polarization transfer	Polarization transfer	Polarization transfer
Hurd and Freeman (1989) PNAS 86: 4402	DQ Coherence transfer	DQ	Selective 1331
Stryjewski et al. (1990) Magn Reson Med 13: 158	Polarization transfer	Polarization transfer Gaussian excitation	Polarization transfer
Trimble et al. (1990) J Magn Reson, p 191	ZQ & DQ coherence transfer	2Q & DQ	Selective 11 on methine

2 Contributions to Metabolite Specific Proton Editing In Vivo

The first methods for spectral editing to allow for observation of lactate in the presence of strong overlapping lipid signal [7] were based on difference spectroscopy using alternate inhibition of J-modulation by double resonance [8]. These methods retain most of the lactate methyl signal, but rely on T_2 discrimination and presaturation for water suppression. The need for double resonance was eliminated [9], by using a binomial pulse [10, 11] on alternate scans to inhibit *J*-modulation, and T_2 discrimination was used to eliminate water. The potential for subtraction error, (caused by motion, or system instability) limits the general application of these *J*-modulation methods to detect lactate, especially in living tissue with large observable lipid signals. Other subtractive methods developed for the detection of lactate in the presence of lipid include a method based on zero quantum coherence, with the advantage of not requiring selective pulses [12], and a method based on selective homonuclear polarization transfer [13]. These methods also require additional water suppression measures such as presaturation, and have reduced detectable signal from the lactate methyl.

Methods developed to avoid subtraction errors include the use of chemical shift selective pulses in concert with homonuclear polarization transfer [14] and in concert with zero quantum coherence transfer [15], and soft COSY [16]. Though these selective methods obviate the need for subtractive editing, they are compromised by the level of desired signal which can be recovered, which is 25% or less of the total. These methods also require additional water suppression such as presaturation or heavy T_2 weighting.

Although double quantum coherence techniques [17, 18] provide excellent water suppression [19, 20], they are also compromised in recovered signal and require further editing of lipid and lactate. One approach to resolving the lactate methyl from the lipid resonance without the use of subtraction is the use of two dimensional NMR methods [21]. Improvements in these methods using selective coherence transfer [22, 23], have addressed the issue of unwanted loss of lactate methyl signal and provide significant lipid suppression.

This approach has been proven to work *in vivo* for both spectroscopy and imaging applications.

Volume localization methods have the added advantage of not picking up the subcutaneous fat present in nearly all these implanted subcutaneous or subdermal tumor models. The importance of the ability to localize small volume of tissue lactate is that by NMR one can achieve the kind of information from volumes similar to those used in chemical analysis (e.g. spectrophotometric analysis) of tissues.

3 Theory of Gradient Enhanced Multiple Quantum Editing

In a magnetic field, an isolated spin 1/2 nucleus has only two energy states, spin quantum states $+1/2$ and $-1/2$ as illustrated in Fig. 1a. Transition between these two states is a single quantum change of $+1$ or -1. A coupled system however, has a more complicated set of energy states generated by spin–spin coupling. An energy level diagram for a simple two-spin system is shown in Fig. 1b. With four energy levels, there are six possible transitions. Four of these are single quantum and are responsible for the four discrete frequencies observed in the NMR spectrum. There is also one double quantum transition and one zero quantum transition. Although all six transitions in this simple spin system can be excited, only single quantum transitions are directly observable, because only these result from transverse magnetization. The zero and double quantum transitions must be observed indirectly. Once generated, the phase of evolving zero and double quantum coherences will be retained once these coherences are converted to observable single quantum signal. Appropriate manipulation of this phase behavior can then be used to edit for zero or double quantum transitions. For this model system, double quantum coherence modulates at the sum, and zero quantum at the difference of the transmitter frequency offsets for the coupled spins.

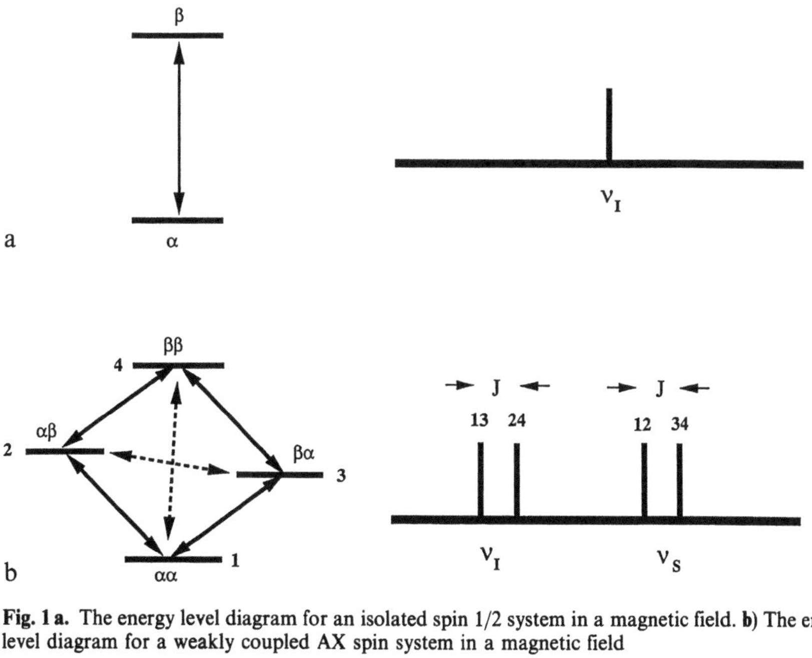

Fig. 1 a. The energy level diagram for an isolated spin 1/2 system in a magnetic field. **b)** The energy level diagram for a weakly coupled AX spin system in a magnetic field

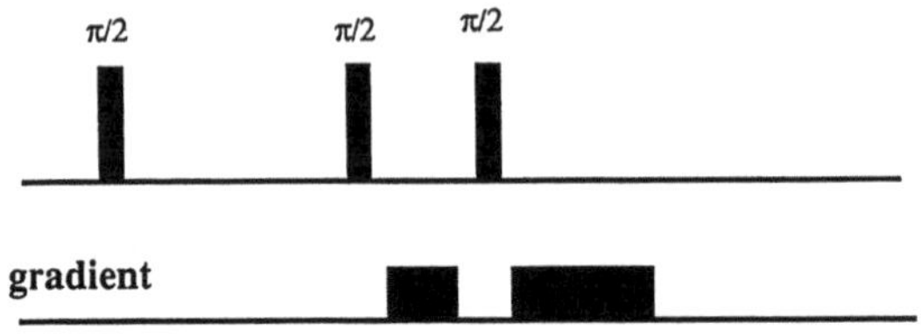

Fig. 2. The simplest sequence for the excitation and indirect detection of double quantum coherence transfer

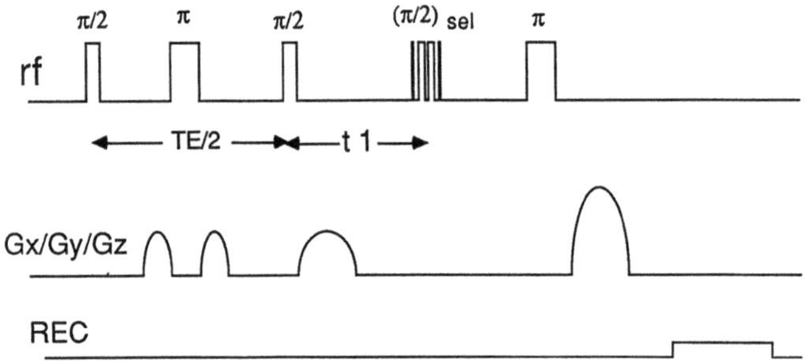

Fig. 3. Selective proton double quantum coherence transfer sequence. Composite gradients in *xy* and *z* are used to select for the desired order of coherence. The use of symmetric excitation and detection, combined with selective coherence transfer provides efficient water and lipid suppression while maintaining sensitivity for the lactate methyl

The use of B_0-field gradients to select for specific coherences in NMR was first described over a decade ago [25–27]. The simplest pulse sequence for generating and selecting double quantum coherence is shown in Fig. 2. For coupled spins, the first two radiofrequency pulses generate double quantum coherence which acquires a phase distribution proportional to the area of the applied gradient (Gt) times twice the gyromagnetic ratio of the excited spins:

$$\mathrm{Gt}\,(\gamma + \gamma). \qquad (1)$$

This coherence is converted to observable single quantum coherence by the final radiofrequency pulse. Since the phase dependence of single quantum coherence is simply proportional to the gradient area multiplied by the gyromagnetic ratio, the single quantum signal, which previously existed as double quantum coherence, can be selectively rephased and detected following application of a gradient with twice the area (2 Gt). Water, which has no scalar coupling, and hence no double quantum coherence is left dephased by the asymmetric gradient pair. More recently this idea has been applied in vivo using a 2D version of the method shown in Fig. 3 to sort out the overlapping signals of lactate and lipid [21].

3.1 *Efficiency of Fat Suppression*

One of the most important subsequent improvements in the double quantum coherence method has been the use of selective coherence transfer to maximize the signal from the lactate methyl and further reduce the lipid signal as illustrated in Fig. 3. As with all of the methods which use chemical shift selective pulses, the signal in a non-selective double quantum coherence sequence varies as: $\sin\alpha\cos^2(\alpha/2)$, where α is the tip angle of the radio-frequency read pulse [28]. This gives a maximum signal intensity of only 0.62. If only the lactate methine is excited with a selective radio-frequency read pulse, then the lactate methyl signal will vary as $\sin\alpha$, which increases the maximum detectable signal to 1.00. In addition, lipid which lacks a significant coupling partner at the lactate methine chemical shift of 4.2 ppm will not generate very much double quantum coherence and will be suppressed [22]. The selective versus non-selective behavior of double quantum coherence transfer is observed experimentally (Fig. 4) and only deviates at larger tip angles; presumably due to B_1 inhomogeneity.

3.2 *Efficiency of Water Suppression*

With the use of composite gradients (simultaneous x, y, and z) to select for double quantum coherence, water suppression has been greatly improved. The

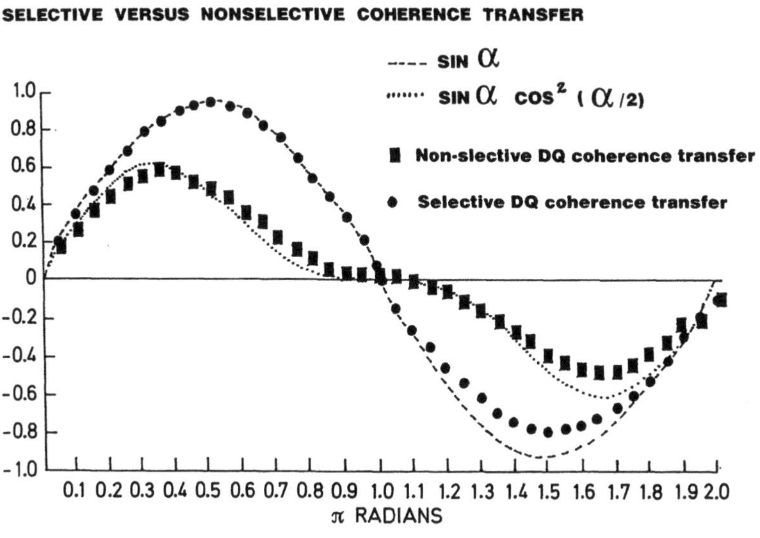

Fig. 4. Selective versus non-selective coherence transfer. Using a $\pi/2$ read pulse, selective coherence transfer should theoretically give a factor of 2.0 improvement over the non-selective method. Using a solution of 100 mM *N*-acetyl alanine this was tested experimentally. The signal in the non-selective version varies as $\sin a\cos^2(\alpha/2)$, where α is the tip angle in π radians for the read pulse. This varies as $\sin\alpha$ in the selective version. Any selective $\pi/2$ pulse which excites the methine and not the methyl will produce similar results

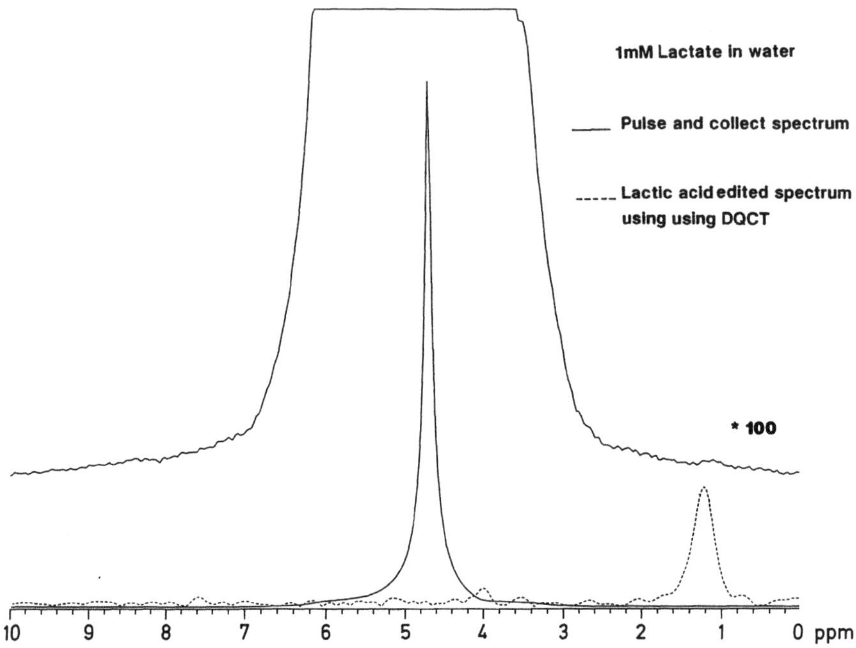

Fig. 5. Proton double quantum coherence spectrum of 1 mM lactic acid in water showing the efficiency of the method for water suppression. The *solid line* is a pulse and acquire spectrum of 1 mM lactate in water. The *dashed plot* shows the lactate methyl using the sequence in Fig. 3, where water is suppressed to better than 100 000, and residual water is essentially undetected

water suppression efficiency is shown in Fig. 5, where 1 millimolar lactic acid dissolved in water can be readily seen using composite gradients for selection of coherence. The efficiency of the water suppression can be attributed to the fact that the water spins remain dephased along the vector sum of the three orthogonal axes, rather than along the primary gradient axes, therefore diminishing the probability of unwanted signal being rephased by spurious gradients during acquisition.

The optimization of the double quantum editing sequence with respect to water suppression and efficiency of editing for the lactic acid methyl is shown in Fig. 6. The modifications were tested on a phantom consisting of 0.5 ml of 100 mM *N*-acetylalanine, (an NMR analogue for lactate). System variables such as spectral width, gain and predelay were kept constant, and the resultant spectrum from one acquisition is presented. Signal to noise measurements are shown to the right.

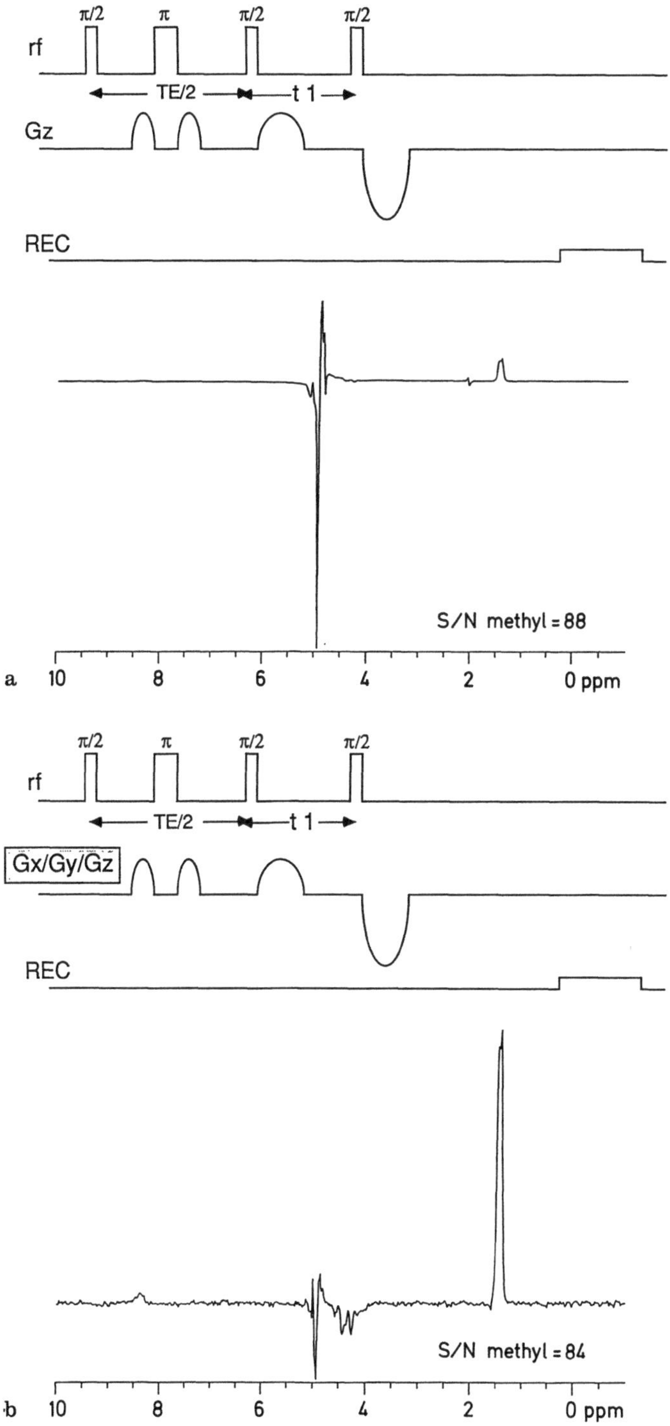
π/2
π
π/2
π/2
rf
TE/2
t 1
Gz
REC
S/N methyl = 88
a
10
8
6
4
2
0 ppm
π/2
π
π/2
π/2
rf
TE/2
t 1
Gx/Gy/Gz
REC
S/N methyl = 84
b
10
8
6
4
2
0 ppm

Fig. 6 a, b

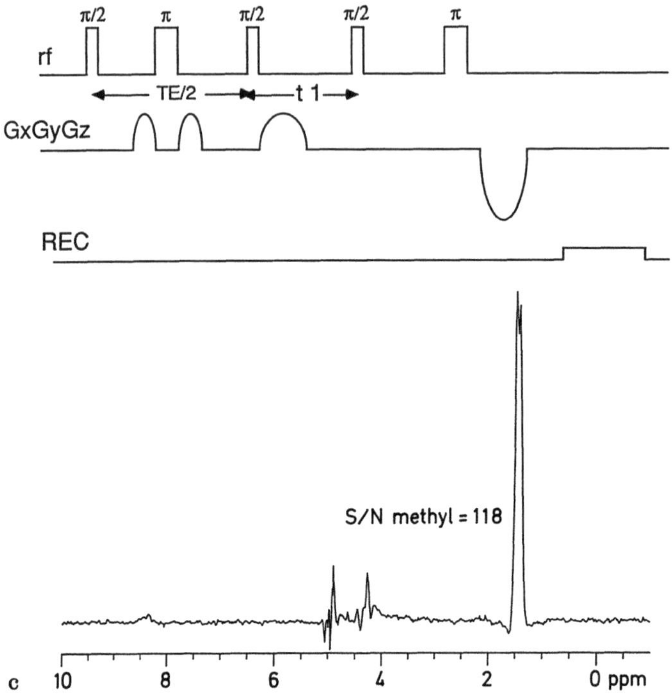
π/2
π
π/2
π/2
π
rf
TE/2
t 1
GxGyGz
REC
S/N methyl = 118
10
8
6
4
2
0 ppm

c

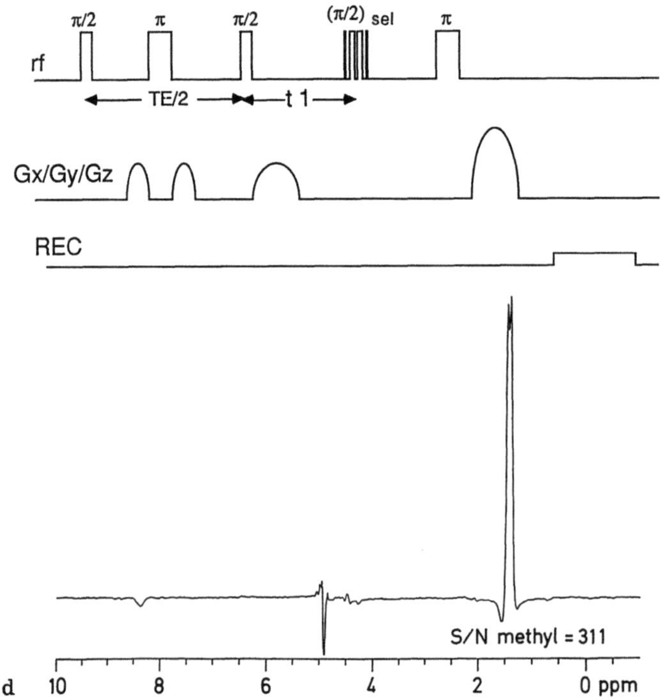
π/2
π
π/2
(π/2) sel
π
rf
TE/2
t 1
Gx/Gy/Gz
REC
S/N methyl = 311
10
8
6
4
2
0 ppm

d

Fig. 6 c, d

4 Metabolite Specific Spectroscopy Using Double Quantum Coherence *In Vivo*

The efficiency of the selective double quantum coherence technique was tested *in vivo* to determine lactate levels in experimental mouse tumors prior to and following treatment with agents known to alter the hypoxic cell fraction and hence the steady state lactic acid levels. Particular attention was paid to mimicking the established clinical protocols for drug administration as well as minimizing examination time in the magnet. Tumors were freeze clamped and extracted in perchloric acid, and assayed enzymatically for comparison with the NMR results.

Mice of the C3H strain, bearing radiation induced fibrosarcomas (RIF-1), were anesthetized, and a four turn radio-frequency solenoid fitted around the tumors. Tumor varied between 0.5 and 1.3 grams in size. Double quantum coherence transfer spectroscopy was carried out at 2.0 T or at 4.7 T on General Electric CSI systems fitted with actively shielded gradients.

Two dimensional double quantum coherence spectra were obtained using a 64 by 2 K data matrix, and incrementing the t_1 dwell time from 8 to 40 ms in 500 μs steps. Spectral width was 2000 Hz in ω_1 dimension and 5000 Hz in ω_2 dimension. For coherence transfer, the transfer was set on the lactate methine at 4.2 ppm, and a 1331 pulse was used to selectively excite this resonance. Under these conditions the double quantum frequency of lactate was 250 Hz at 2.0 T and 580 Hz at 4.7 T.

For the untreated RIF-1 tumors, the advantages gained using selective coherence transfer, and improved water suppression can be clearly seen by comparison of Figs. 7a and 7b. The spectrum in Fig. 7b took only 64 seconds to acquire, since only a single acquisition and a one second recycle time was used, and no phase cycling was required. The one-dimensional spectrum on the chemical shift axis (ω_2) is the first block of the two dimensional array. The spectrum on the double quantum axis (ω_1) was extracted from the two dimensional array at 1.3 ppm. The two dimensional spectrum in Fig. 7a took thirty minutes to acquire since 16 acquisitions were required for each t_1 increment. The result of inefficient water suppression is clearly seen, as well as the observation of significant interfering lipid in the region of the lactic acid methyl. The lactic acid concentration by chemical analysis in this tumor was

Fig. 6 a–d. The optimization of the proton double quantum coherence transfer sequence. The sample used to compare the sequences was 0.5 ml of a 100 mM *N*-acetyl alanine solution in water (as an NMR analogue for lactate AX_3 spin system). For this study all system variables such as gain, spectral width, a single acquisition, were held constant. Pulse sequences, experimental spectra, measured signal to noise ratios (S/N) are shown left to right for each of the four sequences. (**a**) The basic double quantum coherence sequence (**b**) The same as in (a) but with composite gradients in *x*, *y* and *z*. (**c**) The use of symmetric excitation and detection (**d**) The addition of selective coherence transfer

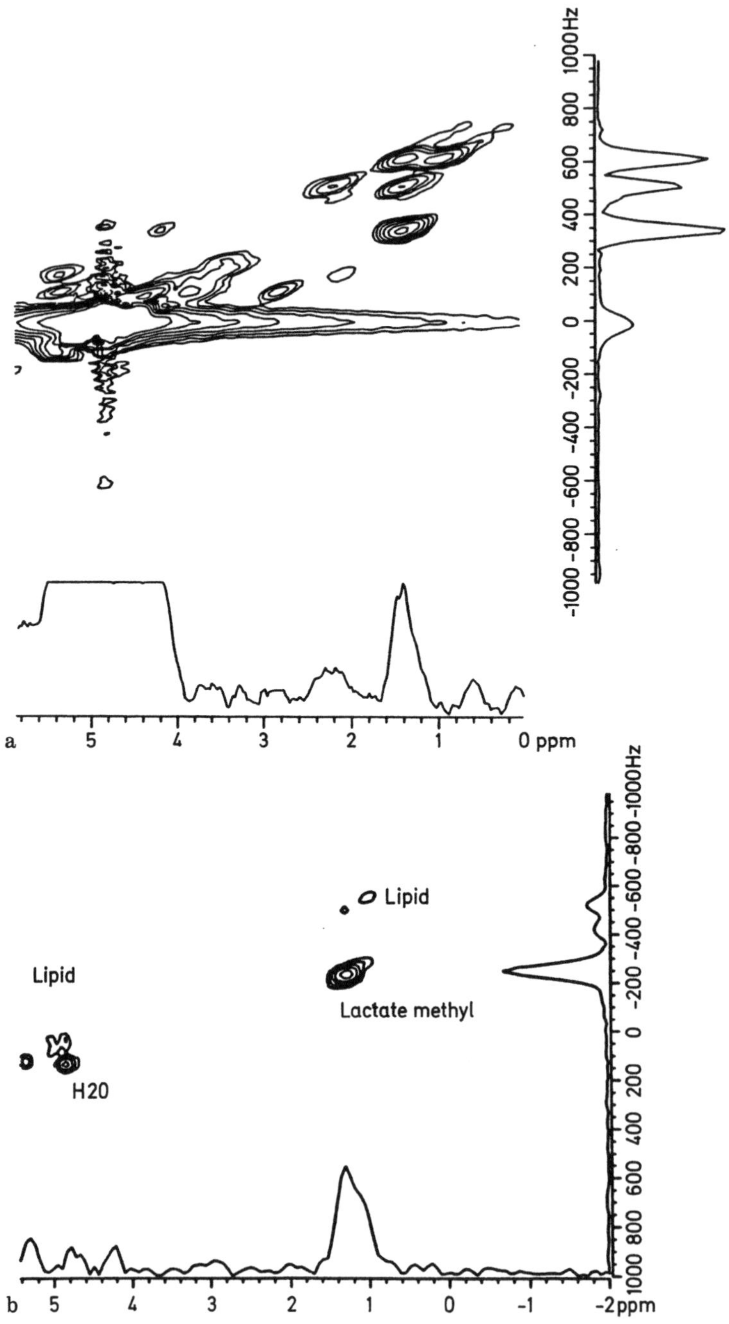

Fig. 7 a. A two dimensional contour plot of the double quantum coherence transfer experiment using the sequence in Fig. 6(a), performed *in vivo* in a RIF-1 tumor implanted on the back of a C3H mouse. The spectrum was acquired in 34 minutes. (**b**) Contour plot of a two dimensional double quantum coherence transfer experiment of a RIF-1 tumor using the sequence in Fig. 6(d). This spectrum was acquired in 64 seconds

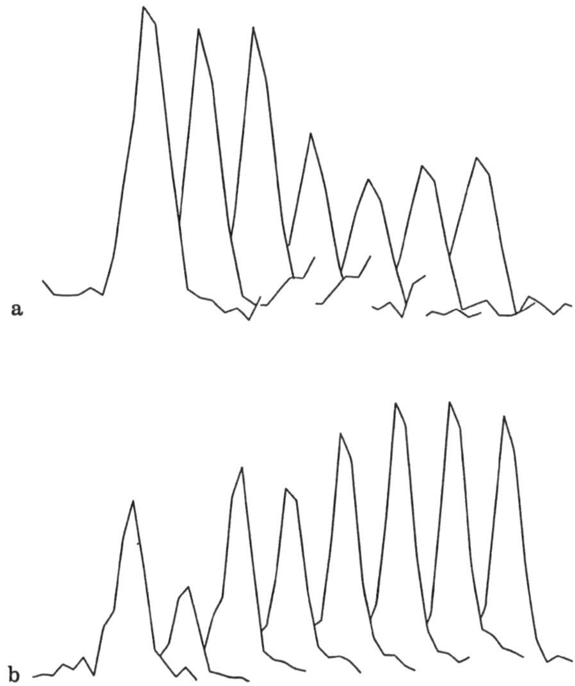

Fig. 8 a, b. The time course of the effect of nicotinamide (a) and hydralazine (b) on the lactic acid level of RIF-1 tumors. In (a) Nicotinamide (1 mg/g) was administered following the acquisition of three control spectra, the stack plot shows the decrease in the lactate methyl resonance comensurate with the increase in tumor oxygenation. Conversly the addition of hydralazine (0.01 mg/g, (b)), results in an increase in the lactate level consistent with an increase in tumor hypoxia

8.17 μmoles/gram wet weight. The lactic acid concentration in the tumor from Fig.7b was 10 μmoles/gram wet weight for a tumor of similar size.

The ability to edit for lactic acid on the order of one minute or less renders the method amenable for use to follow the time course of drug action in these experimental mouse tumor models. The possibility of monitoring regional lactic acid changes in these tumors was tested using two agents known to alter the hypoxic cell fraction of the tumors [29, 30].

Nicotinamide is a radiosensitizer and was administered 1 mg/g body weight intraperitonally. Hydralazine, an agent known to collapse tumor vasculature and slow blood flow was administered at 10 mg/kg body weight. The drugs were given in a manner consistent with that of the established clinical protocols for patients undergoing radiotherapy. Steady state lactic acid levels were monitored prior to and following the drug administration. The time points were taken in accordance with previous studies evaluating the effect of timing of the drug administration and radiotherapy on tumor cell survival.

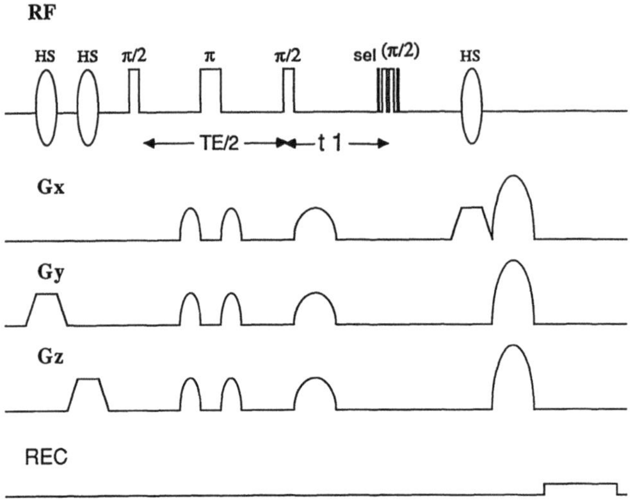

Fig. 9. Volume localized double quantum coherence transfer sequence

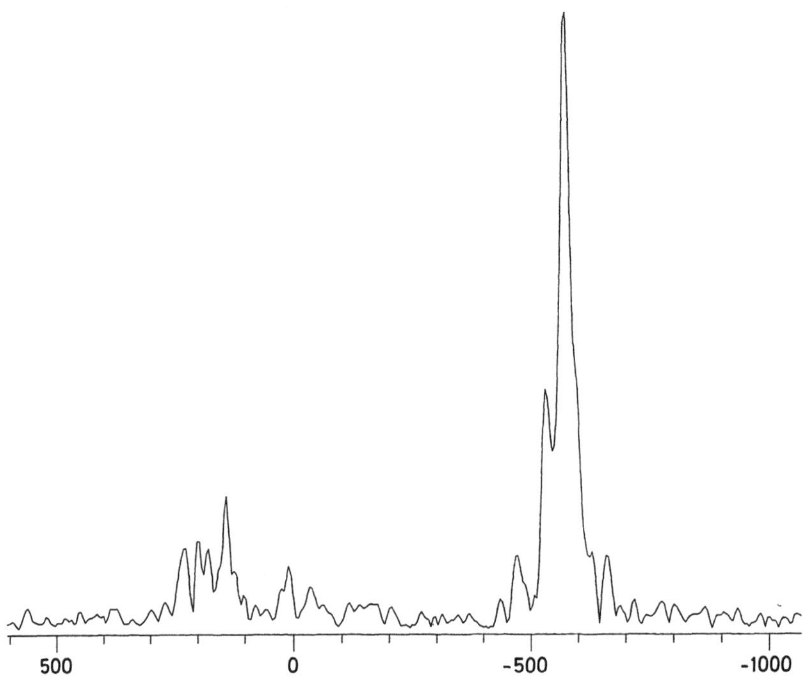

Fig. 10. A spectrum from the volume localized two-dimensional double quantum coherence transfer contour plot obtained from a 5 mm cubed voxel in a RIF-1 tumor at 4.7 T. The total volume represents a 125 μl tissue sample

Addition of nicotinamide (Fig. 8a) resulted in a drop in the lactic acid peak intensity consistent with a drop in the total hypoxic cell fraction in the tumor. Each spectrum represents a one minute acquisition, and the peak is the lactic acid resonance extracted from the two dimensional array at the double quantum frequency for lactate, which at 2.0 T is 250 Hz.

Addition of hydralazine (Fig. 8b) had the expected opposite effect, such that the reduction in blood flow resulted in an increase in the lactate methyl resonance intensity. This represents the first demonstration *in vivo* of the ability to monitor lactic acid levels over a period of time using this method.

The case of combination of this method with localization techniques is shown in Fig. 9, where hyperbolic secant pulses have been implemented to generate an ISIS [31] type volume selection. Figure 10 represents a lactic acid spectrum obtained from a 5 mm cubed voxel (or 125 µl) at 4.7 T acquired in less than one minute from a RIF-1 tumor. The spectrum is extracted from the 2 dimensional contour plot for simplicity.

5 Metabolite Specific Imaging *In Vivo*

It is entirely feasible with the water suppression and efficiency of lactate editing achieved with this type of sequence to convert the method into a metabolite specific imaging technique. Indeed, this has been achieved by adding phase and frequency encoding gradients (Fig. 11) to the standard sequence. The limit of detection of the imaging method was tested using a phantom containing lactate solutions which had quantified chemically prior to incorporation into the phantom. Three different concentration of lactic acid in water 2.5 mM, 5 mM and 10 mM were placed in 5 mm NMR tubes, which in turn were placed in a 12 mm diameter cylinder filled with doped water. As can be seen from the image in Fig. 12, the method can detect 2.5 mM lactic acid in about 30 minutes (see caption for detail of experimental parameters).

In vivo, the metabolite specific imaging technique has been extended to look at the distribution of lactic acid in RIF-1 tumors. The hypoxic cell fractions in these tumors are known to vary with time and oxygen availability to the cells (Sect. 4), as well as with the extent of cell damage due to necrosis. A metabolite specific map of the tumor should help the clinician to plan the fractionated radiotherapy to counteract the resistant hypoxic cells that determine the success of treatment.

An *in vivo* lactate image of a post mortern RIF-1 tumor was obtained using the imaging sequence in Fig. 11. Extracting the lactate image at 250 Hz, a heterogeneous distribution of lactate throughout the volume of the tumor is seen (Fig. 13). The tumor is easily differentiated from the normal tissue (homogeneous lactic acid distribution) as well as the bone marrow (see Fig. 13c) which is the fat image extracted at 0 Hz, in the femur. It is interesting to note that the

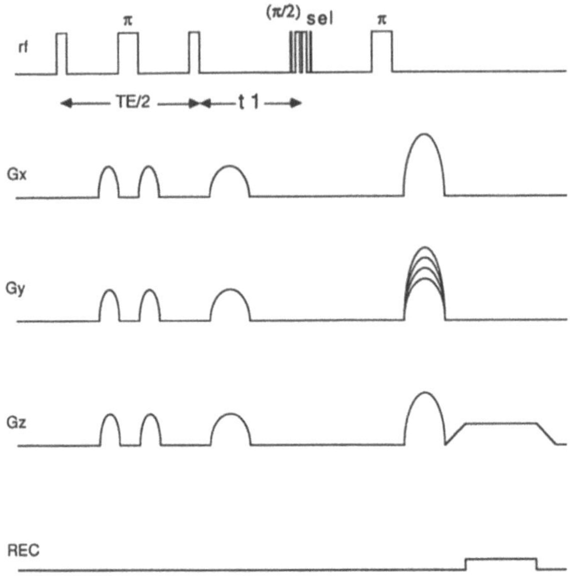

Fig. 11. Proton double quantum coherence transfer imaging sequence for metabolite specific imaging of lactate

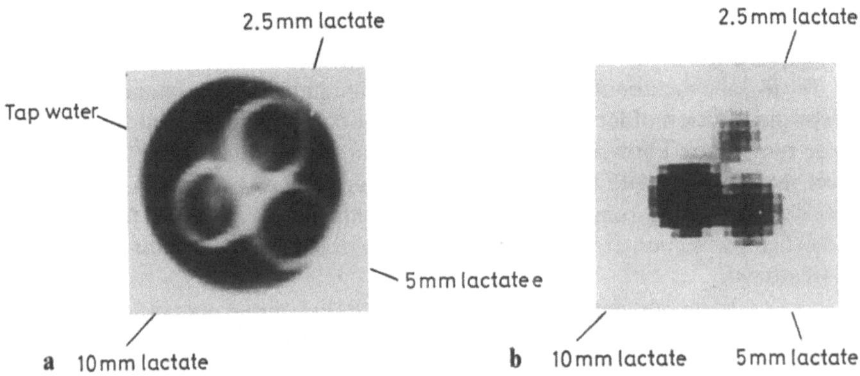

Fig. 12a, b. *In vitro* lactate imaging (**a**) normal proton image with a recycle time of 1.5 s, and echo time 28 ms. (**b**) lactate image obtained using the sequence in Fig. 11, in a phantom of three tubes of 5 mm in diameter containing 10 mM, 5 mM and 2.5 mM lactate placed in a 12 mm diameter cylinder containing tap water

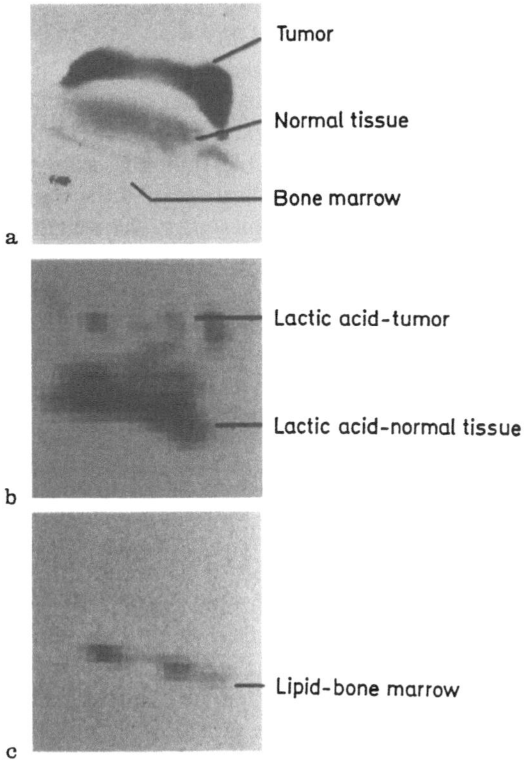

Fig. 13a–c. Proton double quantum images obtained immediately postmortem on a RIF-1 tumor. Data were collected with a total acquisition time of 51 min for a 16 × 64 × 128 array. The images shown are normal T_2 weighted proton images (TR 1.5 s, TE 50 ms) of the tumor (**a**); Lactate image showing heterogeneous distribution of lactate in tumor, and homogeneous distribution of lactate in the surrounding tissue (**b**); Lipid image primarily of bone marrow in the femur (**c**)

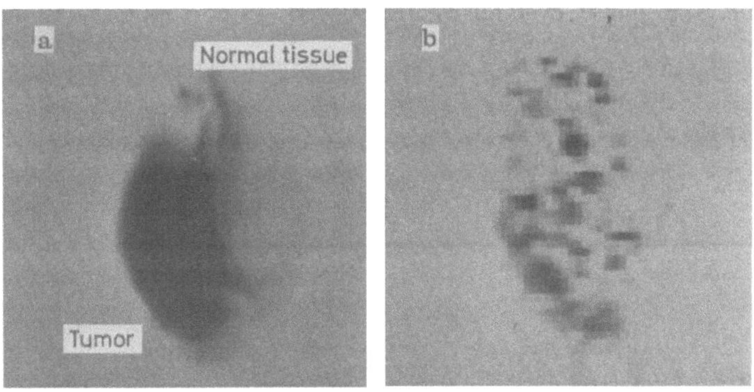

Fig. 14a, b. *In vivo* proton double quantum lactate image of a RIF-1 tumor implanted on the back of a live anesthetized C3H mouse. Data were acquired as described for Fig. 13. The images shown are normal T_2 weighted proton image of the tumor (TR 1.5 s, TE 50 ms), (**a**), and lactate image showing the heterogeneous distribution of the lactate in the tumor (**b**). No lactate was observed in the surrounding tissue technique also performs well in the presence of poor B_1 homogeneity. A T_2 weighted proton image is shown for comparison (Fig. 13a).

Live anaesthetized mice bearing tumors were imaged again at 2.0 T. The results are shown in Fig. 14. Comparing the T_2 weighted water proton image with that obtained from the metabolite specific image (extracted at 250 Hz), the heterogeneous distribution of the lactate throughout the tumor is seen. No lactate signal is observed from the normal tissue. Chemical assay of the tumor following NMR examination gave an average of 12 μmoles/gram wet weight of lactate for this particular tumor. It is therefore possible, to obtain metabolite-specific images on a reasonable timescale, *in vivo*. The possibility exists therefore to follow treatment non-invasively in patients undergoing drug treatment known to alter lactate concentration, in clinical systems.

6 Future Directions

Gradient selection of coherence can be applied to both *in vivo* and *in vitro* work. Recent results [32] reveal some of the benefits of using gradients in high resolution NMR. One method investigated was gradient enhanced, double quantum COSY. In this method, double quantum coherence is selected as described in Sect. 3, using a 1:2 ratio of gradient areas about the final radiofrequency pulse. In addition, to reduce the experiment to a single acquisition per evolution time increment, quadrature in $\omega 1$ is selected using a symmetric gradient pair around the last two radio-frequency pulses to give a final gradient sequence of 1:1:3, shown in Fig. 15.

This method was tested using a sample of 8 mM angiotensin in water. The data shown in Fig. 16 were collected as a 1 K by 1 K matrix at 9.4 T. The result is, that without any phase cycling, or traditional water suppression, the

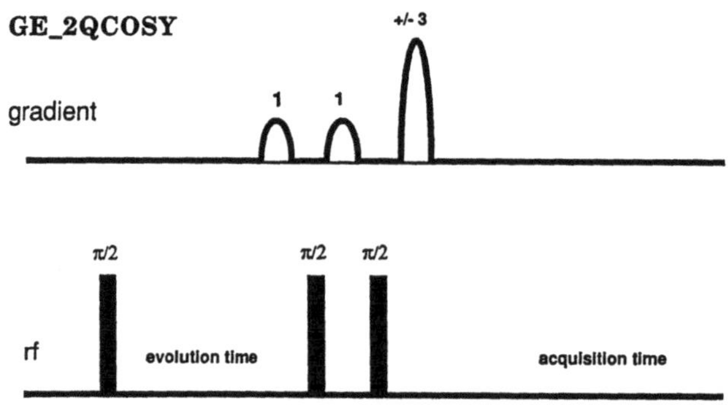

Fig. 15. Gradient enhanced double quantum COSY (GE-2QCOSY) pulse sequence. Evolution time includes the first gradient pulse, and acquisition time starts after the final gradient pulse

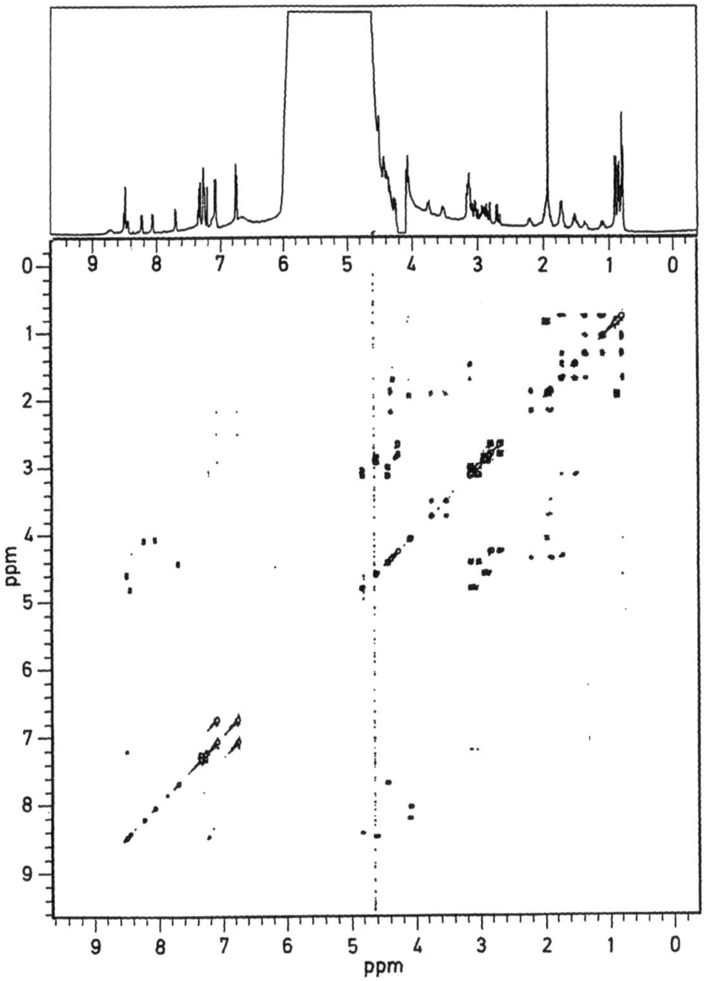

Fig. 16. Contour plot of a 400 MHz proton GE-2QCOSY spectrum of 8 mM angiotensin II in water. There were 1024 t_1 increments collected without phase cycling. Acquisition times of 256 ms provided 4 Hz resolution in both ω_1 and ω_2. One millisecond half sinusoid gradients of 10, 10 and 30 G/cm were used. An 800 fold vertical expansion of the reference one pulse is plotted across the top of the contour plot

correlation of coupled spins is observed even at the water chemical shift. The potential advantage for in vivo applications is the elimination of water signal independant of susceptability and lineshape effects.

One of the most important tools in NMR is proton (reverse) detection of lower gamma nuclei. In classical proton detected carbon experiments, for example, more than 99% of the signal must be subtracted to reveal the protons

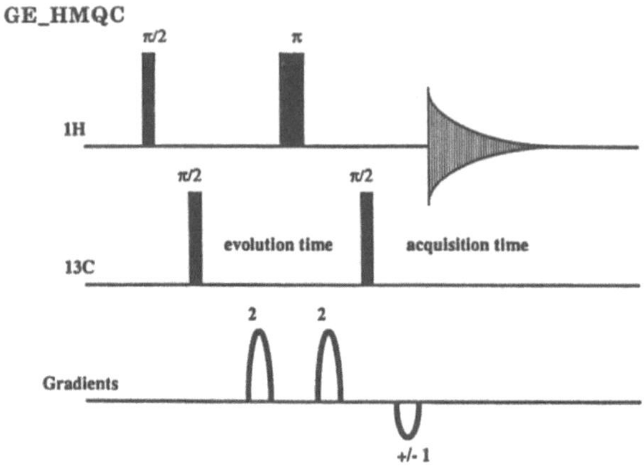

Fig. 17. Gradient enhanced heteronuclear multiple quantum correlation (GE-HMQC) pulse sequence

attached to the 1% abundant ^{13}C nuclei. With gradient selection, protons attached to ^{12}C and from solvents such as water will not be detected, thus avoiding subtraction and dynamic range limitations. This approach has been recently used for indirect detection and imaging of carbon [33–35].

A pulse sequence for gradient enhanced proton detection of carbon is shown in Fig. 17. The effect of gradients on the coherences generated by this sequence are illustrated in Table 2. The {proton pulse, 1/2 J_{C-H} delay, carbon pulse} generates both double and zero quantum coherence only for protons attached to ^{13}C nuclei. For double quantum coherence, the first gradient generates a distribution of phase proportional to the area of the gradient times the sum of the proton and carbon gyromagnetic ratios:

$$sGt(\gamma_{1H} + \gamma_{13C}) \qquad (2)$$

where s is a shape factor and is equal to $2/\pi$ for the half sinusoid shape used here. G is the strength of the gradient field and t is the gradient duration. The proton π pulse inverts the sign of the proton contribution, thus converting the

Table 2. The phase dependence of coherences generated by the GE-HMQC sequence

Before π pulse		After π pulse	Net
$(\gamma_{1H} + \gamma_{13C})$ Gt *DQC*	$\longrightarrow$	$(-\gamma_{1H} + \gamma_{13C})$ Gt *ZQC*	$(2\gamma_{13C})$ Gt $= (0.5\gamma_{1H})$ Gt
$(\gamma_{1H} - \gamma_{13C})$ Gt *ZQC*	$\longrightarrow$	$(-\gamma_{1H} - \gamma_{13C})$ Gt *DQC*	$(-2\gamma_{13C})$ Gt $= (-0.5\gamma_{1H})$ Gt
(γ_{1H}) Gt *SQC*	$\longrightarrow$	$(-\gamma_{1H})$ Gt *SQC*	0

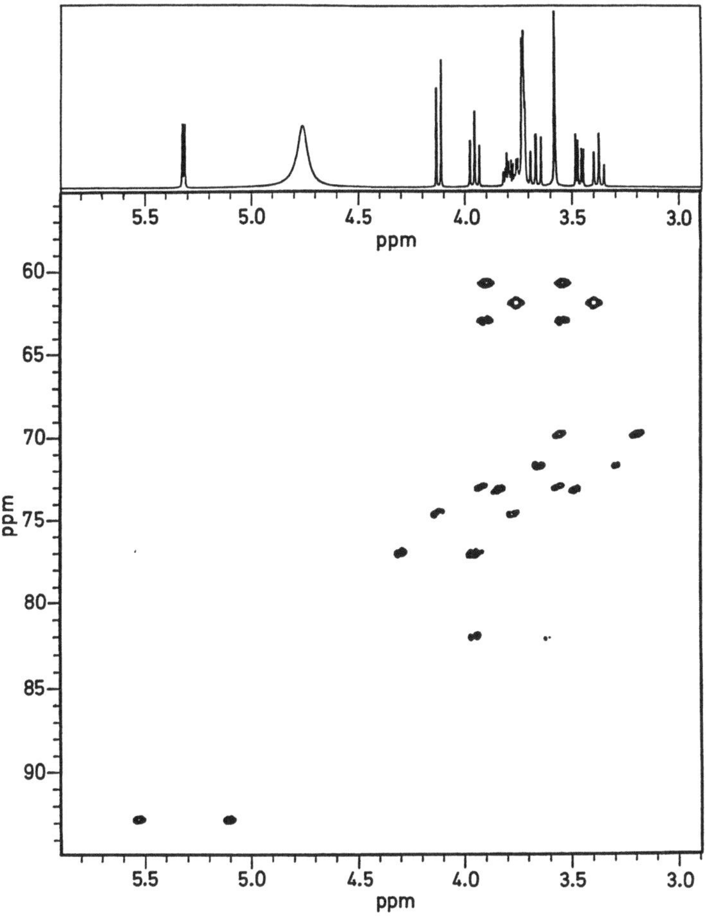

Fig. 18. Contour plot of a 400 MHz proton detected proton carbon correlation of sucrose obtained using the GE-HMQC experiment. A single scan was taken for each of the 128 t_1 increments. A proton acquisition dimension of 4000 Hz was ampled with 2 K complex points. One millisecond, half sinusoid gradients of 28, 28 and −14 G/cm were used to select for protons attached to carbon and provide quadrature detection in ω_1. A one pulse spectrum is shown at the top for reference

double quantum coherence to zero quantum coherence. The effect of the second gradient is to partially rephase this coherence by:

$$sGt(-\gamma_{1H} + \gamma_{13C}). \tag{3}$$

This leaves a net phase factor:

$$sGt(2 \times \gamma_{13C}) \tag{4}$$

or, as expressed as the proton single quantum signal generated by the final

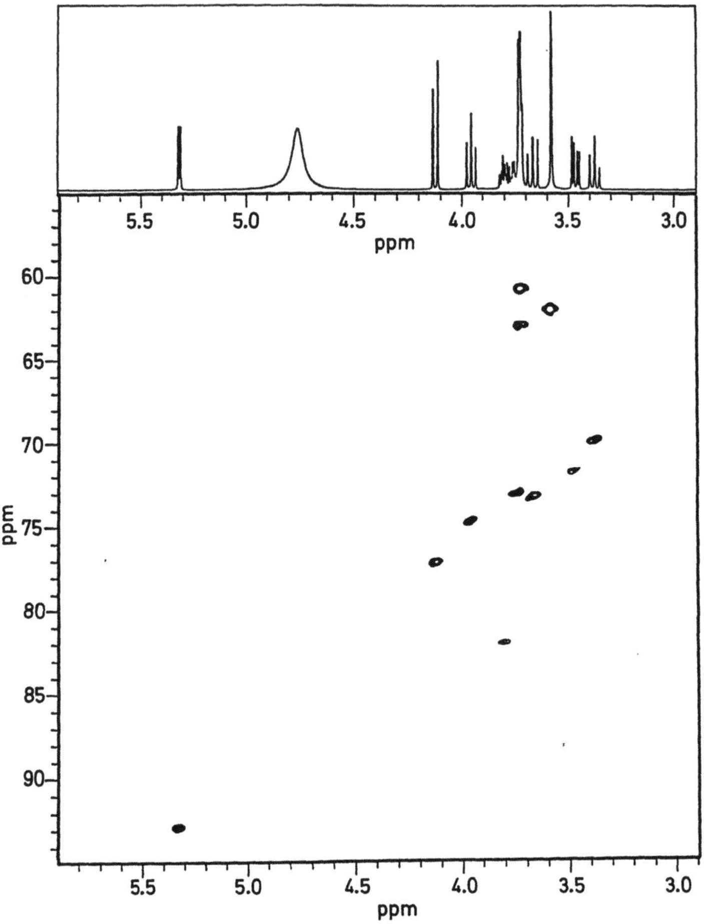

Fig. 19. Contour plot of a 400 MHz proton detected carbon correlation of sucrose acquired as in Fig. 18, except that broadband ^{13}C decoupling has been applied during the proton acquisition time

carbon $\pi/2$ pulse and using a γ_{1H} equal to $4 \times \gamma_{13C}$:

$$sGt(0.5 \times \gamma_{1H}). \tag{5}$$

This means for the observation of protons attached to ^{13}C nuclei, a unit area gradient applied prior to the proton π pulse will generate a distribution of phase for the double quantum coherence proportional to $(1 + 0.25)$ sGt, and that this coherence will then be partially rephased by the unit area gradient applied after the proton π pulse by $(-1 + 0.25)$sGt, to leave a net distribution of phase proportional to 0.5 sGt. Rephasing and detection of the desired signal can be accomplished by conversion of this zero quantum coherence to proton single quantum signal with the final carbon $\pi/2$ pulse, and by applying a gradient with -0.5 times the area of each of the first two gradients surrounding the proton π pulse. Therefore a 2:2:−1 ratio of gradients will rephase this signal. A 2:2:1 ratio will rephase the signal which starts out as zero quantum coherence and is interconverted to double quantum coherence by the proton π pulse. The symmetric 2:2 gradients about the proton π pulse leaves zero phase for the protons not attached to ^{13}C. Therefore any rephasing gradient applied just before aquisition (+1 or −1) will leave these signals dephased and unobserved.

Figure 18 shows a GE-HMQC proton–carbon correlation contour map of sucrose obtained with only a single acquisition per evolution time (ω_1) increment. This spectrum shows the direct proton–carbon correlations without any broadband carbon decoupling. The J_{C-H} splitting remains in the proton dimension (ω_2), and the dominant signals of the protons attached to ^{12}C and from the solvent are suppressed. As illustrated in Fig. 19, broadband ^{13}C decoupling simplifies the spectrum in the proton dimension (ω_2) by removing the effects of proton–carbon coupling during the acquisition period.

7 Conclusions

Gradient enhanced metabolite specific specroscopy and imaging using selective coherence transfer for selection of double quantum coherence provides an efficient method for monitoring changes in steady state lactate levels *in vivo*. Using this method, good spatial and time resolution was established while maintaining sensitivity for the lactate methyl. Metabolite specific editing should provide a sensitive biochemical probe for drug mechanism and action. Drug therapy can be monitored over time since the method is not a difference technique and hence susceptible to subtraction artifacts. It is therefore tempting to speculate that with stable, shielded gradient technology likely to be available on clinical systems in the next few years, gradient enhanced metabolite specific spectroscopy and imaging should provide an excellent addition to existing clinical spectroscopy protocols in patients. Indeed, the majority of high resolution proton and reverse detection NMR experiments are rendered more efficient by the use of gradients, and are now feasible for *in vivo* tissue examination.

Acknowledgements: The authors are grateful to Boban John (GE NMR Instruments), Ray Freeman, Xi-Li Wu, and Helen Geen (University of Cambridge, Cambridge, U.K.) for discussions on the theoretical aspects of the selective double quantum experiment. The authors with to thank Martin Brown and Marilyn Lemmon (Department of Therapeutic Radiology, Stanford University Medical Center) for helpful insights on radiotherapy resistance, and for providing the RIF-1 tumors.

8 References

1. Ross BD, Freeman DM, Chan LC (1986) Kidney Int 29: 131
2. Nunnally RL, Bottomley PA (1981) Science 211: 177
3. Jue T, Shulman R (1989) PNAS 86: 1439
4. Weiner MW (1988) Invest Radiol 23: 253
5. Radda GK (1986) Science 283: 640
6. Bottomley PA (1989) Radiol 170: 1
7. Campbell ID, Dobson CM (1979) Methods Biochem Anal 25: 1
8. Rothman DL, Arias-Mendoza F, Shulman G, Shulman RA (1980) J Magn Reson 60: 430
9. Williams SR, Gadian DG, Proctor E (1986) J Magn Reson 66: 560
10. Plateau P, Gueron M (1982) J Am Chem Soc 104: 7310
11. Hore PJ (1983) J Magn Reson 55: 283
12. Sotak CH Freeman DM (1988) J Magn Reson 77: 382
13. Von Kienlin M, Albrand A, Authier B et al. (1987) J Magn Reson 75: 371
14. Hardy CJ, Dumoulin CL (1987) Magn Reson Med 5: 58
15. Doddrell D, Brereton IM, Leith N et al. (1989) Magn Reson Med 9: 132
16. Stryjewski D, Oschkinat H, Liebfritz D (1990) Magn Reson Med 13: 158
17. McKinnon GC, Boesinger P (1989) Soc Magn Reson Med 222
18. Keller PJ, Schmallbrock F (1989) Soc Magn Reson Med 757
19. Szeverynyi NM, Haacke EM (1986) J Comp Assist Tomog 10(3): 484
20. Dumoulin CL (1985) J Magn Reson 64: 138
21. Sotak CH, Freeman DM, Hurd RE (1988) J Magn Reson 78: 355
22. Hurd RE, Freeman DM (1989) Proc Nat Acad Sci 86: 4402
23. Webb P, Spielman D, Macovski A (1988) Soc Magn Reson Med (2): 758
24. Trimble LA, Shen JF, Wilman AU, Allen PS (1990) J Magn Reson 86: 191
25. Maudsley AA, Wokaun A, Ernst RR (1978) Chem Phys Lett 55: 9
26. Bax A, DeJong PG, Mehlkopf AF, Smidt (1980) Chem Phys Lett 69: 567
27. Barker PB, Freeman R (1985) J Magn Reson 64: 334
28. Mareci TH, Freeman R (1982) J Magn Reson 48: 158
29. Horsman MR, Chaplin DJ, Brown JM (1987) Radiat Res 09: 479
30. Hirst DG, Hezelhurst JD, Brown JM (1985) Int J Radiant Oncol Biol Phys 11: 1349
31. Ordidge RJ, Connelly A, Lohman JAB (1986) 66: 266
32. Hurd RE (1990) J Magn Reson 87: 472
33. Yeung HN, Swanson SD (1989) J Magn Reson 83: 183
34. Hurd RE, John BK (1991) J Magn Reson (in press)
35. Knuttel A, Spohn H, Kimmich R (1990) J Magn Reson 86: 542

Two-Dimensional Correlated Spectroscopy *In Vivo*

Bruce A. Berkowitz

National Institute of Environmental Health Sciences, Research Triangle Park, NC

Table of Contents

The information available from one dimensional NMR spectra *in vivo* is often limited by the problem of severe peak overlap. Two dimensional correlated spectroscopy (COSY) NMR is an important approach for improving spectral resolution. It is shown that the COSY data set can act as a road map for the identification of several metabolites simultaneously in a crowded spectrum *in vivo*, even if these metabolites have resonances in the 1D NMR spectra which are not resolved. In addition, this technique may be useful for elucidating coupling partners which might then be amenable for study by 1D spectral editing techniques. Methodological improvements of the basic COSY experiment have recently been suggested, such as the HOHAHA, SUPERCOSY, and DQF COSY experiments, and their respective advantages and disadvantages for application *in vivo* are discussed.

NMR Basic Principles and Progress, Vol. 27

1 Introduction

The application of one-dimensional NMR to heterogeneous samples in vivo is often limited by the problem of severe peak overlap which makes interpretation of the data difficult. This problem is particularly acute in the application of proton NMR to *in vivo* situations due to the proton spectra having a limited spectral dispersion (≈ 10 ppm) and large dynamic range. The latter problem has, for the most part, been overcome by the application of various combinations of hard and soft pulses and by taking advantage of differences in rates of relaxation [1]. Solutions to the former problem have been limited to spectral editing techniques which isolate a single compound at a time (usually lactate) [2]. However, this approach does not efficiently utilize the information in the free induction decay. In addition, this approach requires prior knowledge about the compound to be studied and this can prevent the observation of unexpected resonances. The issue of poor spectral resolution becomes more pressing as proton NMR is applied as a clinical tool where unequivocal identification is critical.

Two-dimensional NMR is a powerful technique for improving spectral resolution [3]. In particular, the 2D *CO*related *S*pectroscop*Y* (COSY) experiment [3, 4] may be used to identify and monitor metabolites in vivo under certain conditions even when these metabolites are not resolved in the 1D NMR spectrum [5]. Application of the 2D COSY experiment *in vivo* is just beginning and most of the work to-date involves development of methodology for improving the sensitivity, spectral resolution, and spatial localization of the basic COSY experiment.

2 The Basic COSY Pulse Sequence and Related Spin Physics

The simplest COSY pulse sequence is presented in Fig. 1. The COSY experiment is fairly robust with regard to its pulse requirements and although 90° rf pulses will be assumed in this portion of the discussion, lower or spatially varying flip angles (e.g., as found with a surface coil) can also be used (see below) [3, 5]. During the preparation time it is usually desirable to achieve a reproducible initial spin state for the proper functioning of phase cycling routines (*vide infra*) [3]. Following the hard 90° pulse, transverse magnetization from all allowed transitions in a coupled spin system are allowed to evolve. A delay, t_1, is then systematically varied allowing phase and amplitude encoding of the transverse magnetization as a function of t_1. This information will subsequently be used to determine the frequency of the spins following the Fourier transform in the first frequency (F_1) domain. The second rf pulse terminates the evolution period and causes a redistribution, or mixing, of the population among the energy levels

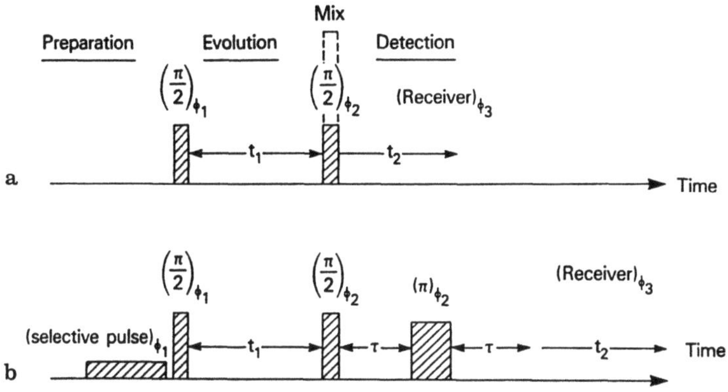

Fig. 1 a, b. Pulse timing diagrams for obtaining 2D COSY [5]. (**a**) Basic COSY sequence used to obtain Fig. 2. (**b**) Water and fat suppression COSY sequence used to obtain Fig. 3. A low power (≈ 3 W) water-selective pulse was gated on for 700 ms and then gated off for 1 ms prior to the first hard 90° pulse. During the detection period a spin echo recovery was used to help suppress broad spectral components and reduce baseline roll. When forming a spin echo with a coil which produces a homogeneous B_1 field a composite 180° pulse (e.g., $90_x°180_y°90_x°$) can be used; if a surface coil is used other approaches may be needed (see text). Typical acquisition parameters for obtaining 2D proton COSY spectra are: spectral width 1500 Hz, 90° pulse width 28 μs (100 W). Data were acquired on a 4.7 T GE CSI system. An external deuterium oxide lock was used to minimize the effects of static field drift. *In vivo* data were collected using a saddle coil tuned for protons (200 MHz) positioned around a surgically exposed rabbit kidney [5]. In addition, data were acquired in quadrature during the t_2 period (on the GE CSI system QP Y = 0, and AB = off) and stored in separate memory locations (AL = 2) in order to obtain phase-sensitive data. In sequences (**a**) and (**b**) the phases were cycled according to Table 1

such that magnetization is exchanged between J-coupled partners. The transverse magnetization components are then detected as a free-induction-decay (FID) during the second time period t_2. This FID contains the frequency of the spins following the second Fourier transform with respect to the second frequency (F_2) axis. Thus, the FIDs collected during the detection period are phase and amplitude encoded with respect to t_1 and frequency encoded with respect to t_2. After collecting a series of FID's with systematic increases in t_1, a 2D Fourier transform on this matrix with respect to t_1 and t_2 produces a 2D spectrum which is a function of the two frequencies F_1 and F_2. F_1 is the frequency of the spins before the homonuclear J-coupled exchange of magnetization and F_2 is the frequency of the spins after exchange. If a spin does not unergo this type of exchange during the second rf pulse then it will have same F_1 and F_2 frequency. However, if homonuclear J-coupled exchange (i.e. coherence transfer) occurred during the second pulse, the spin will have the F_1 frequency during the evolution time and the F_2 frequency of its coupled partner after the mixing pulse. Note that in this experiment the total amount of magnetization exchanged between J-coupled partners is conserved (i.e., no net magnetization is exchanged). An example of a COSY spectrum is seen in Fig. 2, where the homonuclear J-coupling network of ethanol is presented in both stacked and

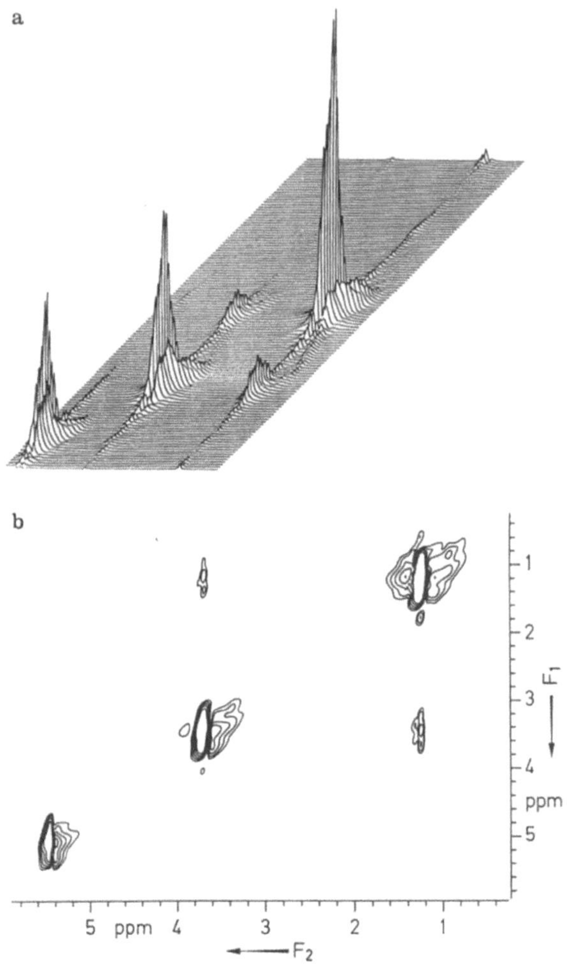

Fig. 2 a, b. Proton COSY spectra of an ethanol phantom obtained using a surface coil [5]. (**a**) Stacked plot display of F_1 vs F_2 which demonstrates the two COSY peak types: the diagonal peaks which represent magnetization which was not exchanged during the mixing pulse and cross peaks which map out the *J*-coupling network and represent magnetization which did exchange during the mixing pulse. (**b**) Contour plot of the same data. The basic COSY sequence (Fig. 1a) was applied. Data were acquired by averaging 16 acquisitions per t_1 step, phase cycling was done according to Table 1, and taking 64 t_1 steps and 1 K t_2 points (or 64 by 1 K). Other data acquisition parameters are given in the legend of Fig. 1. Data processing was performed in the time domain in the following order: Gaussian multiplication; apodization by the first half of a non-shifted sine wave, zero-fill to final matrix size of 1 K by 1 K, Fourier transformation, and presentation as the power spectrum. The data are not symmetrized and demonstrate the feasibility of surface coil 2D experiments. The chemical shift of the CH_2 protons was set to 3.66 ppm. Total data acquisition was 20 min contour plot formats. Details of this experiment are presented in the figure legend. Note that two type of "peaks" are seen: those which have the same F_1 and F_2 (diagonal peaks) and those which have different F_1 and F_2 (crosspeaks). The time evolution of a crosspeak in a simple AX coupled spin system, neglecting relaxation, is given by Ref. [3]:

$$S(t_1, t_2) = M_0 \sin(\pi J t_1) \sin(\pi J t_2) \exp(-iF_1 t_1) \exp(iF_2 t_2) \tag{1}$$

where J is the homonuclear coupling constant, M_0 is the equilibrium magnetization, and F_1, F_2 are the Larmor frequencies of spin A and X, respectively. Thus, after 2D Fourier transformation, the frequency position of the crosspeak, given by (F_1, F_2), provides a visual "road map" of J-coupled partners. In general, in a crowded spectrum, compounds which have unresolved resonances in the 1D spectrum may be readily detected by this method.

As mentioned above, it is often necessary to phase cycle the rf pulses and receiver in order to suppress quadrature artifacts and axial peaks [3, 6]. Axial peaks arise from z magnetization which develops before the second 90° rf pulse due to T_1 relaxation [6]. These resonances contain no new information and are easily eliminated by phase cycling [6]. The rational behind the design of the phase cycling routines has been described elsewhere [7]; a phase cycling scheme found effective on a 4.7 T wide bore (30 cm) General Electric system is given in Table 1 [5]. These phase cycling schemes work best if a reproducible initial spin state is achieved so that unwanted coherences can be canceled prior to each scan. In this regard, motion artifacts must be minimized and this can be achieved by gating the data acquisition to the anesthetized animals'

Table 1. This 16-step sequence provides for quadrature detection in the t_1 domain, removes axial peaks, and corrects for receiver mismatches (CYCLOPS). Odd- and even-numbered steps are stored in different memory locations and analyzed according to States et al. [8]

Phase cycle for COSY experiments

Phase step	ϕ_1	ϕ_2	ϕ_3
1	x	x	x
2	x	y	$-y$
3	x	$-x$	x
4	x	$-y$	$-y$
5	y	y	y
6	y	$-x$	x
7	y	$-y$	y
8	y	x	x
9	$-x$	$-x$	$-x$
10	$-x$	$-y$	y
11	$-x$	x	$-x$
12	$-x$	y	y
13	$-y$	$-y$	$-y$
14	$-y$	x	$-x$
15	$-y$	y	$-y$
16	$-y$	$-x$	$-x$

a

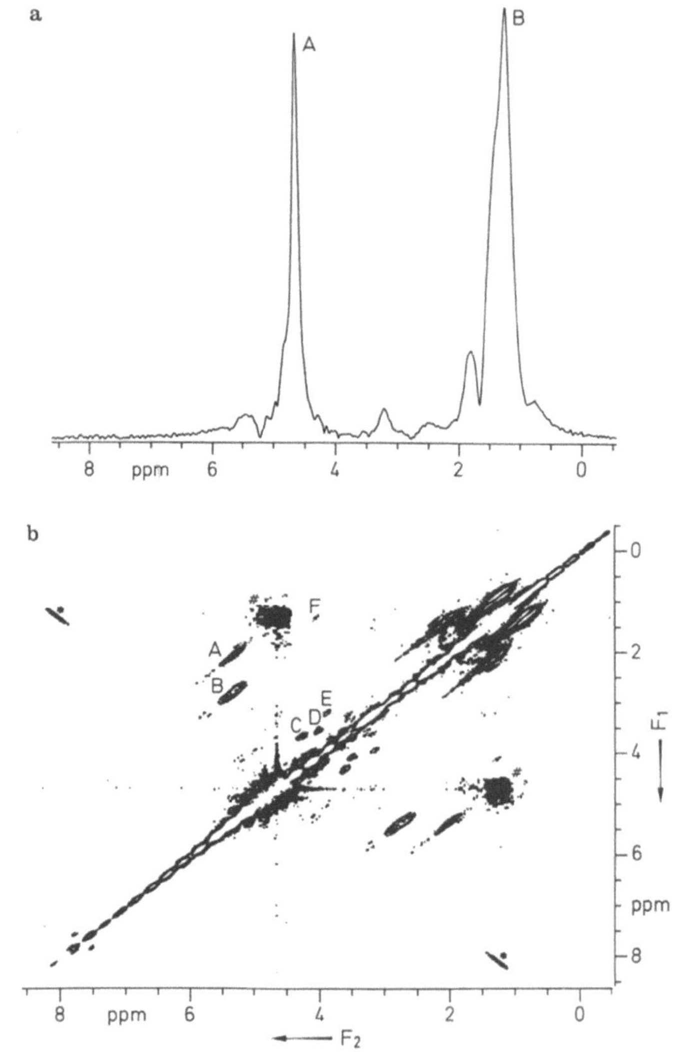

Fig. 3 a–c. Water and fat suppressed proton NMR spectra of a rabbit kidney *in vivo* (5). (**a**) 1D NMR spectrum obtained using the sequence: (selective water pulse, 700 ms, 3 W)-1 ms-90°-τ-(composite 180°)-τ-AQ. For the spin echo $2\tau = 60$ ms. The resultant signal was averaged over 16 acquisitions. Other acquisition and data processing parameters are given in the legends of Figs. 1 and 2. The two dominant peaks are labeled *A* (water) and *B* (fat). Other peak assignments of the 1D spectrum are presented elsewhere [5]. (**b**) Full 1 K by 1 K 2D COSY spectrum obtained using the pulse sequence in Fig. 1b. Symmetrization was used to eliminate t_1 ridges (see text); artifacts from this procedure are marked by a #. Data were acquired in 512 t_1 steps and 1 K t_2 points. Other acquisition and data processing parameters are given in the legends of Figs. 1 and 2. The faint anti-diagonal (marked by an *) is due to incomplete suppression of the anti-echo signal [7]. Six resonances are tentatively identified (see text): *A*: methylene groups adjacent to saturated hydrocarbons (5.4, 2.0 ppm); *B*: methylene protons between two saturated hydrocarbon linkages (5.4, 2.7 ppm), *C*: glycerolphosphorylcholine (GPC) (4.28, 3.61 ppm); *D*: inositol (4.04, 3.53 ppm); *E*: phosphoethanolamine (PEA) (3.91, 3.15 ppm); *F*: lactate (4.08, 1.29 ppm). The chemical shift of

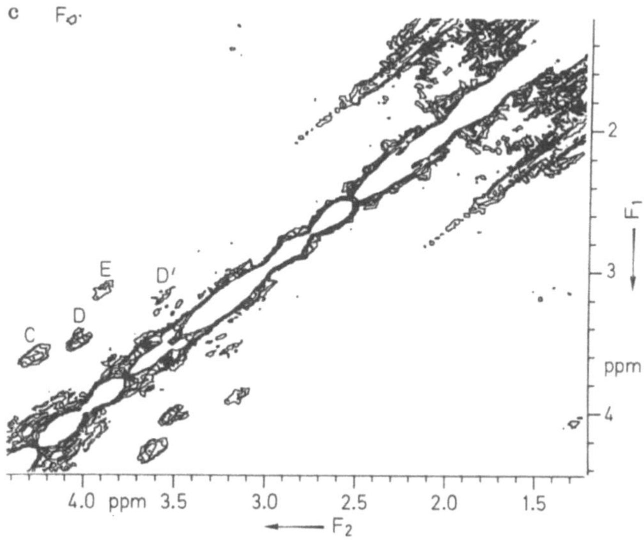

respiration and/or cardiac cycle. Instead of phase cycling, pulsed field gradients may be used (i.e., "crusher gradients") to destroy unwanted coherences and this approach has been described elsewhere [3]. It should be noted that the use of pulsed field gradients can significantly shorten the overall experiment time if signal averaging is not required. In either case, phase sensitive COSY data can be collected although a fundamental 90° phase shift between the cross and diagonal peaks will exist [3, 6]. That is, if the crosspeaks are phased in absorption mode, the diagonal peaks will be in dispersive mode. This is a basic feature of pulse-interrupted free procession experiments [7] and arises from the lack of net magnetization transferred between J-coupled partners. Unfortunately, when the data are presented in this mode the long tails of the dispersive diagonal peaks prevent observing crosspeaks close to the diagonal. Thus, COSY spectra are often presented in absolute value mode after *post facto* "reshaping" of the FID to reduce the long tails presented on the peak after the magnitude calculation [3]. However, it would be desirable to present phase-sensitive data in absorptive mode since this presentation provides significant improvements in spectral resolution compared to the absolute value mode presentation [8]. Methods to accomplish this will be presented below.

For in vivo application to protons the basic COSY sequence in Fig. 1a must be modified to suppress the strong resonances from both water and fat. One method for accomplishing this is presented in Fig. 1b [5]. During the preparation time a water-selective, low-power pulse is applied to suppress the strong water

water was set to 4.66 ppm. Total data acquisition was 3 h. (**c**) Blow up of region (1.2–4.4 ppm). Resonance assignments are as above and in addition peak D' (3.59, 3.24) represents the coupling between the H_5 and H_4, H_6 protons of inositol resonance. During the detection period a spin-echo is performed to suppress the fat resonance(s). Note that the fat suppression technique may introduce asymmetric crosspeak intensities about the diagonal depending on the spin-echo delay that is used. An additional problem, unique to the 2D experiment, is the presence of "t_1 ridges" especially from the suppressed water and fat resonances [5, 7]. In some cases, t_1 ridges can be largely eliminated by post acquisition data manipulations of the 2D matrix (e.g., symmetrization) [7]. Extreme care must be used by these *post facto* data manipulations since they tend to distort the intensity and connectivity information of the spectrum.

Figure 3 is a water/fat suppressed COSY spectrum of a rabbit kidney *in vivo* obtained from a rabbit deprived of water for three days [5]. Details of the experiment are presented in the figure legend. The strong crosspeaks at (5.4, 2.0 ppm) and (5.4, 2.7 ppm) were not seen in the perchloric acid extract of this kidney and probably represent methylene protons in lipids [5]. A blow up of the region between 1.2 and 4.4 ppm is shown in Fig. 3c. Preliminary assignments of glycerolphosphocholine (GPC), inositol, phosphoethanolamine (PEA), and lactate in the 2D COSY spectrum, based on crosspeak chemical shifts, are made in this figure. These chemical shifts agree with published values as well as with shifts of pure compounds obtained in COSY phantom studies. The inositol resonance assignment was supported by the good agreement achieved by comparing the inositol concentration derived from the COSY spectrum (see below) in a perchloric acid extract of the rabbit kidney with the inositol concentration determined in that same extract by HPLC [9].

As mentioned above, the crosspeaks in a COSY spectrum can provide concentration information, as seen in Eq. (1), in the form of M_0. Under the same acquisition conditions the cross peak area is directly related to the compounds concentration [9, 10]. The same requirements for quantitating 1D spectra also apply to the 2D case: the effects of rapid pulsing on the diagonal resonances must be accounted for and an internal or external concentration standard must be used. It should be noted that when each peak in a coupled spin system has a significantly different T_1, fairly rapid pulsing conditions ($TR \leqslant T_1$) will produce intensity differences in the corresponding crosspeaks. Additional details of how to obtain quantitative concentration information from the COSY experiment have been presented by Berkowitz and Balaban [9] and by Alonso et al. [10].

3 Sensitivity Issues

Optimization of the signal-to-noise ratio (SNR) in the COSY experiment will greatly aid in its ability to follow metabolism *in vivo*. The SNR of the COSY experiment compared to that of a 1D experiment on a *per unit time* basis is down by approximately a factor of 2.5 [6]. This loss arises since only a fraction (see Eq. (1)) of magnetization in the diagonal peaks is distributed into a crosspeak and so the SNR of the crosspeak will be down from the SNR of the diagonal peak.

Another loss in this type of experiment involves the cancellation of antiphase components, and thus reduced signal contribution, when the linewidth is much greater than the J value. Therefore it is important that the magnetic field through the tissue of interest be well shimmed. In addition, T_2^* relaxation during the evolution period will also decrease the overall sensitivity of the COSY experiment with the relatively broad lines *in vivo* compared to the high resolution situation. One promising solution to these sensitivity losses issues will be discussed below and involves the net transfer of magnetization through the application of a spin-locking pulse in place of the mixing pulse.

Several factors determine the overall length of the COSY experiment [11]. These factors are interdependent on each other and it is necessary to optimize each one in order to maximize the SNR. For example, from Eq. (1), optimum exchange occurs when the acquisition time in the t_1 domain ($t_{1,max}$) is on the order of or greater than $(2J)^{-1}$. Approximating the J value *in vivo* as 7 Hz yields a $t_{1,max}$ of 72 ms whereas a J value of 2.7 Hz (for inositol, for example) requires a $t_{1,max}$ of 185 ms or a total experiment time 2.6 times greater to obtain a similar SNR assuming equal concentrations. The acquisition time in the t_1 domain can be achieved in the shortest time by making the systematic increment of t_1, or the dwell time, as large as possible (i.e., use the smallest spectral width possible). Using a small spectral width implies that the transmitter is positioned within the spectra and so the appropriate phase cycling must be employed to eliminate quadrature artifacts in the t_1 and t_2 domains. A typical phase cycle which provides quadrature detection in t_1, suppresses axial peaks, and suppresses quadrature errors in t_2 requires 16 steps [5]. However, this might not be a limitation since the NMR detectable concentration limit of metabolites *in vivo* is on the order of mM, some signal averaging is usually necessary to obtain an adequate SNR. Nonetheless, when using the full phase cycling routine the overall experiment time will be increased.

4 Modifications of the Basic COSY Experiment

4.1 *Improved Sensitivity and/or Selectivity*

In general, the basic COSY sequence is a powerful method for improving the spectral resolution *in vivo*. However, its sensitivity and overall experimental time may prevent its general use. Several approaches have recently been developed which directly address these issues. The most promising technique in this regard involves replacing the mixing pulse with a spin-locking pulse. This technique is called either total correlated spectroscopy (TOCSY) [12] or homonuclear Hartman-Hahn (HOHAHA) [13] spectroscopy. The difference between these methods involves the type of spin-locking pulse used. In general the HOHAHA spin-locking sequences [13, 14] are more efficient and robust than those used in the TOCSY experiment. In contrast to the COSY experiment where the minimum $t_{1,max}$ which provides the optimum magnetization transfer is determined, according to Eq. (1), by $(2J)^{-1}$, in the HOHAHA experiment the minimum $t_{1,max}$ only needs to fulfill the resolution requirement of $(\Delta\delta)^{-1}$ where $\Delta\delta$ is the chemical shift difference between two resonances with the minimum needed resolution. The fixed spin-lock period (usually 35 ms) then provides for optimum magnetization transfer. For example, if a 25 Hz resolution is required to resolve two resonances in the F_1 domain, then $t_{1,max}$ is only 40 ms. Thus, losses due to spin–spin relaxation effects during the t_1 time can be minimized. At the end of this delay a spin-locking pulse is applied for a set mixing time. The spin-physics associated with such a pulse is complicated and has been presented elsewhere [12]. In essence, the spin-lock mixing pulse will ideally produce an effective field which will cancel the static field component so that only the J-coupling terms of the Hamiltonian survive. This allows the spins to behave as if they were in a zero-field environment. At zero-field, chemical shift differences between nuclides are negligible so that unlike spins (spins with different frequencies) behave as like spins (spins with the same frequency) and a net periodic oscillation of magnetization will occur between J-coupled spins. In fact, under ideal conditions, almost all of the diagonal magnetization can be transferred into the crosspeak [12]. In larger molecules this magnetization can be effectively relayed throughout the spin-coupling network depending on the length of the spin-locking pulse, and this will correlate the chemical shift of even distant nuclei. In the HOHAHA experiment, the magnetization transferred is in-phase in contrast to the anti-phase magnetization exchanged in the COSY experiment [12]. Therefore, when the HOHAHA spectrum is displayed in phase sensitive mode the spectral resolution between the diagonal and cross peaks will be greatly increased [8]. In addition, sensitivity losses are minimized during the mixing period when a composite spin-locking pulse (e.g., MLEV-17) is used since the spin-locked magnetization will spend some time along the z-axis and so spin–lattice relaxation mechanisms, which are less efficient than spin–spin mechanisms, become important [15].

The improvement in sensitivity expected from using the HOHAHA technique *in vivo* is demonstrated in the work of Alonso et al. on an excised frog gastrocnemius muscle positioned in a high resolution 400 MHz spectrometer [16]. As seen in Fig. 4, a factor of 10 increase in SNR was observed for the GPC cross peak resonance in the HOHAHA experiment as compared to a double quantum filtered (DQF) COSY experiment (*vide infra*) obtained in an equivalent period of time [16]. In addition to an improved sensitivity, the HOHAHA appears to provide new information, in the form of the sign of the cross peak (i.e., positive or negative), with regard to the binding status of a metabolite [16].

Notwithstanding the obvious advantages of the HOHAHA experiment, its implementation on large in vivo samples is not expected to be trivial. Preliminary experiments in this laboratory suggest that both B_1 and B_0 inhomogeneities can produce significant detrimental effects on the HOHAHA experiment when

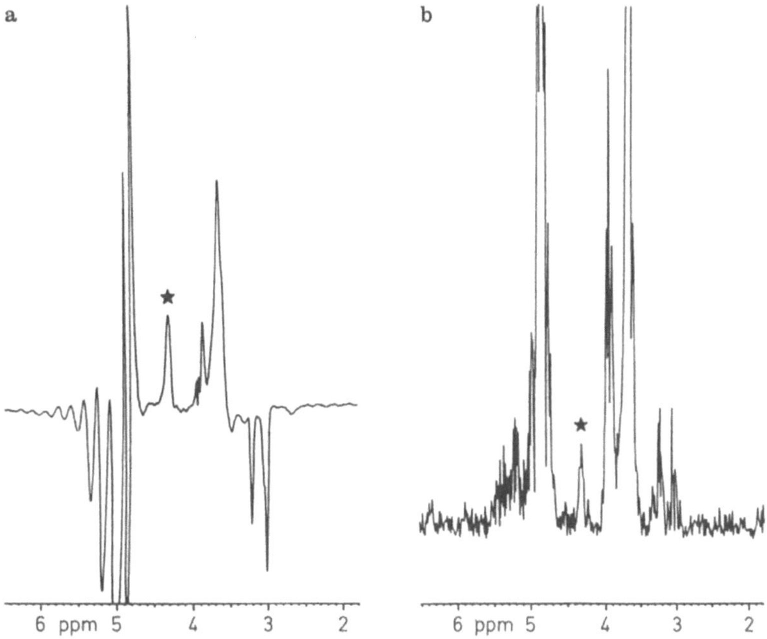

Fig. 4 a, b. SNR comparison of HOHAHA and DQF COSY experiments on excised gastrocnemius frog muscle [17]. Rows containing the maximum intensity of the GPC cross peak (*) taken from (**a**) HOHAHA and (**b**) phase-sensitive DQF-COSY data. The s/n determined for the HOHAHA cross peak is 10 times than that determined for the DQF COSY cross peak. Compared to the basic COSY experiment the s/n of the DQF COSY experiment is down a factor of 2. Thus, the HOHAHA experiment is expected to have a factor of 5 greater SNR than in the basic COSY experiment (see text)

using the original HOHAHA spin locking pulse sequence [13] (data not shown). Evaluation of the effects of inhomogeneities on more broadbanded spin-locking pulses, such as MLEV-17 or WALTZ-16, is necessary in order to determine the general usefulness of the HOHAHA experiment in vivo.

Several other variants of the basic COSY experiment have also been suggested for *in vivo* samples. Typically, these techniques sacrifice spectral sensitivity for gains in spectral resolution. As mentioned above, Alonso et al. used the DQF COSY to study the excised frog gastrocnemius muscle [10, 16]. The DQF COSY is down in sensitivity by a factor of two compared to the COSY experiment since only the double quantum signal, which is half of the original magnetization, is examined at the end of the experiment [7]. The advantage of this technique is that both diagonal and cross peaks in the spectrum can be displayed in absorption mode, thus allowing resonances close to the diagonal to be better studied. In addition, this sequence acts to suppress the singlet water resonance although presaturation may still be required.

Another method which allows the crosspeaks to be displayed with suppressed interference from the diagonal resonances is called SUPERCOSY and has been applied to superfused rat brain slices obtained on a high-resolution 400 MHz system [17]. This pulse sequence is very similar to that in Fig. 1b except that the pulse cluster ($\tau - 180° - \tau$) is placed just before and after the mixing pulse. For a given J value the delay τ can be set so that, for example, the spins which are in antiphase immediately after the mixing pulse are allowed to come in-phase. The 180° pulse simply removes the chemical shift effects of these delays. In addition, since the diagonal resonances which are in phase after the mixing pulse, are allowed to come into anti-phase and so these components tend to cancel leaving a higher crosspeak to diagonal intensity ratio than in the basic COSY spectrum. However, this technique may be somewhat less optimal than COSY for all *in vivo* metabolites. For example, inositol has crosspeaks that are well separated from the diagonal with a J-value of 2.7 Hz, whereas the J value of lactate is 7 Hz. Clearly, delays which are optimal for lactate are not optimal for inositol. In addition, the additional delays added by the SUPERCOSY technique make this experiment more sensitive to T_2^* losses than the basic COSY experiment. Nonetheless, the spin echo experiment is an effective tool for suppressing the fat resonances (*vide supra*) and so a balance between the echo delay for reasonable fat suppression and minimum SNR decrease must be found.

4.2 Spatial Localization

Due to the heterogeneous nature of in vivo samples, some spatial localization is usually needed in order to take full advantage of the improved spectral resolution of the COSY experiment. Although it is possible to isolate the organ of interest surgically (e.g., see Fig. 3), more often less invasive approaches, such as use of a surface coil and/or pulsed field gradients, are needed especially in clinical situations. Recent experiments on phantoms have demonstrated that these less invasive methods can be used in conjunction with the COSY experiment. For example, the basic COSY experiment does not require a homogeneous excitation, and so can be performed using a surface coil (see Fig. 2) [5]. A potential problem with this approach is due to the highly inhomogeneous B_1 field produced by the surface coil which, for example, inhibits effective performance of a spin echo pulse sequence (e.g., for fat suppression or in SUPER COSY). However, this limitation can be partially remedied through the use of adiabatic pulses which can achieve a uniform flip angle over the majority of the surface coils sensitive volume [18]. Another recourse involves performing a crude "depth pulse" by phase cycling the hard 180° pulse according to the EXORCYCLE procedure [19]. In either case the sensitivity of the COSY experiment will be reduced since, in the former case, the adiabatic pulse requires on the order of ms to play out (producing T_2^* losses) or, in the latter case, signals from those regions in space not receiving a "perfect" 180° pulse are suppressed by the phase cycle routine. A somewhat more rewarding approach may be to perform a homogeneous excitation with a larger coil and receive with a surface coil [20]. This method may be expected to maintain the sensitivity of the COSY experiment while localizing with a surface coil at the expense of a more complicated coil arrangement.

Another possibility for obtaining spatially localized COSY spectra involves the use of pulsed field gradients. However, the prospect of obtaining 3 spatial dimensions and 2 frequency dimensions is unappealing in terms of an unreasonable overall experimental time. Recently it has been suggested that experiments which obtain a full 2D data set from either a localized region of space [21, 22] or from a localized plane of spins might be more profitable in this regard [23]. While such localization methods have been successfully demonstrated on phantom systems, successful application *in vivo* remains to be demonstrated. Several inherent problems with these approaches exist. The pulsed field gradients must be on either during the rf pulse or during the evolution time. The former case may be expected to have chemical shift artifact problems while the latter will not be slice selective and obtaining pure phase-sensitive spectra will be difficult since the minimum evolution time is limited by the length of time the gradients are on (typically on the order of ms). The combined use of pulsed field gradients and a surface coil may help alleviate the slice selection problem [23] but will not be useful if phase-sensitive spectra are required. It should be noted that in these types of experiments it is critical to minimize the effects of rapid gradient switching, such as eddy currents, in order to maximize the experiments sensitivity.

5 Conclusions

The 2D COSY experiment is a powerful, non-invasive method which can provide more reliable identification of metabolite resonances *in vivo* than the 1D NMR. However, the sensitivity and spectral resolution of the basic COSY experiment in the presence of the fairly broad lines (compared to the J-coupling value), which are usually observed *in vivo*, are not optimal and this typically necessitates long measurement times.

Several modifications of the COSY experiment have been developed and perhaps the most promising is the HOHAHA experiment. However, it remains to be determined whether or not the HOHAHA experiment can be performed over a large volume or with a surface coil (with the increased B_1 and B_0 inhomogeneities these conditions entail).

6 References

1. Sklenar V, Tschudin R, Bax A (1987) J Magn Reson 75: 352
2. Rothman DL, Behar KL, Hetherington HP, Shulman RG (1983) Proc Natl Acad Sci USA 81: 6330
3. Bax A (1982) Two-dimensional nuclear magnetic resonance in liquids. Reidel, London
4. Jeener J (1971) Ampère International Summer School, Basko Polje, Yugoslavia
5. Berkowitz BA, Wolff SD, Balaban RS (1988) J Magn Reson 79: 547
6. Aue WP, Bartholdi E, Ernst RR (1976) J Chem Phys 64(5): 2229
7. Ernst RR, Bodenhauser G, Wokaun A (1987) Principles of nuclear magn reson in one and two dimensions. Oxford University Press
8. States DJ, Haberkorn RA, Ruben DJ (1982) J Magn Reson 48: 286
9. Berkowitz BA, Balaban RS (1989) in: Oppenheimer NJ, James TL (eds) Methods in Enzymology. Academic Press, San Diego, vol 176
10. Alonso J, Arús C, Westler WM, Markley JL (1989) Magn Reson Med 11: 316
11. Levitt MH, Bodenhausen G, Ernst RR (1984) J Magn Reson 58: 462
12. Braunschweiler L, Ernst RR (1983) J Magn Reson 53: 521
13. Davis DG, Bax A (1985) J Am Chem Soc 107: 2820
14. Bax A, Davis DG (1985) J Magn Reson 65: 355
15. Bax A (1989) in: Oppenheimer NJ, James TL (eds) Methods in enzymology. Academic Press, San Diego, vol 176
16. Alonso J, Arús C, Westler WM, Markley JL (1990) Magn Reson Med 15: 142
17. Gillet B, Mergui S, Beloeil J-C, Champagnat J, Fortin G, Jacquin T (1989) Magn Reson Med 11: 288
18. Bendall MR, Garwood M, Ugurbil K, Pegg DT (1987) *Magn Reson Med* 4: 493
19. Bendall MR, Gordon RE (1983) J Magn Reson 53: 365
20. Crowley MG, Evelhoh JL, Ackerman JJH (1985) J Magn Reson 64: 20
21. Blackband SJ, McGovern KA, McLennan IJ (1988) J Magn Reson 79: 184
22. Desmoulin F, Seelig J (1990) Magn Reson Med 14: 160
23. Cohen Y, Chang L-H, Litt L, James TL (1989) J Magn Reson 85: 203

Two-Dimensional $^{31}P-^{1}H$ Correlation Spectroscopy in Intact Organs and Their Extracts

Gil Navon[1], Tammar Kushnir[2], Nadir Askenasy[1], and Ofer Kaplan[3]

Table of Contents

2D $^{31}P-^{1}H$ correlation spectroscopy is a useful technique for the assignment of ^{31}P-NMR resonances. It allows the separation of overlapping ^{31}P peaks by spreading them in the ^{1}H dimension and gives the chemical shifts of the protons which are spin–spin coupled to the phosphorus atoms. In this chapter we give a short survey of the various 2D heteronuclear techniques with an emphasis on those that might be useful for $^{31}P-^{1}H$ correlation in extracts, intact tissues, and in vivo. Examples are given for such applications in intact pancreases excised from healthy rats and from rats with experimental pancreatitis, in excised pancreatic carcinoma, as well as for extracts from rat pancreas and liver. The potential of 2D $^{31}P-^{1}H$ correlation spectroscopy for the study of isolated and in-vivo systems is discussed.

[1] School of Chemistry, Sackler Faculty of Exact Sciences, Tel Aviv University, Tel Aviv 69978, Israel
[2] Department of Diagnostic Imaging MRI Unit, Sheba Medical Center, Tel Hashomer 52621, Israel
[3] Department of Surgery, Tel Aviv Medical Center, Rokach Hospital, Tel Aviv, Israel

NMR Basic Principles and Progress, Vol. 27

List of Symbols and Abbreviations

ATP—adenosine triphosphate; COSY—2D homonuclear correlation spectroscopy; DPDE—diphosphodiesters; GPC—glycerophosphocholine; GPE—glycerophosphoethanolamine; HETCOSY—2D heteronuclear correlation spectroscopy; HOHAHA—homonuclear Hartmann-Hahn cross polarization; PCh—phosphocholine; PCr—phosphocreatine; PDE—phosphodiesters; PEA—phosphoethanolamine; Pi—inorganic phosphate, PME—phosphomonoesters; Pur 3′—purine 3′-phosphate; Pur 2′—purine 2′-phosphate; Pyr 3′—pyrimidine 3′-phosphate; Pyr 2′—pyrimidine 2′-phosphate,

1 Introduction

Assignment of NMR resonances in biological tissues is a problem of particular importance. Most of the analytical techniques are capable to detect and measure only those compounds that the analysis is designed to find. In contrast, the NMR spectrum displays at the same time resonances from all the compounds carrying the measured nucleus, provided that their levels are high enough to produce signals with sufficient signal-to-noise ratio. This gives us the opportunity to observe new and unexpected metabolites, that escape detection by other methods, simply because we have no prior knowledge about their presence in the sample. Recent examples are the lecithin-taurocholate micelles in experimental pancreatitis [1], cyclic inositol phosphate in tumors [2], and fructose-3-phosphate and sorbitol-3-phosphate in lenses of diabetic animals [3].

The assignment of the various resonances in the ^{31}P-NMR spectra of tissues is commonly done via the analysis of their extracts [4]. Obvious advantages of the extract method are the improved resolution, particularly after eliminating paramagnetic impurities, and the possibility to vary the pH. Many of the phosphorus metabolites exhibit pH titration of their chemical shifts. For these metabolites, the pH variation enables separation of overlapping peaks. This is a valuable tool for assignment since one can compare the entire pH titration curves with those of known genuine compounds.

Assignment of the tissue spectrum using extracts has few serious drawbacks which may lead to errors: substances can be eliminated during the extraction, while others can be produced. In some cases, compounds are present in the tissue but do not appear in the extract, either because of limited solubility in the extracting solution or because they undergo a chemical modification during the extraction procedure. At the same time, new compounds may appear in the extracts due to degradation of metabolites or macromolecules.

The use of two-dimensional (2D) NMR techniques in intact tissues provides a means to assign the varius resonances without preparing extracts and thereby avoiding the above mentioned problems. Furthermore, these techniques are potentially useful in assigning NMR signals in vivo, including human patients, and thus avoiding the need to take biopsies for extracts. The technique of 2D ^{31}P–^{1}H correlation spectroscopy is particularly useful for the assignment of ^{31}P-NMR spectra. The peaks which appear in the 2D map originate only from those protons and phosphorus nuclei that are *J*-coupled with each other, leaving all other signals undetectable. This leads to a dramatic simplification of the spectrum, particularly in the ^{1}H dimension. Only few protons out of hundreds contribute to the signals in the 2D ^{31}P–^{1}H map. Noted in its absence is the water signal which usually masks many proton signals which have chemical shifts close to it, such as sugar protons. In the present chapter we will present and discuss a few examples where the application of this technique has been useful in the assignment of ^{31}P resonances in extracts and in intact tissues.

2 Methods of 2D Heteronuclear Correlation Spectroscopy

In this section we describe the basic pulse sequences of 2D heteronuclear correlation spectroscopy (HETCOSY) methods, with an emphasis on those that are most suitable for $^{31}P-^{1}H$ correlation. For simplicity we use the specific notation of ^{1}H for the sensitive nucleus and X for the insensitive nucleus. For a more detailed account of these methods and their numerous variations and modifications the reader is referred to other sources [5–8].

HETCOSY in its original version [9] is a simple extension of its homonuclear analogue (COSY). As was first suggested by Maudsley and Ernst [9], the second nonselective 90° pulse in the COSY experiment is replaced by simultaneous 90° pulses for the two nuclei (Fig. 1). This version of HETCOSY is the simplest of all other versions. It does not require prior knowledge of the J coupling constant and spans the shortest time between the first excitation pulse and acquisition. Therefore this is the method of choice for systems with short T_2. Maudsley and Ernst in their original work introduced also a saturating random pulse train prior to the first 90° pulse, in order to obtain the small NOE enhancement due to transfer of magnetization from the X nucleus to ^{1}H. However, this saturating random train has not been used in later works. Bolton and Bodenhausen [10] were the first to use this pulse program for $^{31}P-^{1}H$ correlation in organic compounds. They used it for two cellular phosphates: pyridoxal phosphate and 2′-guanosine monophosphate.

The main disadvantages of this method stem from the fact that peaks belonging to the same multiplet may have opposite phases. This prohibits the possibility of broad band ^{1}H decoupling during acquisition. Also, for systems with J_{XH} smaller than the linewidth, partial cancellation of the signal will occur. In order to bring all the signals of a multiplet into the same phase, a few pulse sequences have been suggested [11–14]. The one that has been most widely used [14] is depicted in Fig. 2a. In this sequence the frequency labeling of the ^{1}H magnetization is being performed during t_1; a X-nucleus 180° pulse is applied in the middle of this period in order to cancel the J_{HX} modulation of the protons. A period $\Delta_1 = 1/2\,J_{HX}$ is then added in order to bring the two components of the

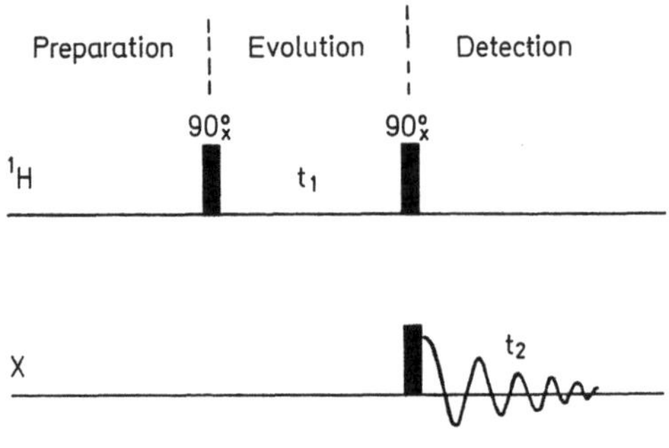

Fig. 1. The basic 2D heteronuclear correlation sequence suggested by Maudsley and Ernst [9]

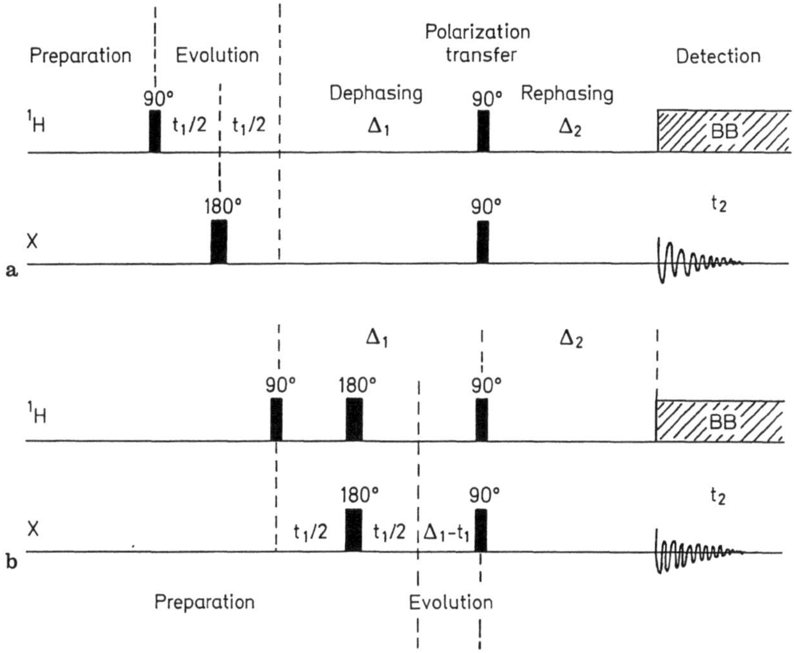

Fig. 2a,b. A comparison between two HETCOSY experiments which allow spin decoupling. **a)** Freeman and Morris's pulse sequence [14]. **b)** Kessler's COLOC sequence [16]

^{1}H doublet into a 180° phase difference. This is a necessary condition for the $^{1}H \rightarrow X$ polarization transfer to take place during the simultaneous X and ^{1}H 90° pulses that follow this delay. During the delay Δ_2 the components of the XH_n multiplet of the X nucleus are brought into phase so that ^{1}H broad band decoupling (BB) could be applied. The optimal length of the Δ_2 delay is $0.5/J_{HX}$, $0.25/J_{HX}$, or $0.196/J_{HX}$ for XH, XH_2, or XH_3 spin systems, respectively [6]. In order to cover all these possibilities, an average value of $\Delta_2 = 0.3/J_{HX}$ is commonly used. Pardi et al. [15] were the first to apply this method for the assignment of ^{31}P signals in oligonucleotides.

Numerous modifications to the Freeman and Morris pulse sequence (Fig. 2a) have been suggested in the literature. Of particular relevance to ^{31}P–^{1}H correlation of cellular metabolites are those that are designed to detect small HX coupling constants (for a review see Ref. [7]). However, most of them deal with the problem of eliminating interferences from strong coupling constants—a problem that does not occur in cellular phosphates. A very useful modification of the Freeman and Morris pulse sequence, for systems with small HX coupling constants, is the COLOC (correlation spectroscopy via long-range coupling) sequence devised by Kessler et al. [16, 17] (Fig. 2b). In this sequence the variable evolution time t_1 is incorporated inside the fixed delay Δ_1. Thus, the duration of the whole sequence is significantly reduced, a fact which is helpful in systems with short T_2. Another advantage of the COLOC method is that the proton spectrum appears as a fully decoupled spectrum. This is a direct consequence from the fact that the interval between the first pulse and the polarization transfer step is fixed and therefore the homonuclear coupling does not modulate the signal as a function of t_1. Thus, the final 2D map is fully decoupled in both the X and the ^{1}H dimensions, resulting in improved chemical shift resolution and signal-to-noise ratio. The incorporation of the evolution time t_1 inside the Δ_1 delay poses some limitations on the digital resolution in the F_1 dimension. Since t_1 must be smaller than Δ_1 which is equal to $1/2\,J_{HX}$ and the maximum value of t_1 is equal to the number of increments (TD_1) times the dwell time ($DW_1 = 1/2\,SW_1$), it follows that $SW_1/TD_1 \geqq J_{HX}$, i.e., the maximum digital resolution (neglecting zero filling) is equal to J_{HX}. Modifications of the COLOC sequence, providing pure absorption phase constant time HETCOSY maps and eliminating the antiphase modulation due to proton–proton coupling constants, were successfully applied by Gorenstein et al. to the study of oligonucleotides [18–20].

An additional aid in assignment of phosphorus metabolites may be obtained by RELAYed HETCOSY experiments [21–27]. In this technique correlation is obtained between remote protons and the X nucleus through a neighboring proton, spin coupled to the X nucleus. This method is useful in cases where the neighboring proton resonates in a very crowded spectral region and X cannot be assigned by its one-bond correlation with this proton. The assignment is thus performed by correlation with remote protons. Following the first experiments performed by Bolton and Bodenhausen [21, 22], a variety of schemes have been proposed in order to achieve this goal [23–27]. One of these methods, overcoming the problems of antiphase peaks and giving pure absorption spectra [27], is depicted in Fig. 3. The net magnetization transfer among protons is achieved via homonuclear Hartmann–Hahn (HOHAHA) type cross polarization [28, 29]. A modification of this procedure has been used by Perlman et al. [30] for the assignment of phosphorus metabolites in liver extracts (see below).

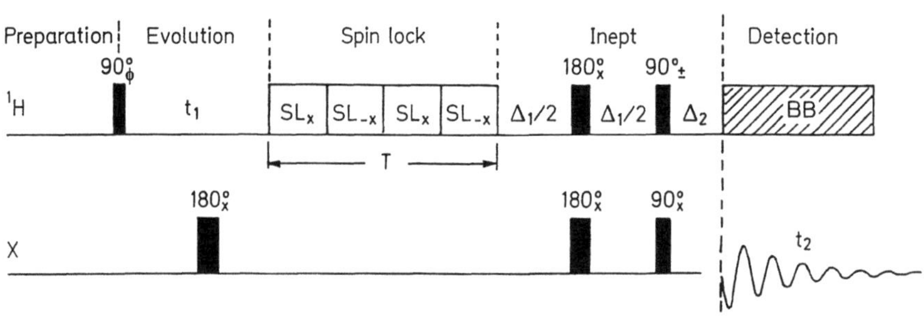

Fig. 3. The RELAY sequence employing Hartmann–Hahn cross polarization [27]. The phase ϕ of the first ^{1}H pulse is cycled $x, y, -x, -y$. Data in odd and even scans are stored separately. The result is a 2D pure absorption spectrum

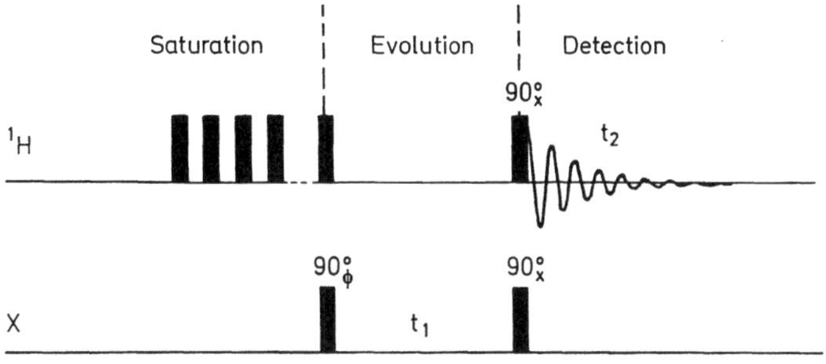

Fig. 4. The ^{1}H-detected HETCOSY sequence suggested by Sklenar et al. [31]. The phase ϕ of the first X pulse is cycled *x*, *y*, -*x*, -*y* and the receiver is cycled *x*, -*x*, *x*, -*x*. Data in odd and even scans are stored separately. The result is a 2D pure absorption spectrum

One should note that the sensitivity of the RELAY experiments is significantly lower than for X–H direct correlation experiments, due to an inefficient transfer of magnetization between the protons.

Significant improvements of the sensitivity of the heteronuclear correlation methods may be achieved by ^{1}H-detected ("inverse") sequences. There are several approaches to such experiments. The simplest of these [31] is an analog of the Maudsley and Ernst's original HETCOSY [9] and is depicted in Fig 4. A number of other ^{1}H-detected HETCOSY methods [32–38] including direct and RELAYed experiments are routinely used. Among them are the ^{1}H-detected heteronuclear multiple-quantum coherence (HMQC) [32–34], and heteronuclear multi-bond coherence (HMBC) [35] 2D correlation spectroscopy experiments. Byrd et al. [39] were the first to apply the HMQC method for ^{31}P NMR using it for the assignment of phosphorus signals in oligonucleotides. Although in these types of experiments the generation and the observation of the coherences are made with the ^{1}H frequency, one does not recover the full ^{1}H sensitivity because the intermediate steps are sometimes very inefficient. One should be aware that running the ^{1}H-detected "inverse" sequences requires special hardware modifications that at the present time are not available on many NMR spectrometers. However, since the gyromagnetic ratio of the ^{31}P nucleus is relatively large, the advantage of using the inverse detection methods is less significant than for other X nuclei such as ^{13}C and ^{15}N. Also, it is expected that difficulties to get rid of a residual water signal might be encountered, especially when intact tissues are being measured. At the present moment we are not aware of any published ^{1}H-detected HETCOSY experiment on intact tissues.

3 2D ^{31}P–^{1}H Correlation Studies of Intact Organs and Tissues

3.1 Experimental Pancreatitis

Acute pancreatitis is a most important pathology of the pancreas [40, 41]. It is an inflammatory process which may progress into hemorrhagic necrosis with a high fatality rate. A ^{31}P-NMR study of experimental pancreatitis [42], induced in rats by direct intraparenchymal injections of bile salts, demonstrated the appearance of a prominent peak in the phosphodiester region, $\delta_p = -0.18$ ppm, immediately after the induction of the disease (peak "X" in Fig. 5a). ATP levels decreased concomitantly with the progression of the disease in the live animal

a

b

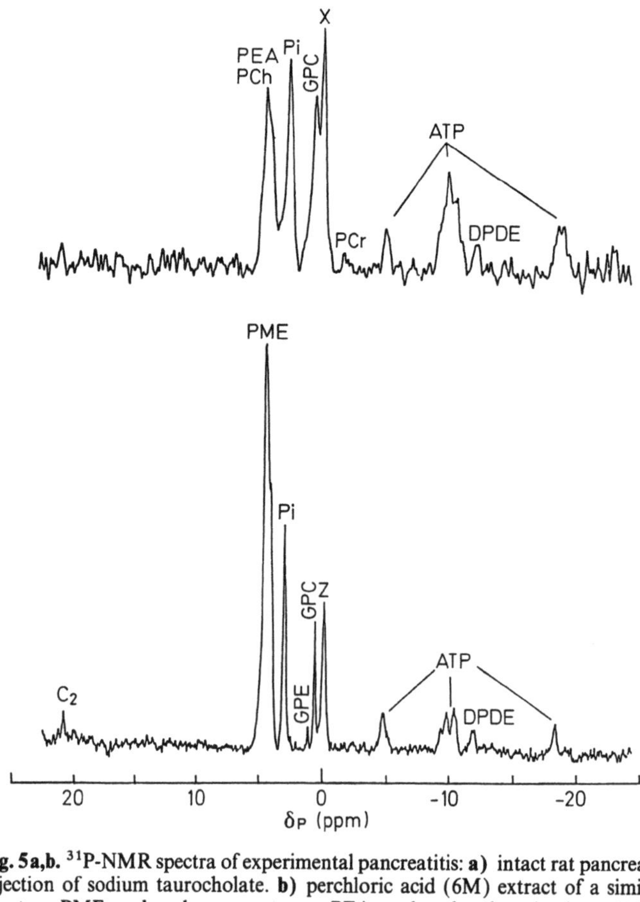

Fig. 5a,b. ^{31}P-NMR spectra of experimental pancreatitis: **a)** intact rat pancreas excised 30 min after injection of sodium taurocholate. **b)** perchloric acid (6M) extract of a similar pancreas. Assignments: PME—phosphomonoesters, PEA—phosphoethanolamine, PCh—phosphocholine, Pi—inorganic phosphate, GPE—glycerophosphoethanolamine, GPC—glycerophosphocholine, PCr—phosphocreatine, ATP—adenosine triphosphate, DPDE—diphosphodiesters, X—lecithin-taurocholate complex, Z—phosphodiesters of RNA oligonuleotides [44]

and were found to be correlated with the severity of the histologic damage. The new peak, which was absent in the ^{31}P spectrum of the intact healthy pancreas, reached a maximal intensity at a short time after induction of the disease and gradually decayed thereafter in the live animal, becoming almost undetectable after about 2h. However, in excised pancreases, kept in an oxygenated physiologic medium at 4°C, the presence of the peak was prolonged to periods longer than 20h.

In order to define the properties of the "X" signal its relaxation times were measured in the intact organ. The observed values: $T_1 = 950$ ms and $T_2 = 30$ ms are characteristic of molecules with long rotational correlation time, i.e., either molecules with high molecular weight or any kind of molecular assembly. In order to identify the characteristic signal of the disease, additional measurements were performed on tissue extracts. These measurements included ^{1}H, ^{13}C, and ^{31}P spectroscopy, as well as thin-layer and partition chromatography. These revealed no obvious difference between the healthy and diseased organs. The identification of signal "X" was enabled by the 2D ^{31}P–^{1}H correlation spectroscopy of the intact pancreas.

The 2D ^{31}P–^{1}H correlation spectrum of a diseased pancreas is shown in Fig. 6 [1]. The method used in this experiment was the simple HETCOSY described in Fig. 1 (Sect. 2). The reasons for using this method were the short T_2 of the signal and the absence of a priori information concerning the coupling constants. The reference used in these experiments for both the proton and phosphorus domains was GPC; this is a natural constituent of tissues and its chemical shift is independent of the pH. The 2D map consists of two major peaks in the phosphomonoester (PME) region: phosphocholine (PCh, $\delta_P = 3.8$,

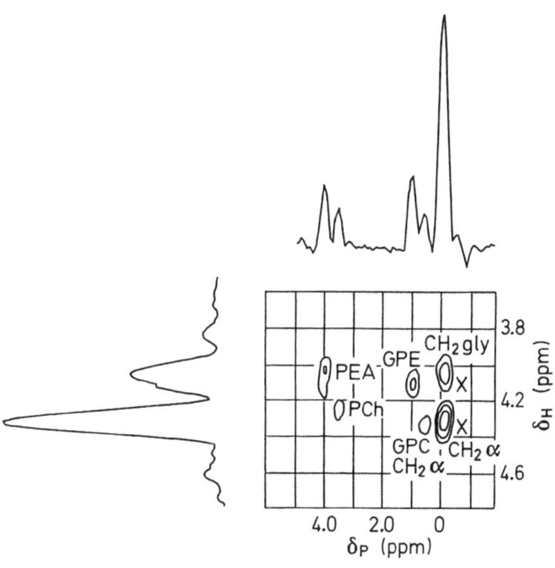

Fig. 6. 2D ^{31}P–^{1}H correlation spectrum of acute pancreatitis. The intact pancreas was excised 30 min after induction of the disease by sodium taurocholate injection. Assignment: X—a micellar complex of lecithin and taurocholae; CH_2, CH_2gly—methylene groups of the choline and glycerol residues of lecithin, respectively [1]

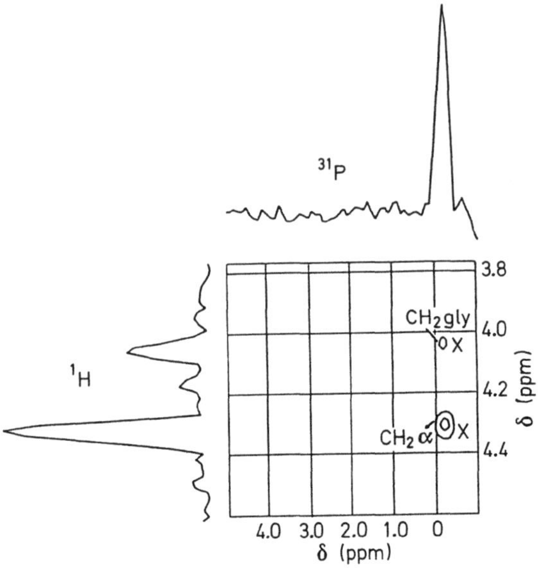

Fig. 7. 2D $^{31}P-^{1}H$ correlation spectrum of lecithin-taurocholate complex at a ratio of 0.66. (For assignments see Fig. 6) [1]

$\delta_H = 4.15$ ppm) and phosphoethanolamine (PEA, $\delta_P = 4.3$, $\delta_H = 3.96$ ppm) and in the phosphodiester (PDE) region: glycerophosphocholine (GPC, $\delta_P = 0.49$, $\delta_H = 4.33$ ppm) and glycerophosphoethanolamine (GPE, $\delta_P = 1.03$, $\delta_H = 4.06$ ppm). The observed proton chemical shifts correspond to the O–CH_2 protons of the choline and ethanolamine residues of GPC and GPE, respectively. These compounds are common to the 2D maps of healthy and diseased pancreases. The 2D map of a diseased pancreas (Fig. 6) contained two additional cross peaks (X), both resonating at $\delta_P = -0.18$ ppm in the phosphorus domain. The proton chemical shift of one of them was identical to that of the choline residue of GPC, ($\delta_H = 4.33$ ppm). The chemical shift of the other ($\delta_H = 4.02$ ppm) was similar to that of the methylene of the GPC glycerol residue. The combined data of the 2D experiment on the intact organ and the knowledge of the 1D chemical shifts of different compounds in the same spectral region led to the unambiguous assignment of the X peak as taurocholate-lecithin micelles. A further support for the assignment was obtained from the 2D $^{31}P-^{1}H$ correlation map of a mixture of lecithin and taurocholate (Fig. 7) which generated the same two cross peaks. This assignment explains the fact that the signal X was missing in the extracts, as the micelles are destroyed by the treatment with perchloric acid. The standard method of comparing ^{31}P chemical shifts with known genuine compounds is not applicable in this system, because the ^{31}P chemical shift of lecithin is highly dependent on its dispersion state. For instance, in the form of micelles with taurocholate it varies from $\delta_P = -0.18$ to $\delta_P = -0.52$ ppm when the ratio of lecithin to taurocholate changes from 0.66 to 0.33.

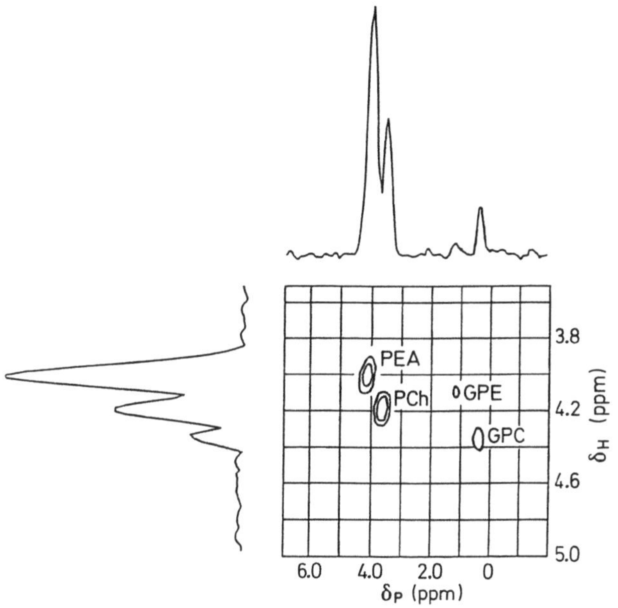

Fig. 8. 2D ^{31}P–^{1}H correlation spectrum of an intact pancreatic adenocarcinoma

3.2 Pancreatic Carcinoma

2D ^{31}P–^{1}H correlation spectroscopy has been used also in the identification of the PME signals of pancreatic cancer [43]. Tumors were developed by subcutaneous implantation of human pancreatic adenocarcinoma cells in hamsters. The intact solid tumors were excised and kept at 4°C and the external medium was bubbled with oxygen during the NMR measurements. ^{31}P-NMR spectra of the tumors showed an unresolved superposition of peaks in the PME region. The chemical shifts of these peaks (average $\delta_P = 4.04$ ppm) differed from the PME chemical shifts of healthy pancreases (average $\delta_P = 4.17$ ppm), while the pH was identical (pH = 7.15 ± 0.1). The assignment of the PME peaks could be easily achieved by the spread in the proton domain of the 2D map (Fig. 8). The PME signals in both healthy pancreases and pancreatic adenocarcinoma were identified as phosphocholine ($\delta_P = 3.8$ ppm) and phosphoethanolamine ($\delta_P = 4.3$ ppm).

4 2D ^{31}P–^{1}H Correlation Studies of Tissue Extracts

2D ^{31}P–^{1}H correlation spectroscopy in tissue extracts provides chemical shifts of both protons and phosphorus and thereby direct information on the chemical nature of the molecule under consideration. The identification process is much shorter than in the 1D ^{31}P spectra of extracts, where comparison with genuine compounds is essentially a trial-and-error procedure. 2D $^{31}P-^{1}H$ correlation spectroscopy has been particularly useful in discovering compounds which were artifacts of the extraction procedure. Two examples of cases of this kind are given below.

4.1 Experimental Pancreatitis

In taurocholate-induced experimental pancreatitis a characteristic signal ("X") appeared in the PDE region of the ^{31}P-NMR spectra of the intact organ (see Sect. 3.1). In order to identify this signal, perchloric acid extracts were prepared [44]. In ^{31}P spectra of these extracts a signal with a similar chemical shift was observed (peak "Z" in Fig. 5b). However, the same signal was found to appear also in extracts of healthy pancreases. A comparison between the 2D $^{31}P-^{1}H$ correlation spectra of the extracts from either healthy or diseased pancreas (Fig. 9) and that of the intact diseased pancreas (Fig. 6) clearly demonstrated that the signal Z in the extracts and the signal X in the intact organ originate from totally different chemical species. The chemical shifts of the phosphorus-coupled protons of Z, obtained from the 2D map of the extracts, have led to the assignment of this peak to the phosphodiester residue of oligonucleotides. This

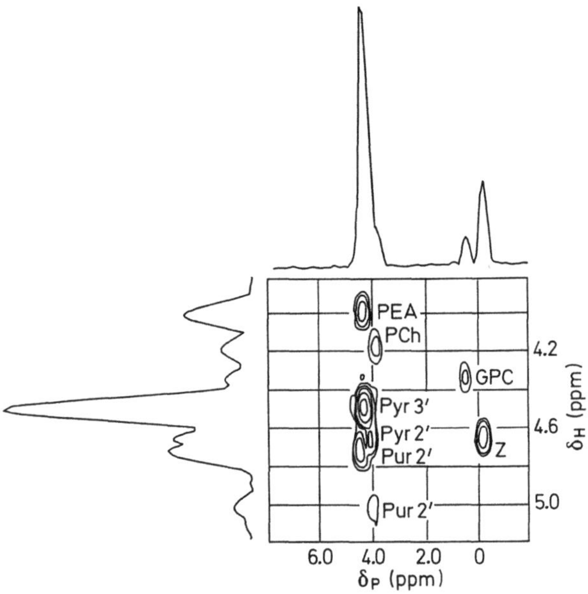

Fig. 9. 2D $^{31}P-^{1}H$ correlation spectrum of a perchloric acid (6M) extract of a healthy pancreas. Assignments: Pyr 3'—pyrimidine 3'-phosphate, Pyr 2'—pyrimidine 2'-phosphate, Pur 3'—purine 3'-phosphate, Pur 2'—purine 2'-phosphate; Z—phosphodiesters of the RNA oligonucleotides [44]

assignment was supported by the appearance of additional degradation products of nucleic acids in the PME region, i.e. pyrimidine and purine mononucleotides (Fig. 9). All these signals belong to degradation products of RNA. In fact, RNA that underwent the same extraction procedure, gave a very similar 2D $^{31}P-^{1}H$ correlation spectrum (Fig. 10). It may be noted that the 5′-mononucleotides that must also be present in the extract solutions do not appear in the 2D map. This is probably due to a particular value of $^{1}H-^{1}H$ spin–spin coupling constants causing an accidental cancellation of the cross peaks [44]. Practically no hydrolysis products of DNA are present in the extracts of the pancreas. DNA that underwent a similar extraction procedure yielded PME and PDE signals with lower intensities and with an entirely different 2D $^{31}P-^{1}H$ correlation spectrum (Fig. 11). The reason for the higher resistance of DNA to acid hydrolysis is presumably the lack of oxygen in the 2′ position so that the 2′–3′-anhydride intermediate cannot be formed. Thus it was concluded that RNA hydrolysis products are important artifacts of acid extracts of tissues. By measuring the yields of the RNA hydrolysis products as a function of the acid concentration used in the extraction procedure it was found that the $HClO_4$ concentration of 0.5 M gave a minimum yield of these products [44].

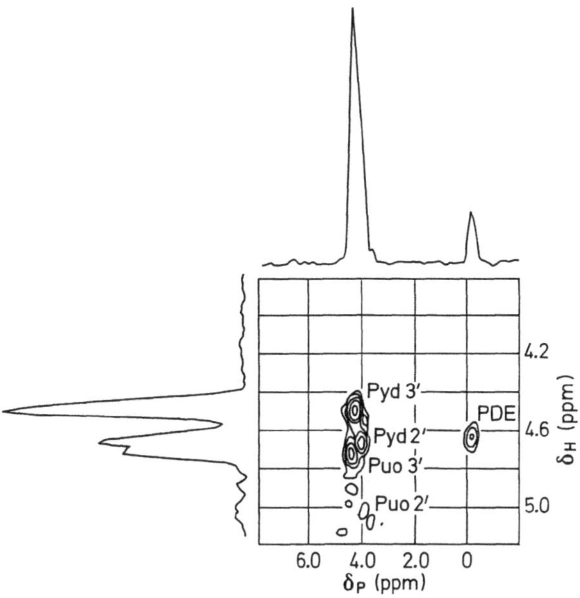

Fig. 10. 2D $^{31}P-^{1}H$ correlation spectrum of a perchloric acid (3 M) extract of RNA. RNA oligonucleotides are represented by the phosphodiester cross peak (PDE). The mononucleotides are 2′- and 3′-monophosphates [44]

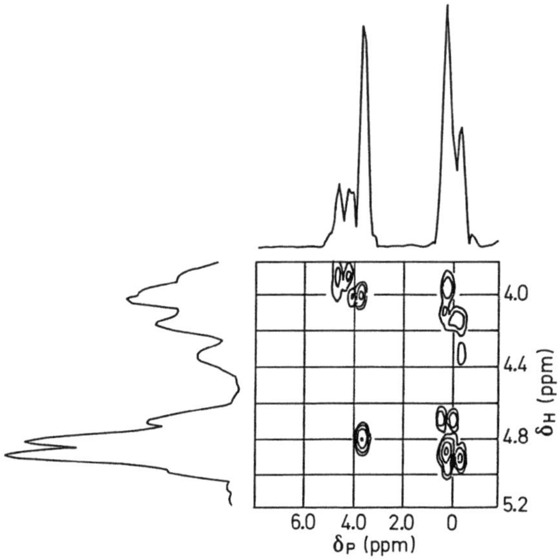

Fig. 11. 2D $^{31}P-^{1}H$ correlation spectrum of a perchloric acid (3 M) extract of DNA. Assignments: $\delta_H = 4.0 \pm 0.2$ ppm region contains the 5′-methylenes, the $\delta_H = 4.8 \pm 0.2$ region consists of the 3′-methylenes

4.2 Liver Extracts

2D $^{31}P-^{1}H$ correlation spectra of liver extracts from hexosamine-treated rats were reported by Perlman et al. [30]. The method used in this work was the phase-sensitive 2D $^{31}P-^{1}H$ relayed correlation spectroscopy, originally proposed by Bax et al. [27] (see Sect. 2). In order to reduce the sensitivity loss due to T_2, they omitted the last refocusing ^{31}P and ^{1}H 180° pulses with the associated delays. An example of their 2D $^{31}P-^{1}H$ correlation spectrum of an extract of a liver taken from a rat treated with glucosamine is given in Fig. 12. In addition to the assignment of various uridine diphosphosugars, they also reported the presence of compounds which were artifacts of the extraction procedure. The preparation of perchloric acid extracts include a neutralization step which is usually performed either with KOH or with K_2CO_3. In this way, high salt concentration in the extracts is avoided since the slightly soluble $KClO_4$ is precipitated and removed by centrifugation. Neutralization with KOH must be done very carefully since the endpoint may be easily missed and the resulting alkaline pH may lead to hydrolysis of some phosphate esters. On the other hand, it was found by Perlman et al. [30] that neutralization with K_2CO_3 led to the appearance of new ^{31}P signals (Fig. 13). On the basis of the ^{1}H chemical shifts obtained in the 2D $^{31}P-^{1}H$ correlation spectrum they assigned these signals to the formation of carbamates.

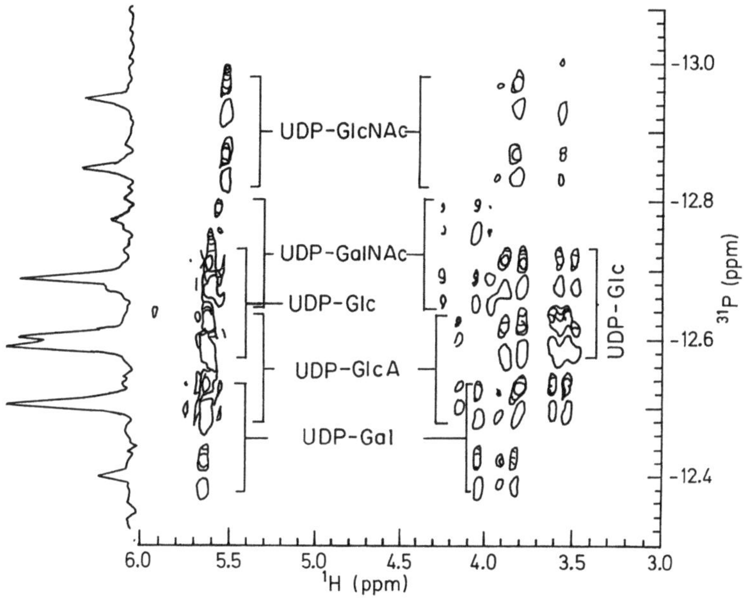

Fig. 12. The UDP-sugar region of 2D HOHAHA relay ^{31}P–^{1}H correlation spectrum of a perchloric acid extract of a rat liver [30]

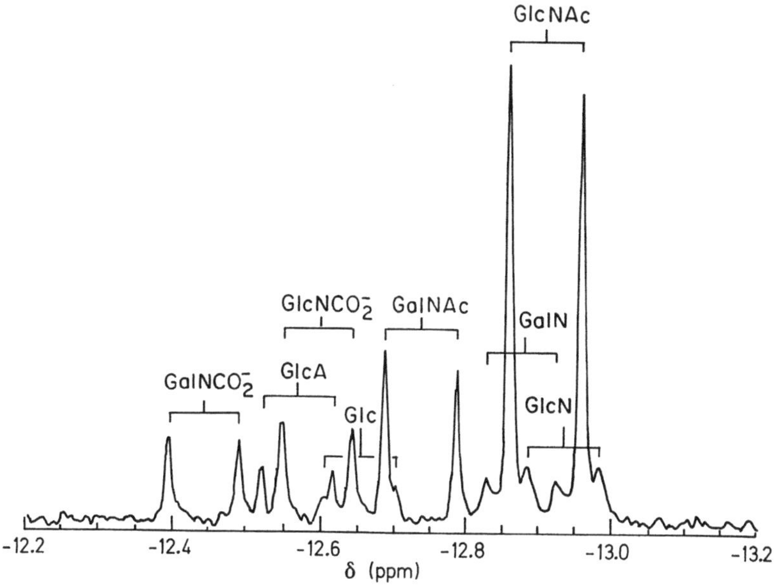

Fig. 13. The UDP-sugar region of ^{31}P-NMR spectrum of a liver extract from rats treated with galactosamine. The neutralization of the perchloric acid extract was done with K_2CO_3 [30]

5 The Potential of 2D $^{31}P-^{1}H$ Correlation Spectroscopy to Studies of Intact Tissues and to In-Vivo Measurements

^{31}P-NMR spectroscopy is a very valuable tool for studying physiology and pathophysiology of tissues. Using this technique it is possible to monitor a variety of biochemical processes, including energy metabolism, pH balance, and phospholipid pathways. The focus of these studies has been mainly on the first two subjects, while the phospholipid metabolites have received less attention. The commonly occurring phospholipid metabolites can be classified as either phosphomonoesters (PME) or phosphodiesters (PDE). Historically, the signals in the lower field region (PME) were labelled as sugar phosphates [45]. Subsequently it was found that in many tissues, the strongest signals in this region belong to the phospholipid pathways precursors, phosphocholine (PCh) and phosphoethanolamine (PEA) [4, 46, 47]. Examples are cancer cells and solid tumors [4, 46–55], brain [56–61], liver [62, 63], testis [64, 65], and pancreas [42]. Changes in these peaks were noted during maturation and differentiation [56–58, 66–68]. There are numerous publications concerning changes in PME and PDE levels in tumors, following chemotherapy, hyperthermia, and irradiation, which are outside the scope of this review. In muscles, the signals in the PME region are normally weak. They do become more apparent in stimulated muscles and in various pathological situations. For instance, elevated levels of unidentified PME signals were found upon stimulation of muscles of rabbits [69] and iodoacetate treated rats [70, 71]. A strong PME signal was found to appear in functional electrical stimulated quadriceps muscle of paraplegic patients [72]; two well-resolved peaks were observed in human forearm muscle after exercise [73]. From the in-vivo pH variations they were able to assign one of them as glucose-6-phosphate. However, the titration of the other peak was consistent with a variety of metabolites and therefore remained unidentified. Examples of elevated PME peaks in pathologic muscles are canine and human phosphofructokinase deficiencies [74, 75]. In some cases, uncommon phosphorus metabolites may be responsible for PME signals. One example is the accumulation of myo-inositol-1-phosphate in agonist-stimulated cat brain [76]. In all of these cases 2D $^{31}P-^{1}H$ correlation may prove to be very useful in the assignment of the specific PME signal for each case.

The PDE region in ^{31}P spectra is governed by glycerophosphocholine (GPC) and glycerophosphoethanolamine (GPE) which are degradation products of phospholipids. GPC was first identified in muscles [77] and subsequently, together with GPE [4], was found to be most pronounced in cancer cells [4, 46, 47], in tumors (for reviews see [48, 53]), and in the epididymis [64]. Presently it is not clear whether the intracellular levels of GPC and GPE have a physiological significance. It has been postulated that they are involved in mechanisms of membranes composition control, cell differentiation, and activation [55, 78, 79]. Their levels in cancer cells change following therapy and when the cells acquire multi-drug resistance [80]. However, the PDE region may include signals of other compounds, such as membrane particles and fragments of RNA [81] and compounds characteristic of pathological processes (e.g. acute pancreatitis—Sect. 3.1). In chicken tissues the PDE signal originates from serine ethanolamine phosphodiester (SEP) instead of GPC [82–85] and in frog lenses a signal of threonine ethanolamine phosphodiester (TEP) was present in addition to SEP [86].

Many in-vivo NMR studies dealt with the PME/PDE ratio. Attempts were made to use this ratio as an indicator of malignant processes, the efficacy of specific treatment and also to evaluate rejection reactions following organ transplantation. However, the significance of the results is often obscure since in many instances the compounds in the PME and PDE regions were not identified. It should be noted that mobile phospholipids give in some cases relatively sharp signals, that may lead to their incorrect assignment as GPC [87–90]. Whenever new compounds are anticipated, especially in deranged metabolic status of diseases, it is necessary to obtain an unequivocal assignments of the signals, and it appears that 2D $^{31}P-^{1}H$ correlation spectroscopy can provide these data.

Acknowledgment: The Authors are grateful to Dr. Hadassah Shinar for helpful discussions.

References

1. Kushnir T, Kaplan O, Askenasy N, Navon G (1989) Magn, Res. Med. 10:119
2. Szwergold BS, Brown TR, Freed JJ (1989) J. Cell. Physiol. 138:227
3. Graham RA, Meyer RA, Szwergold BS, Brown TR (1987) J. Biol. Chem. 262:35
4. Navon G, Ogawa S, Shulman RG, Yamane T (1977) Proc. Natl. Acad. Sci. USA 74:87
5. Ernst RR, Bodenhausen G, Wokaun A (1987) Principles of nuclear magnetic resonance in one and two dimensions. Clarendon Press, Oxford
6. Derome AE (1987) Modern NMR techniques for chemistry research. Pergamon Press, Oxford
7. Martin GE, Zektzer AS (1988) Magn. Reson. Chem. 26:631
8. Brey WS (ed) (1988) Pulse methods in 1D and 2D liquid-phase NMR. Acad. Press, New York
9. Maudsley AA, Ernst RR (1977) Chem. Phys. Letters 50:368
10. Bolton PH, Bodenhausen G (1979) J. Am. Chem. Soc. 101:1080
11. Maudsley AA, Muller L, Ernst RR (1977) J. Magn. Reson. 28:463
12. Bodenhausen G, Freeman R (1977) J. Magn. Reson. 28:471
13. Bodenhausen G, Freeman R (1978) J. Am. Chem. Soc. 100:320
14. Freeman R, Morris GA (1978) J. Chem. Soc. Chem. Commun. :684
15. Pardi A, Walker R, Rapoport H, Wider G, Wuthrich K (1983) J. Am. Chem. Soc. 105:1652
16. Kessler H, Griesinger C, Zarbock J, Loosli HR (1984) J. Magn. Reson. 57:331
17. Kessler H, Griesinger C, Lautz (1984) Angew. Chem., Intl. Ed. Engl. 23:444
18. Gorenstein DG, Schroeder SA, Miyasaki M, Fu JM, Roongta V, Abuaf P, Metz JT, Jones CR (1986) Bull. Magn. Reson. 8:137
19. Jones CR, Schroeder SA, Gorenstein DG (1988) J. Magn. Reson. 80:370
20. Fu JM, Schroeder SA, Jones CR, Santini R, Gorenstein DG (1988) J. Magan. Reson. 77: 577
21. Bolton PH (1982) J. Magn. Reson. 48:336
22. Bolton PH, Bodenhausen G (1982) Chem. Phys. Lett. 89:139

23. Kessler H, Bernd M, Kogler H, Zarbock J, Sørensen OW, Bodenhausen G, Ernst RR (1983) J. Am. Chem. Soc. 105:6944
24. Bax A (1983) J. Magn. Reson. 53:149
25. Bolton PH (1983) J. Magn. Reson. 54:333
26. Sørensen OW, Ernst RR (1983) J. Magn. Reson. 55: 338
27. Bax A, Davis DG, Sarkar SK (1985) J. Magn. Reson. 63:230
28. Hartmann SR, Hahn EL (1962) Phys. Rev. 128:2042
29. Pines A, Gibby MG, Wangh JS (1973) J. Chem. Phys. 59:563
30. Perlman ME, Davis DG, Gabel SA, London RE (1990) Biochemistry 29:4318
31. Sklenar V, Miyashiro H, Zon G, Miles HT, Bax A (1986) FEBS Lett. 208:94
32. Müller L (1979) J. Am. Chem. Soc. 101: 4481
33. Bax A, Griffey RH, Hawkins BL (1983) J. Magn. Reson. 55:301
34. Bendall MR, Pegg DT, Doddrell DM (1983) J. Magn. Reson. 52:81
35. Bax A, Summers MF (1986) J. Am. Chem. Soc. 108: 2093
36. Frey MH, Wagner G, Vasak M, Neuhaus D, Worogotter E, Kagi JHR, Ernst RR, Wuthrich K (1985) J. Am. Chem. Soc. 107:6847
37. Field LD, Messerle BA (1986) J. Magn. Reson. 66:483
38. Lerner L, Bax A (1986) J. Magn. Reson. 69:375
39. Byrd RA, Summers MF, Zon G, Spellmeyer Fouts C, Marzilli LG (1986) J. Am. Chem. Soc. 108:504
40. Robbins SL, Cortan RS, Kumar V (1984) Pathologic basis of disease, 3rd ed. Philadelphia, Saunders, p 963
41. Beger HG, Buchler M (1987) Acute pancreatitis. Springer-Verlag, Berlin, p 25
42. Kaplan O, Kushnir T, Sandbank U, Navon G (1987) J. Surg. Res. 43:172
43. Kushnir T, Kaplan O, Askenasy N, Navon G (1989) Biophys. J. 55:448a
44. Askenasy N, Kushnir T, Navon G, Kaplan O (1990) NMR Biomed. 3:220
45. Gadian DG (1982) Nuclear magnetic resonance and its applications to living systems. Oxford Univ Press, Oxford
46. Evans FE, Kaplan NO (1977) Proc. Natl. Acad. Sci. USA 74:4909
47. Navon G, Navon R, Shulman RG, Yamane T (1978) Proc. Natl. Acad. Sci. USA 75:891
48. Evanochko WT, Ng TC, Glickson JD (1984) Magn. Reson. Med. 1:508
49. Desmoulin F, Galons JP, Canioni P, Marvaldi J, Cozzone PJ (1986) Cancer Res. 46:3768
50. Degani H, Horowitz A, Itzchak Y (1986) Radiology 161:53
51. Podo F, Carpinelli G, DiVito M, Giannini M, Proietti E, Fiers W, Gresser I, Belardelli F (1987) Cancer Res. 47:6481
52. Daly PF, Lyon RC, Faustino PJ, Cohen JS (1987) J. Biol. Chem. 262:14875
53. Daly PF, Cohen JS (1989) Cancer Res. 49:770
54. Kuesel AC, Graschew G, Hull WE, Lorenz W, Thielmann HW (1990) NMR Biomed. 3:78
55. Kaplan O, Aebersold P, Cohen JS (1989) FEBS Lett. 258:55
56. Cady EB, Costello AM, Dawson M, Delpy DT, Hope E, Reynolds EOR, Tofts P, Wilkie DR (1983) Lancet I:1059
57. Tofts P, Wray S (1985) J. Physiol. 359:417
58. Gyulai L, Bolinger L, Leigh JS, Chance B (1984) FEBS Lett. 178:137
59. Brenton DP, Garrod PJ, Krywawych S, Reynolds EOR, Bachelard HS, Cox DW, Morris PG (1985) Lancet I:115
60. Pettegrew JW, Kopp SJ, Dadok J, Minshew NJ, Feliksik JM, Glonek T, Cohen MM (1986) J. Magn. Reson. 67:443
61. Maris JM, Evans AE, McLaughlin AC, D'Angio GJ, Bolinger L, Manos H, Chance B (1985) New Eng. J. Med. 312:1500
62. Cohen SM (1983) Hepatology 3:738
63. Oberhaensli RD, Rajagopalan B, Taylor DJ, Radda GK, Collins JE, Leonard JV (1988) Pediat. Res. 23:375
64. Navon G, Gogol E, Weissenberg R (1985) Arch. Androl. 15:153
65. Bretan PN Jr, Vigneron DB, Hricak H, McClure RD, Yen TS, Moseley M, Tanagho EA, James TL (1987) Radiology 162:867
66. Schmidt HC, Gooding CA, James TL, Gonzalez-Mendez R, James JL (1986) Pediatr. Radiol. 16:144
67. Baldassari AM, Giomini M, Giuliani AM, Tritta E, Boicelli CA (1987) Physiol. Chem. Phys. Med. NMR 19:95

68. Boesch C, Gruetter R, Martin E, Due G, Wuthrich K (1989) Radiology 172:197
69. Burt CT, Pluskal MG, Sreter FA (1982) Biochem. Biophys. Acta 721:492
70. Kuwabara T, Yuasa T, Miyatake T (1986) Muscle & Nerve 9:138
71. Argov Z, Nagle D, Giger U, Leigh JS Jr. (1989) Eur. J. Appl. Physiol. 58: 808
72. Levy M, Kushnir T, Shporer M, Mizrahi J, Itzchak Y (1989) Soc. Magn. Reson. Med. 8th Ann. Mtg. (Abst.): 1088
73. Pan JW, Hamm JR, Rothman DL, Shulman RG (1989) Soc. Magn. Reson. Med. 8th Ann. Mtg. (Abst.): 541
74. Argov Z, Bank WJ, Maris J, Leigh JS, Chance B (1987) Ann. Neurol. 22:46
75. Giger U, Argov Z, Schnall M, Bank WJ, Chance B (1988) Muscle & Nerve 11:1260
76. Renshaw PF, Schnall MD, Leigh JS Jr. (1987) Magn. Reson. Med. 4:221
77. Burt CT, Glonek T, Barany M (1976) Biochemistry 15:4850
78. Burt CT, Ribolow HJ (1984) Biochem. Med. 31:21
79. Proietti E, Carpinelli G, DiVito M, Belardelli F, Gresser I, Podo F, (1986) Cancer Res. 46:2849
80. Cohen JS, Lyon RC, Chen C, Faustino JC, Batist G, Shoemaker M, Rubalcaba E, Cowan KH (1986) Cancer Res. 46: 4087
81. Guidoni L, Mariutti G, Rampelli GM, Rosi A, Viti T (1987) Magn. Reson. Med. 5:578
82. Chalovich JM, Burt CT, Cohen SM, Glonek T, Barany M (1977) Arch. Biochim. Biophys. 182:683
83. Burt CT, Chalovich JM (1978) Biochim. Biophys. Acta 529:186
84. Pettegrew J, Minshew N, Feit H (1984) Muscle & Nerve 7:442
85. Burt CT (1985) Trends Biochem. Sci. 10:404
86. Burt CT, Levine JF (1990) Biophys. Acta 1033:189
87. Cerdan S, Harihara Subramanian V, Hilberman M, Cone J, Egan J, Chance B, Williamson JR (1986) Magn. Reson. Med. 3:432
88. Bates TE, Williams SR, Gadian DG (1989) Magn. Reson. Med. 12:145
89. Murphy EJ, Rajagabalan B. Brindle KM, Radda GK (1989) Magn. Reson. Med. 12:282
90. Karczmar GS, Tavares NJ, Weiner NW (1989) Magn. Reson. Med. 9:8

Measurement of Reaction Rates In Vivo Using Magnetization Transfer Techniques

Markus Rudin and Andre Sauter

Preclinical Research, Sandoz Pharma Ltd., CH-4002 Basel, Switzerland

Table of Contents

NMR Basic Principles and Progress, Vol. 27

The knowledge of reaction rates, in addition to the absolute concentrations of metabolites, provides important information for a deeper understanding of physiological and pathological processes. In vivo NMR offers the unique possibility to determine, by magnetization transfer, reaction rates and fluxes noninvasively and without disturbing the chemical equilibrium. Several magnetization transfer methods, such as saturation transfer, inversion transfer, inversion recovery and two-dimensional exchange spectroscopy, have been developed to date and are dealt with, both from a theoretical as well as practical point of view. A general mathematical solution for a first-order reaction system with multiple exchanging sites is formulated. The hardware requirements, problems associated with inhomogeneous radiofrequency excitations, localization techniques and extraction of kinetic parameters from experimental data sets are discussed. The different magnetization transfer methods are compared in terms of measuring time, quality of information obtained, applicability, etc. Finally, the potential, limits, and problems in interpreting the experimental data are disclosed on the basis of selected examples, such as phosphocreatine kinase, F_1F_0 ATP synthase, adenylate kinase, membrane transport, etc. It is concluded that important information concerning the metabolism and function of an organ may be obtained by magnetization transfer methods in a most direct way.

1 Introduction

The determination of reaction rates and fluxes constitutes an important step in the characterization of metabolic processes. The principle of such measurements is to generate a non-equilibrium state of the system under study by changing one or several parameters and to monitor its adaptation to the new conditions. NMR techniques may be used for both, creating a disequilibration of the system and for the recording of the ensuing concentration changes.

A system may be disturbed with respect to the chemical equilibrium by changing the concentrations of a reactant. For example, the tissue levels of high energy phosphates (phosphocreatine and ATP) can be reduced by reducing the nutritional blood flow to the tissue, resulting in an impairment of their synthesis. The system will adapt to this new steady-state condition. ^{31}P NMR has been used extensively to monitor such processes in vivo, as well as the recovery of the tissue following reperfusion. A typical experiment is illustrated in Fig. 1, which shows the recovery of phosphocreatine (PCr) and inorganic phosphate signals after global ischemia of rat skeletal muscle. From such measurements the net flux reflecting the difference between PCr synthesis and consumption, rather than a specific reaction flux, can be determined. Assuming a PCr concentration of 35 mM in normally perfused skeletal muscle [1], the net rate of PCr synthesis then calculates to 0.35 mM/s. The time resolution of this experiment is dictated by the time required to record a spectrum, and typically is of the order of a few seconds to minutes. Hence only net 'rate constant' $k < 0.1\,s^{-1}$ can be measured in this way.

Similar studies may be carried out by disturbing the isotope equilibrium of a system by introducing an isotopically labelled substrate. Using ^{13}C as label the formation of labelled metabolites can then be followed by ^{13}C-NMR spectroscopy. Well known examples are formation of labelled glycogen in liver or heart after an infusion of ^{13}C-labelled glucose [2, 3]. A drawback of such studies is the necessity, due to the inherent low sensitivity of NMR, to introduce relatively large quantities of labelled substrates, which may also significantly affect the chemical equilibrium of the system.

A third possibility makes use of the magnetic properties of the system exclusively, without affecting the chemical equilibrium. The magnetic equilibrium is disturbed by suitable radiofrequency excitation producing longitudinal non-equilibrium magnetization of a spin system of reactant A, which during the chemical reaction is transferred from A to B resulting in an observable magnetization exchange between the two species. For example, the terminal phosphate group of ATP (γ-ATP) may be selectively excited, yielding, when transferred to creatine via the creatine kinase reaction, magnetically labelled PCr. This technique allows the measurement of considerably faster reaction rates than those previously mentioned, because the rates accessible do not depend on the time required to record a spectrum, but rather on the lifetime of the magnetically labelled states, which is characterized by the spin lattice

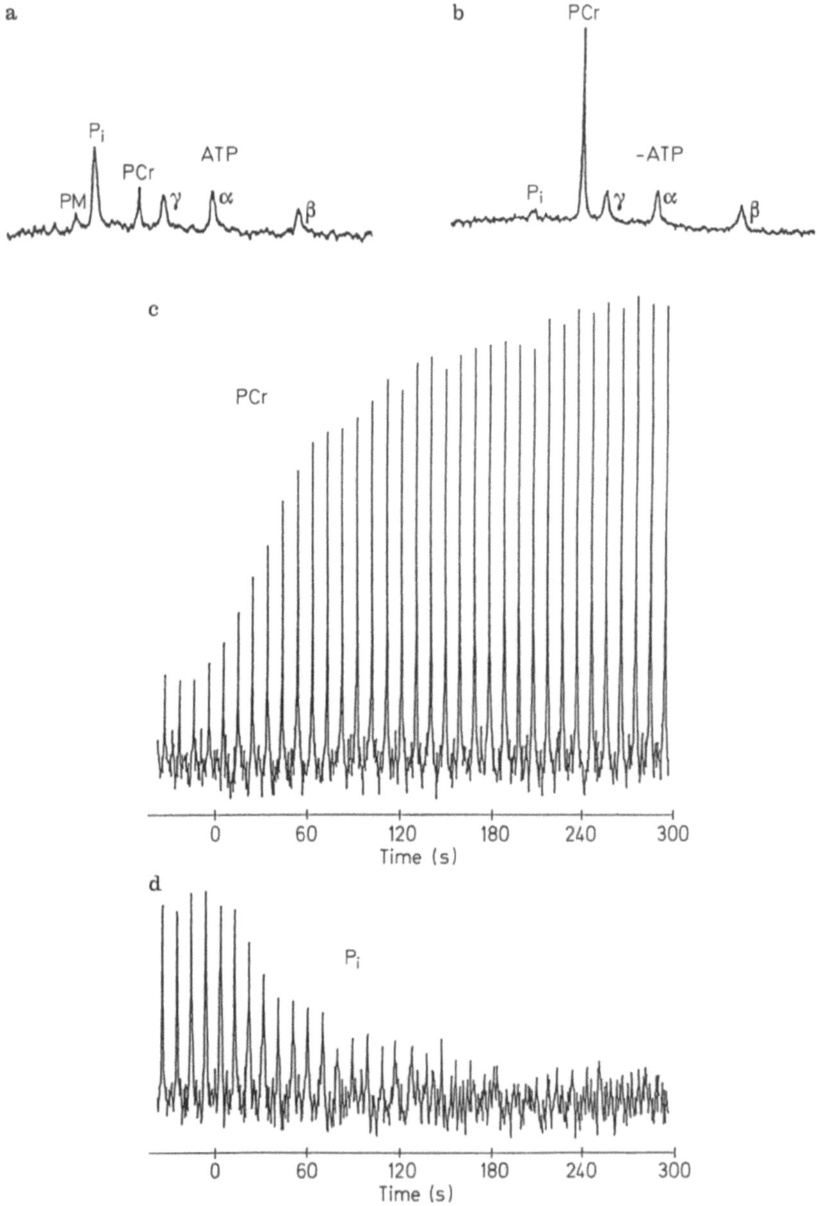

Fig. 1 a–d. Recovery of phosphocreatine (PCr) and inorganic phosphate P_i signals in the rat gastrocnemius muscle after global ischemia. Forty minutes after onset of ischemia spectrum (**a**) was recorded. The resonances correspond to phosphomonoesters (PM), inorganic phosphate (P_i), phosphocreatine (PCr), and the γ-, α-, and β-resonances of ATP. When blood flow was released both PCr (**c**) and P_i (**d**) recovered to their preischemic equilibrium values, represented in spectrum (**b**). Regression analysis, assuming single exponentials, yielded recovery rate constants of $k(PCr) = 0.0098 \pm 0.0006\,s^{-1}$ and $k(P_i) = 0.0110 \pm 0.0019\,s^{-1}$. The net PCr synthesis rate during recovery therefore is 0.34 ± 0.02 mM/s, assuming a PCr concentration in normoxic muscle of 35 mM. ^{31}P spectra were recorded at 81 MHz, the time resolution in (**c, d**) is 10 s

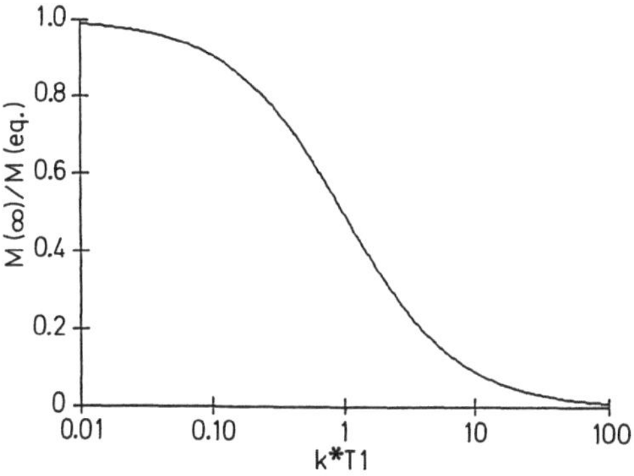

Fig. 2. Sensitive range of saturation transfer experiments for the determination of rate constants. The steady-state signal $M_z(\infty)/M_{z,eq}$ (Eq. (17)), where $M_z(\infty)$ denotes the magnetization after a long saturating pulse and $M_{z,eq}$ the corresponding equilibrium value, is plotted as a function of the product $k*T_1$. Since the longitudinal relaxation T_1 for the metabolites of interest in vivo is of the order of seconds, the sensitive range typically is $0.1\,s^{-1} < k < 50\,s^{-1}$

relaxation time T_1. This is illustrated in Fig. 2 for the saturation transfer experiment (see below). The relative steady-state magnetization $M_z(\infty)/M_{z,eq}$ is shown as a function of the rate constant for a two-site exchange reaction. This diagram demonstrates that the sensitive range of rate constants assessible in saturation transfer (or more general magnetization transfer) experiments is of the order of typically $0.1 < kT_1 < 10$, corresponding to the situation of intermediate exchange with respect to the spin–lattice relaxation time. For tissues in vivo typically having T_1 values of the order of seconds, this yields k values in the range of $0.1–50\,s^{-1}$.

Another great advantage of magnetization transfer experiments is that they allow the measurement of *unidirectional* rate constants and fluxes. However, for complex reaction systems the extraction of the individual rate constants may become tedious or even impossible.

Magnetization transfer experiments have first been described by Forsén and Hoffman [4–6]. The first application to a living system was the study of ATPase kinetics in aerobic *E. coli* cells [7]. Since then many in vivo magnetization transfer experiments, in whole animals as well as perfused organs, have been reported, as reviewed in references [8–11]. The purpose of the present chapter is to describe the principles of the magnetization transfer experiment including the mathematical analysis. Practical aspects such as hardware requirements etc., will also be considered. Finally the problems in the interpretation of the MR-derived rate constants and fluxes will be discussed on the basis of selected examples.

2 Magnetization Transfer Experiments: Principle

The general magnetization transfer experiment is schematically shown in Fig. 3. It may be described using the four-phase concept commonly applied in two-dimensional NMR spectroscopy [12]. During the preparation period, the non-equilibrium longitudinal magnetization is generated, which in 2D exchange spectroscopy [13, 14] is allowed to evolve. During the mixing period, incoherent transfer of magnetization between species A to B due to the chemical reaction or other exchange processes (cross polarization, spin diffusion) occurs. At the end of the mixing phase a read pulse is applied to generate detectable transverse magnetization. The observed NMR signals then contain the kinetic encoding obtained during the mixing period.

All types of magnetization transfer may be described using this scheme since they differ only with respect to the preparation and mixing processes. Hence, for the theoretical description of these experiments, we may formulate a general solution and then adapt it to the various magnetization transfer techniques by inserting their specific initial conditions.

Figure 4 shows several magnetization transfer experiments, which may be classified into two categories: type I without and type II with selective saturation of one or more resonances during the mixing period. Inversion recovery, inversion transfer, and two-dimensional exchange spectroscopy belong to the first category, saturation transfer or inversion recovery, with selective irradiation between the nominal π and $\pi/2$ pulses, to the second.

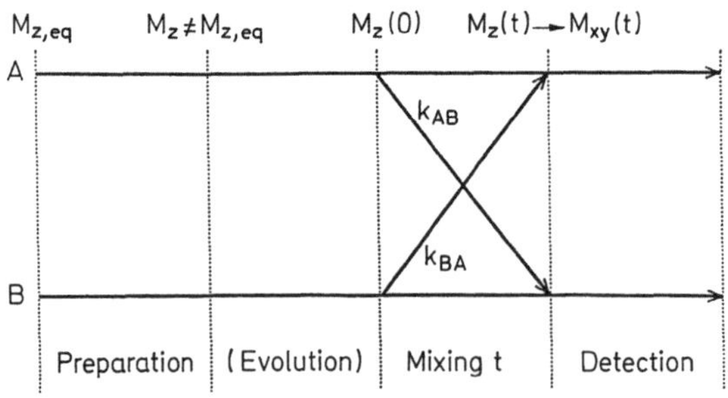

Fig. 3. Schematic representation of the magnetization transfer experiment in the two-site system. The system, which is originally in equilibrium ($M_{z,eq}$), is prepared such that at the beginning of the mixing interval there is non-equilibrium longitudinal magnetization $M_z(0) \neq M_{z,eq}$. During the mixing period, this magnetization is exchanged between the reaction partners A and B in an incoherent process, e.g., a chemical reaction, yielding $M_z(t)$. Thereafter a read-out pulse is applied to generate observable transverse magnetization $M_{xy}(t)$. In the two-dimensional exchange and related experiments, the preparation period is followed by an evolution phase

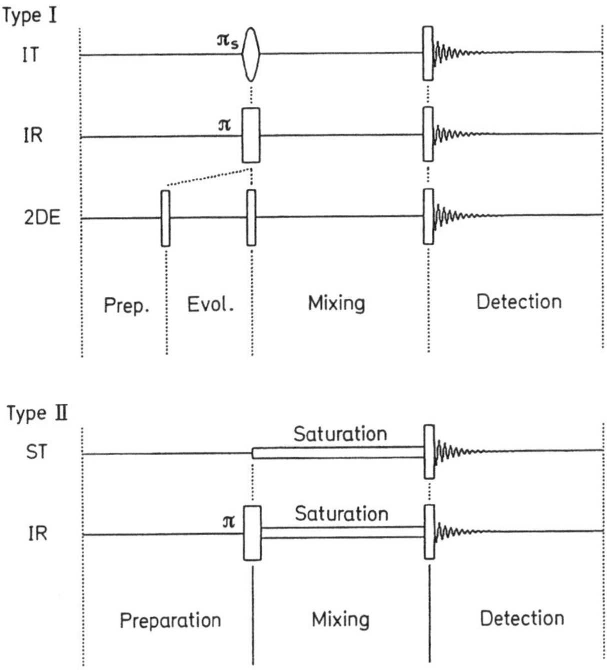

Fig. 4. Experimental realization of magnetization transfer experiments. Two types of experiments may be distinguished: (*I*) without and (*II*) with selective irradiation during the mixing period. Typical type-I experiments are inversion transfer (IT), inversion recovery (IR), and two-dimensional exchange spectroscopy (2DE). Type-II representatives are saturation transfer (ST) and inversion recovery (IR) with selective irradiation during the mixing phase. Unless indicated otherwise, all pulses are of nominal flip angle $\pi/2$; the subscript s indicates a frequency selective pulse

3 Magnetization Transfer: Theoretical Description

3.1 Magnetization Transfer Without Irradiation During the Mixing Period (Type I)

Magnetization Transfer in a Multisite Pseudo-First Order System

Consider a chemical network involving *N* exchange sites. All reactions shall be of pseudo-first order. The rate equations may then be consistently formulated using matrix algebra:

$$d/dt\mathbf{A} = \mathbf{K}\mathbf{A}, \tag{1}$$

where **K** is the kinetic matrix comprising the rate constants k_{ij}, and **A** the *N*-dimensional vector containing the individual concentrations.

The effect of the exchange processes on the longitudinal magnetizations may be described in an analogous way [15]:

$$d/dt\mathbf{M}_z = \mathbf{K}\mathbf{M}_z + (\boldsymbol{\delta}_{ij}T_{1i})[\mathbf{M}_z - \mathbf{M}_{z,eq}]. \tag{2}$$

The concentrations in Eq. (1) have been replaced by z-magnetizations $\mathbf{M}_z = (M_{z1}, \ldots, M_{zN})$ and $(\boldsymbol{\delta}_{ij}T_{1i})$ is the diagonal matrix containing the individual T_1 relaxation times. McConnell [16, 17] was the first to derive these equations.

For a complete two-sided system Eq. (2) may be transformed to become a homogeneous differential equation, substituting M_{zi} by $\delta M_{zi} = M_{zi} - M_{zi,eq}$, yielding:

$$d/dt\boldsymbol{\delta}\mathbf{M}_z = (\mathbf{K} - \boldsymbol{\delta}_{ij}T_{1i})\boldsymbol{\delta}\mathbf{M}_z. \tag{3}$$

In deriving Eq. (3), the principle of micro-reversibility, $k_{ij}M_{zi,eq} - k_{ji}M_{zj,eq} = 0$, has been applied.

In order to solve Eq. (3) we have to find the eigenvalues and eigenvectors of the matrix $\mathbf{K} - \boldsymbol{\delta}_{ij}T_{1i}$, i.e., a transformation matrix $\mathbf{U}$ such that:

$$\mathbf{U}(\mathbf{K} - \boldsymbol{\delta}_{ij}T_{1i})\mathbf{U}^{-1} = (\boldsymbol{\delta}_{ij}a_i). \tag{4}$$

In Eq. (4) $\mathbf{U}^{-1}$ represents the inverse of $\mathbf{U}$ and $(\delta_{ij}a_i)$ the diagonal matrix containing the eigenvalues. The general solution of Eq. (3) thus becomes:

$$\boldsymbol{\delta}\mathbf{M}_z(t) = \mathbf{U}^{-1}(\boldsymbol{\delta}_{ij}\exp(a_it))\mathbf{U}\boldsymbol{\delta}\mathbf{M}_z(0), \tag{5}$$

or expressed in terms of z-magnetizations rather than the differences from the equilibrium values:

$$\mathbf{M}_z(t) = \mathbf{M}_{z,eq} - \mathbf{U}^{-1}(\boldsymbol{\delta}_{ij}\exp(a_it))\mathbf{U}(\mathbf{M}_z(0) - \mathbf{M}_{z,eq}). \tag{6}$$

Since $a_i < 0$, Eq. (6) implies that for a long mixing time $t \to \infty$ the system will approach the equilibrium state. This holds for all type-I experiments.

Explicit analytical solutions for the individual z-magnetizations may only be derived for two-site or pseudo-two-site kinetic systems, i.e., systems which are reduced to the two-site problem after suitable manipulation of the kinetic scheme (e.g., saturation transfer in a system involving three exchanging sites).

In the following we shall discuss chemical exchange in the two- and three-site system.

Magnetization Transfer in a Two-Site System

We consider the two-site system $A \rightleftharpoons B$ with k_{AB} and k_{BA} being the rate constant for the forward and reverse reaction, respectively. The kinetic matrix then becomes:

$$\mathbf{K} - \boldsymbol{\delta}_{ij}T_{1i} = \begin{bmatrix} -R_A & k_{BA} \\ k_{AB} & -R_B \end{bmatrix}, \tag{7}$$

with the eigenvalues:

$$a_{1,2} = -(R_A + R_B)/2 \pm 1/2[(R_A - R_B)^2 + k_{AB}k_{BA}]^{1/2}, \tag{8}$$

where $R_A = k_{AB} + 1/T_{1A}$ and $R_B = k_{BA} + 1/T_{1B}$. The transformation matrix **U** then becomes:

$$\mathbf{U} = \begin{bmatrix} k_{BA}/[(R_A + a_1)^2 + k_{BA}^2]^{1/2} & (R_A + a_1)/[(R_A + a_1)^2 + k_{BA}^2]^{1/2} \\ k_{BA}/[(R_A + a_2)^2 + k_{BA}^2]^{1/2} & (R_A + a_2)/[(R_A + a_2)^2 + k_{BA}^2]^{1/2} \end{bmatrix}. \tag{9}$$

This yields the general solution for the two-site exchange process:

$$\begin{aligned} M_{zA}(t)/M_{zA,eq} &= 1 - (1 - M_{zA}(0)/M_{zA,eq})[c_1 \exp(a_2 t) + c_2 \exp(a_1 t)] \\ &\quad - c_3[(M_{zB}(0) - M_{zB,eq})/M_{zA,eq}][\exp(a_1 t) - \exp(a_2 t)], \\ M_{zB}(t)/M_{zB,eq} &= 1 - c_4[(M_{zA}(0) - M_{zA,eq})/M_{zB,eq}][\exp(a_1 t) - \exp(a_2 t)] \\ &\quad - (1 - M_{zB}(0)/M_{zB,eq})[c_1 \exp(a_1 t) + c_2 \exp(a_2 t)], \end{aligned} \tag{10}$$

with coefficients:

$$\begin{aligned} c_1 &= (R_A + a_2)/(a_2 - a_1), \\ c_2 &= -(R_A + a_1)/(a_2 - a_1), \\ c_3 &= k_{BA}/(a_2 - a_1), \\ c_4 &= k_{AB}/(a_2 - a_1). \end{aligned} \tag{10a}$$

The solution for the individual magnetization transfer experiments is obtained by inserting the appropriate initial conditions $M_{zA}(0)$ and $M_{zB}(0)$, respectively. As an example the explicit solution for the *inversion transfer* experiment are given.

We assume the resonance of B to be selectively inverted. The initial conditions thus become $M_{zA}(0) = M_{zA,eq}$ and $M_{zB}(0) = \beta M_{zB,eq}$, where $-1 < \beta < 1$ accounts for the inhomogeneity in the *rf* excitation ($\beta = -1$ for complete inversion). The time-dependent z-magnetizations for A and B then become:

$$\begin{aligned} M_{zA}(t)/M_{zA,eq} &= 1 - (\beta - 1)k_{BA}(M_{zB,eq}/M_{zA,eq}) \\ &\quad \cdot [\exp(a_1 t) - \exp(a_2 t)]/(a_2 - a_1) \end{aligned} \tag{11}$$

and

$$\begin{aligned} M_{zB}(t)/M_{zB,eq} &= 1 + (\beta - 1)[(R_A + a_2)\exp(a_2 t) \\ &\quad - (R_A + a_1)\exp(a_1 t)]/(a_2 - a_1). \end{aligned}$$

It is important to realize that the efficiency of the magnetization transfer $B \rightarrow A$, characterized by $M_{zA}(t)/M_{zA,eq}$, depends on the equilibrium concentration of the exchange partner, while the selective inversion recovery of B is independent of $M_{zA,eq}$. Therefore, if the equilibrium concentration of B is much smaller than that of A, the dynamic range for inversion transfer becomes very small and the

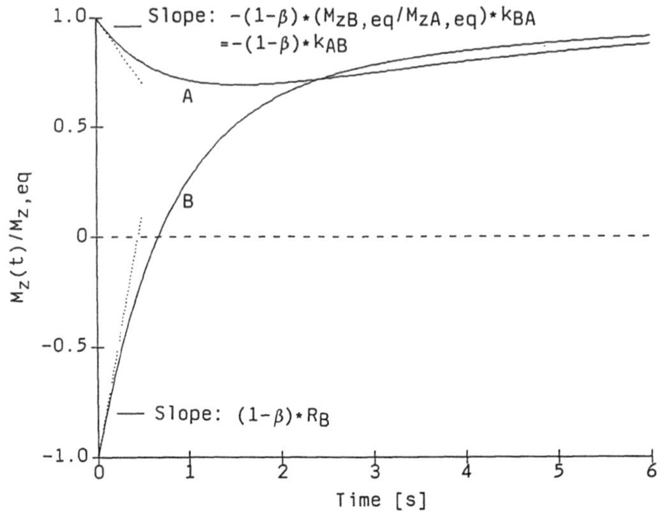

Fig. 5. Inversion transfer in the two-site system A⇌B. At $t = 0$ the signal of B is selectively inverted and the recovers as a function of the mixing time. The intensity of A originally decreases due to the exchange process and then, for long mixing times, recovers to its equilibrium value due to the dominance of the relaxation term. The initial slopes of A and B are $dM_{zA}/dt(t=0) = -(1-\beta)k_{BA}M_{zB,eq}/M_{zA,eq}$ and $dM_{zB}/dt(t=0) = (1-\beta)R_B$. It is easy to show that the rate constant k_{AB} may be derived from the minimum of the curve M_{zA} versus t according to $k_{AB} = -(M_{zA}^{min} - M_{zA,eq})/[T_{1A}(M_{zA}^{min} - M_{zB}^{min}M_{zA,eq}/M_{zB,eq})]$, provided T_{1A} is known from an independent experiment. The upper index min indicates the signals of A and B at t_{min}. The following parameters have been used for the calculation: $k_{AB} = 0.3\,s^{-1}$, $k_{BA} = 0.6\,s^{-1}$, $T_{1A} = 6.0\,s$, $T_{1B} = 2.0\,s$, intensity ratio A:B = 2:1

transfer may become difficult or even impossible to detect. Figure 5 shows the dependence of $M_{zA}(t)/M_{zA,eq}$ and $M_{zB}(t)/M_{zB,eq}$ on the mixing time.

The initial conditions of various type-I magnetization transfer experiments are given in Table 1, together with the explicit solutions for the z-magnetization for the two-site exchange A⇌B.

3.2 *Magnetization Transfer with Selective Irradiation During the Mixing Period (Type II)*

Magnetization Transfer in a Multisite Pseudo-First Order System

A second type of magnetization transfer experiments (type II, Fig. 4) involves permanent selective irradiation (on resonance 1) during the mixing period. Under these circumstances $M_{z1}(t)$ becomes zero for all mixing times t, and the differential equation Eq. (1) has to be changed to:

$$d/dt(\delta\mathbf{M}_{z}\text{-}) = (\mathbf{K}_{-} - \delta_{ij}T_{1i,-})(\delta\mathbf{M}_{z}\text{-}) + \mathbf{K}'_{-}(\mathbf{M}_{z,eq}\text{-}) \qquad (12)$$

where the subscript "−" indicates that row 1 and column 1 have been omitted

Table 1. Initial conditions and time-dependent longitudinal magnetizations (in the two-site system $A \rightleftharpoons B$) for the different magnetization transfer experiments shown in Fig. 2. The coefficients c_1 to c_4 are given in the text (Eq. (10a)). The ratio of the equilibrium concentrations is given by $K = M_{zB,eq}/M_{zA,eq}$. The abbreviations indicate IR: inversion recovery, IT: inversion transfer, SR: saturation recovery, 2DE: 2D exchange spectroscopy, and ST: saturation transfer

Type	Exp	$M_{zi}(0)/M_{zi,eq}$ A	$M_{zi}(0)/M_{zi,eq}$ B	Time-dependent z-magnetization $M_{zi}(t)/M_{zi,eq}$
I	IR	−1	−1	A: $1-2[c_1\exp(a_2t)+c_2\exp(a_1t)]+2Kc_3[\exp(a_1t)-\exp(a_2t)]$ B: $1-2[c_1\exp(a_1t)+c_2\exp(a_2t)]+2K^{-1}c_4[\exp(a_1t)-\exp(a_2t)]$
I	IT	1	−1	A: $1+2Kc_3[\exp(a_1t)-\exp(a_2t)]$ B: $1-2[c_1\exp(a_1t)+c_2\exp(a_2t)]$
I	PS	0	0	A: $1-[c_1\exp(a_2t)+c_2\exp(a_1t)]+Kc_3[\exp(a_1t)-\exp(a_2t)]$ B: $1-[c_1\exp(a_1t)+c_2\exp(a_2t)]+K^{-1}c_4[\exp(a_1t)-\exp(a_2t)]$
I	2DE[a]	1[b]	1[b]	AB: $c_3K^{1/2}[\exp(a_1t)-\exp(a_2t)]$ BA: $c_4K^{-1/2}[\exp(a_1t)-\exp(a_2t)]$ AA: $1-[c_1\exp(a_2t)+c_2\exp(a_1t)]$
II	IR	−1	0[c]	A: $1/(R_AT_{1A})-[1+1/(R_AT_{1A})]\exp(-R_At)$
	ST	1	0[c]	A: $1/(R_AT_{1A})+[1-1/(R_AT_{1A})]\exp(-R_At)$
	PS	0	0[c]	A: $1/(R_AT_{1A})[1-\exp(-R_At)]$

[a] AB: cross-peak intensity divided by $(M_{zA,eq}M_{zB,eq})^{1/2}$. Note that AB = BA for a two-site system AA: diagonal peak intensity.
[b] Amplitude of the modulation terms $(1-\cos\Omega_At_1)$ and $(1-\cos\Omega_At_1)$.
[c] Saturation of B during mixing period.

in the respective matrices and vectors. $\mathbf{K}'_-$ accounts for all exchange processes involving the site 1. In Eq. (12) we have made use of the fact that $k_{ij}M_{zi,eq} = k_{ji}M_{zj,eq}$. The solution is obtained as described above, yielding:

$$\mathbf{M}_{z^-} = \mathbf{M}_{z,eq^-} + \mathbf{U}^{-1}(\delta_{ij}a_i^{-1})\mathbf{U}\mathbf{K}'_-\mathbf{M}_{z,eq^-} + \mathbf{U}^{-1}(\delta_{ij}\exp(a_it))\mathbf{U}[\mathbf{M}_{z^-}(0) - \mathbf{M}_{z,eq^-} - \mathbf{U}^{-1}(\delta_{ij}a_i^{-1})\mathbf{U}\mathbf{K}'_-\mathbf{M}_{z,eq^-}], \quad (13)$$

where a_1 and a_2 are the eigenvalues of the homogenous part of the differential equation and $\mathbf{U}$ the transformation matrix containing the eigenvectors. Note that Eq. (13) contains a term which is independent of the mixing time t. Hence $M_{z^-}(t \to \infty) \neq M_{z,eq^-}$ in contrast to Eq. (6), i.e., upon increasing the mixing time t the system tends to a steady state, which is different from magnetic equilibrium.

In the following we discuss two special cases, namely magnetization transfer in a two- and three-site system.

Magnetization Transfer in a Two-Site System

Consider a two-site system $A \rightleftharpoons B$ with resonance B being saturated during the mixing period. The matrices $\mathbf{K}_- - \delta_{ij}T_{1i,-}$ and $\mathbf{K}'_-$ consist of one element, $-R_A$

and k_{AB}, respectively, and Eq. (12) then becomes:

$$d/dt\delta M_{zA} = -R_A \delta M_{zA} + k_{AB} M_{zA,eq} \tag{14}$$

with the solution:

$$M_{zA}(t)/M_{zA,eq} = 1/(R_A T_{1A}) + [M_{zA}(0)/M_{zA,eq} - 1/(R_A T_{1A})] \exp(-R_A t), \tag{15}$$

with $R_A = k_{AB} + 1/T_{1A}$. Again the explicit solutions for the different type II experiments shown in Fig. 4 are obtained by introducing the appropriate initial conditions (see Table 1). For the conventional *saturation transfer* experiment $M_{zA}(0) = M_{zA,eq}$, yielding:

$$M_{zA}(t)/M_{zA,eq} = 1/(R_A T_{1A}) + [1 - 1/(R_A T_{1A})] \exp(-R_A t). \tag{16}$$

In contrast to the type-I experiments, the z-magnetization of A does not depend on the equilibrium magnetization of the exchange partner, but only on the rate constant k_{AB} and the relaxation time T_{1A}. This is an advantage if $M_{zB,eq}$ is very small. Moreover, for long mixing times t, Eqs. (15) and (16) approach the limit

$$M_{zA}(\infty)/M_{zA,eq} = 1/(1 + k_{AB} T_{1A}). \tag{17}$$

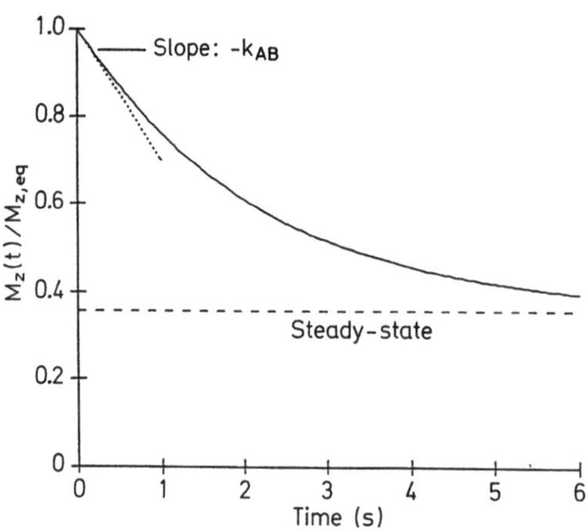

Fig. 6. Saturation transfer for a two-site exchange A⇌B. Upon saturation of B the signal of A decays exponentially as a function of the mixing time according to Eq. (16). The initial slope is $dM_{zA}/dt(t=0) = -k_{AB}$. Similarly, the rate constant may be derived from the steady-state value $M_{zA}(\infty)/M_{zA,eq} = 1/(1 + k_{AB} T_{1A})$ (see Eq. (17)). Parameters are the same as in Fig. 5

If T_{1A} is known from a different set of experiments, the rate constant k_{AB} may be derived from a single experiment with a long mixing time, otherwise a series of experiments with varying mixing times has to be recorded. The saturation transfer experiment is depicted in Fig. 6.

Initial conditions for other type-II experiments are given in Table 1 together with the corresponding solutions for the z-magnetizations.

Saturation Transfer in the Three-Site System

We consider saturation transfer in the three-site system $A \rightleftharpoons B \rightleftharpoons C$, where there is permanent irradiation at the C resonance during the mixing period. The rate equations for M_{zA} and M_{zB} is given in Eq. (13) with $(\delta\mathbf{M}_{z^-}) = (M_{zA} - M_{zA,eq}, M_{zB} - M_{zB,eq})$:

$$\mathbf{K}_- - \delta_{ij} T_{1i,-} = \begin{bmatrix} -R_A & k_{BA} \\ k_{AB} & -R_B \end{bmatrix} \quad \text{and} \quad \mathbf{K}'_- = \begin{bmatrix} 0 & 0 \\ 0 & k_{BC} \end{bmatrix} \tag{18}$$

The relaxation rates become $R_A = k_{AB} + 1/T_{1A}$ and $R_B = k_{BA} + k_{BC} + 1/T_{1B}$. The time dependencies of the individual z-magnetizations after saturation of the resonances A, B, and C are shown in Fig. 7. It is important to realize that the

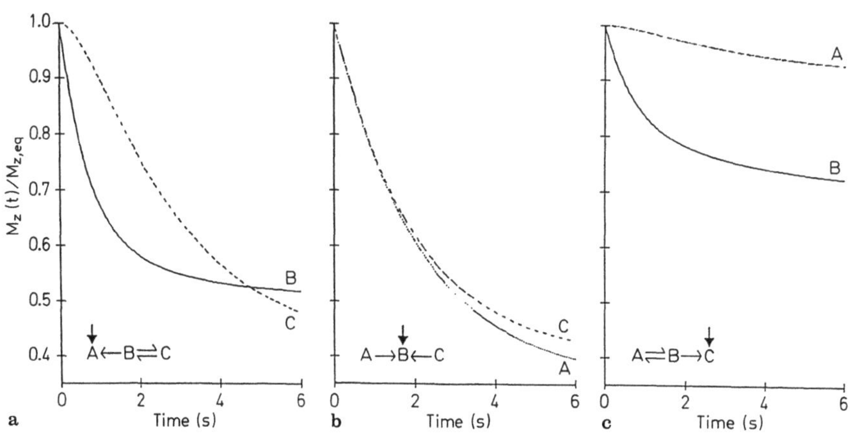

Fig. 7. Saturation transfer in the three-site system. The rate constants $k_{AB} = 0.3\,s^{-1}$, $k_{BA} = 0.6\,s^{-1}$, $k_{BC} = 0.3\,s^{-1}$, relaxation times $T_{1A} = 6\,s$, $T_{1B} = 2\,s$, $T_{1C} = 5\,s$, and an intensity ratio of A:B:C of 4:1:0.5 have been used for the calculations. Selective irradiation at resonances A or C, leads to double exponential dependence of the observed signals (B, C or A, B) on the mixing time. When B is saturated, the three-site system becomes decoupled and may be considered as two independent two-site systems, which may be analyzed in a straightforward way (Eqs. (16) and (17))

signals decay biexponentially when A or C are saturated, while saturation of B causes uncoupling of the system yielding single exponential signal behavior according to Eq. (16).

An alternative, experimentally more sophisticated approach to handle three- (or multisite) systems is the multiple saturation transfer experment proposed by Ugurbil [18]. If in the three-site system the resonance C is constantly saturated, it is reduced to an apparent two-site system described by the differential equations:

$$d/dt\mathbf{M}_{z^-} = \mathbf{K}_-(\mathbf{M}_{z^-} - \mathbf{M}^*_{z^-}) \tag{19}$$

where the symbols have been defined above. Note that the equilibrium magnetizations M_{z,eq^-} have been replaced by the new steady-state magnetizations $M^*_{z^-}$ for saturated C. The condition for the micro-reversibility has to be replaced by:

$$\mathrm{k_{AB}} M^*_{zA} = \sigma \mathrm{k_{BA}} M^*_{zB} \tag{20}$$

with

$$\sigma = (M^*_{zA}/M^*_{zB})(M_{zA,eq}/M_{zB,eq})^{-1}. \tag{21}$$

If A is saturated in addition to C, we obtain for the z-magnetization of species B:

$$M_{zB}(t)/M^*_{zB} = (1 - \mathrm{k_{BA}}\sigma/R_B) + (\mathrm{k_{BA}}\sigma/R_B)\exp(-R_B t). \tag{22}$$

Equation (22) is formally identical to Eq. (16) with the exception of the appearance of σ, which can be determined experimentally by measuring both the equilibrium magnetization $M_{z,eq}$ and the steady-state magnetization M^*_z when C is saturated. It is important to realize that the rate constants $\mathrm{k_{BA}}$ and $\mathrm{k_{BC}}$ may not be derived from a single experiment. Either the spin–lattice relaxation rate R_B has to be determined from a separate experiment (inversion recovery), while resonances A and C are saturated, or the saturation transfer experiment has to be carried out for several mixing times.

3.3 Nonlinear Chemical Reactions

The magnetization transfer experiment involves only linear operators, M_z being proportional to $\Sigma\gamma\hbar\langle Im|I_z|Im\rangle$. Therefore, the rate equations describing the time dependence of the longitudinal magnetizations are linear, always producing apparent first-order rate constants.

For example consider the reaction $2A \rightleftharpoons B$, wih the corresponding rate equation for substrate A:

$$dA/dt = -k_1[A]^2 + \mathrm{k}_{-1}[B]. \tag{23}$$

The rate constants derived from the magnetization transfer experiment are $\mathrm{k_f} = \mathrm{k_1}[A]$ and $\mathrm{k_r} = \mathrm{k_{-1}}$ for the forward and reverse reaction, respectively. Only if the reaction is in steady-state, the forward rate constant is truly a constant. In this case, knowledge of the steady-state concentration allows calculation of second order rate constant k_1.

It is important to realize that magnetization transfer experiments, yielding only pseudo-first order rate constants, are not suited for the determination of the exact mechanism and the order of unknown reactions. However, they may provide a qualitative proof that an exchange is actually taking place.

3.4 *Interpretation of the Experimentally Derived Rate Constants of Enzymatic Reactions*

Consider the simplest case of an enzymatic reaction involving the formation of the enzyme-substrate complex and a product forming step:

$$E + S \overset{K_S}{\rightleftharpoons} ES \overset{k_1}{\rightarrow} E + P \tag{I}$$

where K_S is the equilibrium constant of the enzyme substrate complex ES, and k_1 the rate constant for the product formation step. The forward reaction rate is $k_1[ES]$, yielding for the pseudo-first order rate constant obtained upon saturation of P:

$$k_{for} = k_1[E]_{tot}(1 + [S]/K_S)^{-1}. \tag{25}$$

In Eq. (25) the mass conservation for the enzyme E, $[E]_{tot} = [E] + [ES]$, and the equilibrium constant for the enzyme-substrate complex, $K_S = [ES]/[E][S]$, have been used. The reaction rate correlates with the total enzyme concentration. This has been demonstrated by Bittl et al. [1], who compared the creatine kinase reaction rates in brain, heart, and skeletal muscle.

Looking at an enzymatic conversion of S to P with two intermediate steps, we may extend the reaction scheme (I) to:

$$E + S \overset{K_S}{\rightleftharpoons} ES \underset{k_{-1}}{\overset{k_1}{\rightleftharpoons}} EP \overset{K_P}{\rightleftharpoons} E + P. \tag{II}$$

The pseudo-first order forward an reverse rate constants become:

$$k_{for} = k_1[E]_{tot}(K_S D)^{-1} \quad \text{and} \quad k_{rev} = k_{-1}[E]_{tot}(K_P D)^{-1}, \tag{26}$$

with

$$D = 1 + [S]/K_S + [P]/K_P.$$

Again the rate constant is directly proportional to the total tissue activity of the enzyme. Scheme (II) corresponds, for example, to an intramolecular rearrangement reaction (e.g., *cis/trans* isomerization) with the formation of a binary substrate-enzyme complex.

Many enzymatic processes involve ternary enzyme-substrate complexes. Consider a scheme, which formally corresponds to the creatine kinase:

$$A + B \rightleftharpoons P + Q.$$

Formulating the individual complex formation and dissociation steps we obtain:

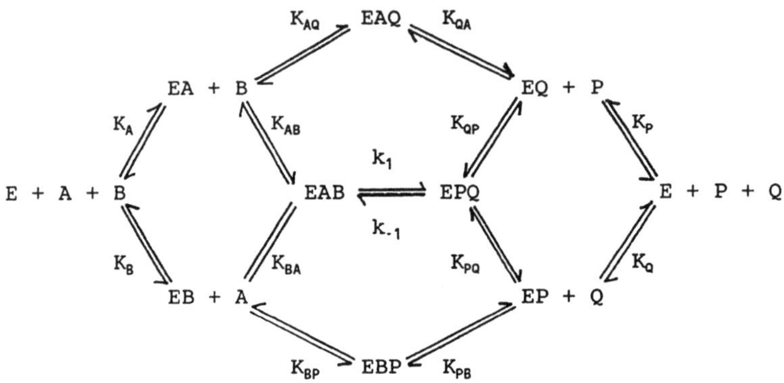

(III)

For the creatine kinase (CK) reaction we have the identifications A = adenosin diphosphate, B = phosphocreatine, P = adenosin triphosphate and Q = creatine. The two ternary complexes EAQ and EBP are so-called dead-end complexes, since they do not lead to a reaction. For CK the complex EBP (= E PCr ATP) has not been observed [19]. The two rate constants then become [19, 20]:

$$k_{for} = k_1[E]_{tot}[A](K_A K_{AB} D)^{-1} \text{ and } k_{rev} = k_{-1}[E]_{tot}[Q](K_P K_{PQ} D)^{-1}, \tag{27}$$

with

$$D = (1 + [A]/K_A[B]/K_B[P]/K_P + [Q]/K_Q + [A][B]/K_{AB}K_A + [P][Q]/K_{PQ}K_P + [A][Q]/K_{AQ}K_A).$$

A general conclusion based on the three reaction schemes discussed is that the measured rate constants should depend on the total tissue enzyme activity. They are global rate constants, involving all steps in the enzymatic reaction mechanism. Therefore, the individual elementary rate constants k_1 and k_{-1} may not be derived from the experimental data without knowledge of the dissociation constants for the individual enzyme-substrate complexes and the respective substrate concentrations. The relative values of the dissociation constants and substrate concentrations determine, which of the substrates is rate-limiting. For the CK reaction it has been shown that the rate-limiting substrate is ADP for the forward, and creatine for the reverse reaction [1].

Rate constants and reaction fluxes derived from magnetization transfer and analogous experiments depend on the total activity of the individual enzymes involved. The activity of a single enzyme, however, can only be derived, if the exchange process under consideration is exclusively mediated by this enzyme. This is the case, e.g., for creatine kinase, which exclusively catalyzes the transfer of a phosphate group between PCr and ATP.

In general, however, the experimentally determined flux is the result of more than one reaction step. Only in the special case, where one step is completely rate-limiting, can the reaction rate be derived simply from the measurement of the overall flux through the pathway. More common is the situation where flux control is exerted by several enzymes. In this case the flux through an individual enzymatic step may be determined by modulating the enzyme activity [21, 22].

Similarly, the influence of parallel enzymatic reactions may be eliminated by selective inhibition. This has been demonstrated for ATP synthase, where, in order to obtain meaningful values for the mitochondrial oxidative phosphorylation, the glycolytic contribution to ATP synthesis had to be inhibited [23, 24].

An additional complication in the interpretation of the experimentally determined reaction fluxes arises from compartmentation. Substrate transport across membranes influences the overall reaction rates [24].

In order to correctly interpret the experimentally determined rate constants, it is necessary to test criteria which allow these different possibilities to be discriminated. For example, if the $P_i + ATP$ fluxes assessed by magnetization transfer reflect the net rates of oxidative phosphorylation, then the relation flux $(P_i \rightarrow ATP) = \alpha \cdot MVO_2$ should be fulfilled, where α is the molar ratio between the ATP synthetized and oxygen consumed, and MVO_2 the oxygen consumption. Additionally, in steady state the rate of ATP utilization, which may be determined from multiple saturation transfer experiments [18], should be equal to the rate of ATP synthesis [24].

4 Practical Aspects

4.1 *Spectrometer Requirements for Magnetization Transfer Experiments*

The spectrometer hardware required for magnetization transfer studies depends on the specific experiment to be carried out. Type-I experiments, in contrast to type-II, do not involve extended time periods of irradiation, and hence can be run on any standard single channel spectrometer.

Selective irradiation may be achieved using shaped pulses (e.g., Gaussian amplitude modulation) or by using DANTE pulse trains [25] (see below). Type-II experiments involve a period of low-power irradiation for selective saturation of a resonance. The rf power required is of the order of 1 mG (for $1/T_1T_2 = 100\,s^{-2}$). There are two possibilities to fulfill this criterion; either to use a second channel with a low-power amplifier (500 mW), which must be operative at high duty cycles (up to 90%), or to use frequency selective high-power pulse trains (100 to 500 W) of long duration but low duty cycles (<0.1%).

When a separate low-power channel is used it has to be inserted into the signal path using a unidirectional power combiner. The second channel should be fully software controlled in order to guarantee the on-resonance condition for the desired resonance and to switch the frequency rapidly back and forth for simultaneous recording of the transfer and control spectrum (see below). For an inhomogeneously broadened line it is more convenient to saturate it by sweeping the saturation frequency through the resonance. This guarantees homogeneous saturation throughout the region of interest. An important point when two amplifiers are used is to move the low-power frequency outside the observation bandwidth during the data acquisition. This avoids leakage to the signal channel, which may give rise to undesirable carrier glitches, which may hide any residual signal, making it difficult to judge whether the saturation has been complete or not.

Using a single channel system, selective saturation may be achieved using DANTE pulse trains. The excitation profile of a DANTE pulse is shown in Fig. 8a, b. In going from Fig. 8a to b the duration of the pulse has been increased ten-fold. As a consequence, the selection profile becomes sharper and the

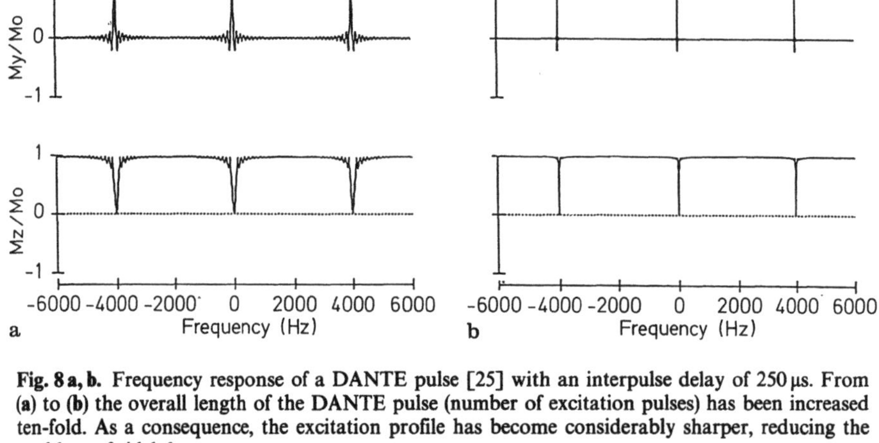

Fig. 8 a, b. Frequency response of a DANTE pulse [25] with an interpulse delay of 250 μs. From (**a**) to (**b**) the overall length of the DANTE pulse (number of excitation pulses) has been increased ten-fold. As a consequence, the excitation profile has become considerably sharper, reducing the problem of sidelobes problems of the sidelobes less severe. An important feature of the DANTE pulse is that multiple frequencies with a separation of $1/T$ are excited, where T is the time interval between the individual low-angle excitation pulses. The timing has to be chosen such that there is no interference of the sidebands. Figure 9 demonstrates this problem. Figure 9b shows the desired experiment with selective saturation of the γ-ATP resonance, while in Fig. 9c the spacing of the individual pulse was set to $1/(f_{\gamma\text{-ATP}} - f_{\beta\text{-ATP}})$, therefore both the γ- and β-ATP resonances are simultaneously saturated.

This last experiment shows a potential application of the DANTE pulses for multifrequency excitation, which may be achieved by setting the appropriate spacings of the short excitation pulses or better by combination of two DANTE pulse trains [25].

A typical duty cycle of a DANTE pulse is $<1/500$ (corresponding to 4 kHz spacing between sidebands and 0.5 μs excitation pulses), giving a power deposition of 10 mW at 5 W incident rf power. For clinical application DANTE pulses may not be suited due to the values of the specific absorption rate exceeding regulations.

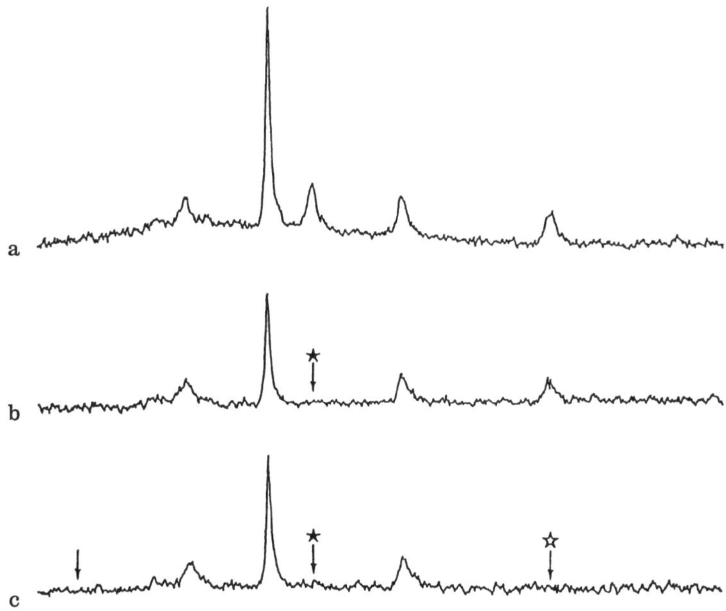

Fig. 9 a–c. Performance of a DANTE pulse for selective saturation of the γ-ATP resonance for rat gastrocnemius muscle. (**a**) shows the control spectrum. In (**b**) the γ-ATP resonance has been saturated using a DANTE pulse with 250 μs interpulse delay. In (**c**) the delay was set to 908 μs, corresponding to the reciprocal chemical shift difference of the γ- and β-ATP signal. As a consequence both signals are simultaneously saturated. The arrows represent the positions of selective irradiation (central frequency on γ-ATP and sidelobes), the filled asterisk the desired, the open asterisk the undesired saturation

4.2 Magnetization Transfer with Inhomogeneous Excitation

Due to their simplicity, transmitter/receiver surface coils are frequently used for in vivo MRS [26]. However, the inherent problem associated with surface coils as transmitting devices is the inhomogeneity of their radiofrequency field distribution [26, 27], which affects MRS experiments requiring accurate excitation pulse angle (e.g., the inversion transfer/recovery or the 2D exchange). The use of excitation and inversion pulses with flip angles, which are largely independent of the B_1 field strength, reduces this problem to a great extent [28, 29]. The problem of rf field inhomogeneity is less severe for experiments requiring saturation of resonances. A rather homogeneous saturation throughout the sample may be achieved, at least for long mixing periods. Consequently, the saturation transfer sequence works reasonably well with a single transmitter/receiver surface coil.

The problems of inhomogeneous excitation may be circumvented by using separate coils for excitation and detection [30]. The desired characteristics for the transmitter coil would be a homogeneous rf field, while the receiver coil should be optimized with respect to sensitivity and localization. This two-coil setup requires appropriate insulation of the two radiofrequency circuits.

4.3 Combination with Localization Techniques

Combination of magnetization transfer and localization methods is readily possible [31–35]. The basic sequence may be described with the four phases:

PREPARATION—MAGNETIZATION TRANSFER—LOCALIZATION—DETECTION

All spatially resolved NMR flux measurements published to date make use of saturation transfer for deriving the kinetic data. However, for localization different approaches have been reportd.

Challiss et al. [31] and Blackledge et al. [32] use the phase-modulated rotating frame technique, i.e., the localizing step consists of the sequence $n\Theta_{\pm x} - \lambda_y$, λ_y being the phase encoding pulse and Θ the frequency encoding pulse increment.

Alternatively, Bottomley and Hardy [33] have combined phase-encoded spectroscopic imaging of the human brain with saturation transfer. Images of the ratio $M_z(\text{PCr}; \gamma\text{-ATP})/M_z(\text{PCr}; \text{Ctl})$, i.e., the image obtained by saturating the γ-ATP resonance divided by the image obtained with irradiation at the control position, were calculated. This ratio is related to the CK forward rate constant according to Eq. (17). The image resolution in this study was $2 \times 2\,\text{cm}^2$ at a slice thickness of 5 cm. In order to speed up data acquisition, a repetition time shorter than $T_{1,\text{PCr}}$ has been chosen and a saturation factor R_{PCr} has been introduced to account for incomplete relaxation:

$$M_z(\text{PCr}; \gamma\text{-ATP})/M_z(\text{PCr}; \text{Ctl}) = [1 + k_f T_{1,\text{PCr}}]^{-1} R_{\text{PCr}}. \tag{28}$$

Hsieh and Balaban [34] used a spin-echo imaging technique for spatially resolving the creatine kinase kinetics in rabbit skeletal muscle. Due to the short T_2 relaxation time of ATP, the only signal detected was that of PCr. A saturation transfer (irradiation of γ-ATP resonance) and control image (irradiation on symmetrical position with respect to PCr) was recorded, and the ratio of the two images calculated as mentioned above.

Mora et al. [35] have applied a slightly modified procedure of the one mentioned above to obtain images of PCr in the presence of control and γ-ATP irradiation. PCr was selectively excited using a frequency-selective $\pi/2$ sinc pulse (duration 10 ms) followed by a conventional spin-echo sequence with 30 ms echo time. Thereby the problem of spectral deconvolution has been avoided.

Wolf and Balaban [36] have studied the exchange between the "free" tissue water and that in a pool with restricted motion (macromolecules) using saturation transfer in combination with a spin-warp imaging sequence. Images of the rabbit kidney were recorded with selective irradiation on the broad resonance due to "bound" water (mixing time 3.5 s, power 0.4 W, 5 kHz downfield to "free" water resonance) immediately before slice selection.

There are several potential pitfalls associated with localized magnetization (saturation) transfer studies. Due to variations of the local magnetic susceptibility of tissue (e.g., in the head), the resonance frequency becomes position dependent. It may therefore be impossible to maintain complete saturation throughout the region of interest. This may lead to misinterpretation of the results. Similarly, T_1 may vary throughout the region of interest. It is then not possible to derive the kinetic information from an experiment with one mixing time. Instead, images have to be recorded for several mixing times, leading to long data acquisition times. The kinetic information must then be obtained from exponential regression with the parameters k and T_1 according to Eq. (16) for each voxel. A further consequence of T_1 variations is that the saturation factor R has to be determined for each voxel individually. Finally, poor signal/noise ratio may prevent the accurate determination of rate constants and fluxes for the individual voxels.

4.4 Evaluations of Rate Constants from Experimental Data Sets

The determination of the kinetic and relaxation parameters from experimental data sets may be quite tedious and pitfalls are easily possible. Low signal/noise ratios, as frequently encountered in in vivo NMR, may be the cause of erroneous results. If the parameters are obtained from a regression analysis, cross-correlation among the individual parameters have to be checked very carefully, otherwise, their actual value may be wrong. In the case of more than two exchanging sites, multiexponential signal behavior will ensue. However, the experimental data sets will hardly be sufficient, both as far as signal/noise and number of time points (mixing times) recorded are concerned, to allow a reliable determination of the two time constants, if they are of comparable order of magnitude.

Potential problems are identified on the basis of model calculations on a two-site system $A \rightleftharpoons B$. The rate constants are $k_{AB} = 0.3\,s^{-1}$, $k_{BA} = 0.6\,s^{-1}$, the corresponding spin–lattice relaxation times $T_{1A} = 6$ s, $T_{1B} = 2$ s, and the relative signal intensities 2:1. These values approximately correspond to the CK reaction in heart or brain (see below). Based on these values, the z-magnetizations for mixing times $t = 0.25$, 0.5, 1, 2, 4, and 8 s were calculated for the saturation transfer experiment. Then different noise levels were added, and a nonlinear regression analysis was carried out in order to redetermine the parameters. Results for the saturation transfer are given in Table 2. Obviously as signal/noise decreases, the uncertainty in the fitted parameters becomes larger. Interestingly, the T_1-values are more susceptible to errors than are the rate constants. Rydzy et al. [37] comparing the accuracy of several type-II magnetization transfer experiments (saturation transfer, inversion and saturation recovery) obtained similar results carrying out two- and three-parameters regression analysis.

Considerable correlations among fit parameters may occur. This is especially the case for inversion transfer and recovery, where each signal (M_A, M_B) is governed by both rate constants and relaxation times [38]. The determination of the individual rate constants from a single dependence is only possible if the two eigenvalues a_1 and a_2 differ considerably. For the model system described above the values are -0.25 and -1.31. Led and Gesmar [39, 40] have pointed out that in the case of comparable eigenvalues complementary experiments have to be carried out in order to obtain robust kinetic parameters. For inversion transfer this means that all four curves $M_{zA}(A)$, $M_{zA}(B)$, $M_{zB}(A)$, and $M_{zB}(B)$, where the parenthesis indicates the selectively inverted signal, have to be analyzed simultaneously. In saturation transfer the problem of parameter correlation is less severe.

Table 2. Model calculation of saturation transfer in a two-site system $A \rightleftharpoons B$ with the rate constants $k_{AB} = 0.3\,s^{-1}$, $k_{BA} = 0.6\,s^{-1}$ and relaxation times $T_{1A} = 6$ s, $T_{1B} = 2$ s. The relative concentrations are A = 2 and B = 1. Different noise levels were added to the calculated signals and the kinetic parameters determined from least square regression analysis. Twenty analyses were carried out for each 'experiment' (saturation of either A and B for a given noise level). The values in the table represent mean ± SD

Method	Noise	A		B	
		k_{AB}	T_{1A}	k_{BA}	T_{1B}
ST	0.1	0.30±0.01	6.1±0.7	0.60±0.06	2.0±0.3
(Eq. (16))	0.2	0.31±0.03	6.5±1.9	0.66±0.20	2.1±1.3
	0.3	0.31±0.06	7.0±3.8	0.66±0.27	2.7±1.6
ST ss	0.1	0.30±0.03		0.60±0.06	
(Eq. (17))	0.2	0.30±0.04		0.60±0.17	
	0.3	0.31±0.07		0.60±0.29	

4.5 Comparison of the Different Magnetization Transfer Experiments

We would like to compare from a pragmatic point of view four commonly used magnetization transfer techniques by studying the CK reaction in rat skeletal muscle. The corresponding kinetic parameters have been reported by several laboratories [1, 41–43]. Our own experiments were carried out on a Bruker Biospec BMT 47/15 spectrometer. A saddle coil tuned to 81 MHz was used as radiofrequency transmitter/receiver. DANTE pulses [25] with pulse lengths of 0.5 μs and interpulse delays of 250 μs were used for selective irradiation. Spectra were recorded from the gastrocnemius muscle of anesthetized rats.

Figure 10a shows a representative *inversion transfer* experiment. At $t = 0$ the PCr resonance is selectively inverted. The initial PCr magnetization is 70% of its equilibrium value corresponding to $\beta = -0.7$ in Eq. (11). Due to the CK reaction, magnetization is transferred from PCr to the γ-ATP signal, which decreases and reaches a minimum at a mixing time of approximately 1.5 s. For longer mixing times it recovers to its equilibrium value. The other resonances are not affected by the exchange process. The spectra represent an average of 32 scans, with a repetition delay of 10 s, corresponding to a total acquisition time of 5.5 min. The duration of the total inversion transfer experiment was approximately 60 min. The rate constants are derived from multiparameter/multifunction regression analysis, i.e., simultaneous regression of the $M_{zPCr}(t)$ and $M_{zATP}(t)$ for inversion of either the ATP- or PCr-resonance.

The *dynamic version of the saturation transfer* experiment with the irradiation frequency set to the γ-ATP resonance is shown in Fig. 10b. The total experimental time was approximately 60 min. The PCr signal decays exponentially with increasing mixing time. The decrease of the β-ATP signal is due to an exchange process, which is catalyzed by all enzymes that interconvert ADP and ATP, since saturation of the γ-ATP resonance also results in saturation of the β-ADP signal, which is less than 1 ppm upfield [44]. The origin of the decrease of the α-ATP resonance upon saturation of the γ-ATP signal is not clear. The rate constants are obtained from regression analysis with Eq. (16).

The corresponding *steady-state saturation transfer* experiment is depicted in Fig. 10c. The mixing time was 6 s and the repetition delay 10 s. The individual traces represent saturation transfer (Sat), control (Ctl), and difference spectrum Ctl–Sat (Dif). The total experimental time is 20 min. The rate constants can be evaluated according to Eq. (17).

Figure 11 finally shows the corresponding *two-dimensional* (2D) *exchange* spectrum of the rat gastrocnemius muscle at rest. The mixing time was set to 600 ms, t_1 was incremented in 128 steps from 0.2 to 25.6 ms, the total acquisition time was 1.8 h. Axial peaks were reduced with suitable phase cycling [45]. The only observed crosspeaks are those between PCr and γ-ATP.

Table 3 summarizes the important advantages and disadvantages of several commonly used mgnetization transfer techniques.

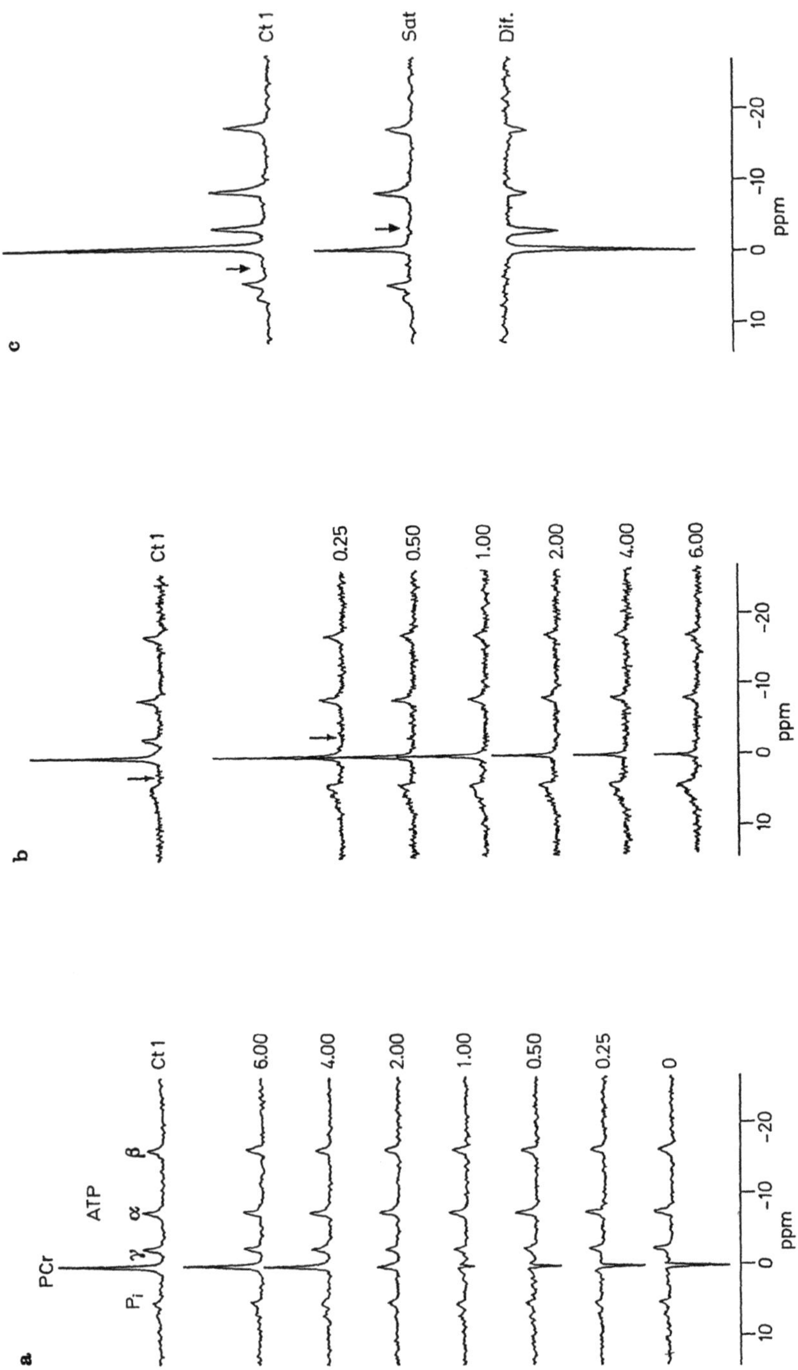
a
PCr
P_i
ATP
γ
α
β
Ct 1
6.00
4.00
2.00
1.00
0.50
0.25
0
10
0
-10
-20
ppm
b
Ct 1
0.25
0.50
1.00
2.00
4.00
6.00
10
0
-10
-20
ppm
c
Ct 1
Sat
Dif.
10
0
-10
-20
ppm

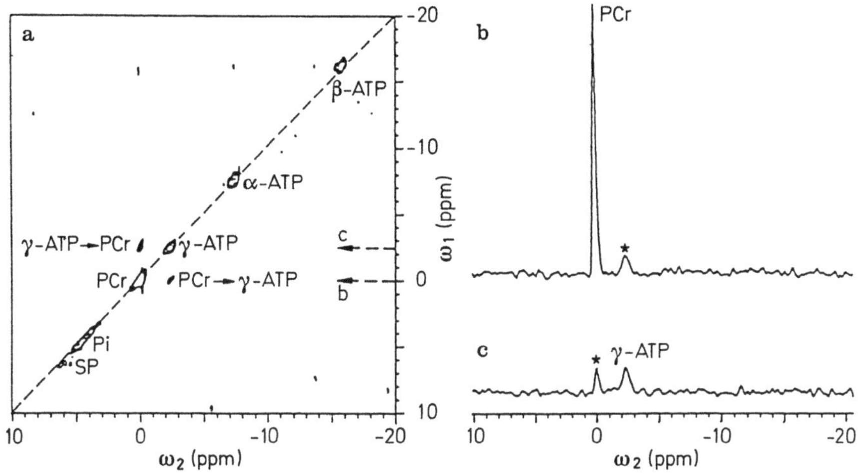

Fig. 11 a–c. Two-dimensional ^{31}P-exchange spectrum of rat gastrocnemius muscle. (**a**) shows contourplot with the one-dimensional spectrum along the diagonal and crosspeaks between PCr and γ-ATP, indicating the respective forward and backward exchange. Two crossections at the positions indicated in (**a**) show the intensity of the exchange peaks PCr→γ-ATP (**b**) and γ-ATP→PCr (**c**)

The measurement of the time-dependent *z*-magnetizations of all substrates involved in the exchange provides a sound data basis for regression analysis. This holds true for inversion and saturation transfer as well as two-dimensional exchange spectroscopy. With nonselective recovery techniques (e.g., inversion recovery) strong correlations among the individual regression parameters may occur. A definite disadvantage of type-I experiments is that measurements with different mixing times have to be carried out, which is time consuming. This may be prohibitive for in vivo studies because of the instability of the preparation or because of the temporal resolution required.

The strength of the 2D exchange technique is that several exchange reactions can be studied simultaneously. The occurrence of a crosspeak (f_i, f_j) unambiguously proves exchange between the species *i* and *j*. The disadvantage of 2D exchange is the experimental time required; spectra have to be recorded for several mixing times. Even with frequency selective excitation pulses in combination with a small number of t_1 steps (evolution period), measurement times of the order of hours will ensue. Methods to speed up 2D exchange

◄

Fig. 10 a–c. One-dimensional ^{31}P-magnetization transfer experiments on creatine kinase in rat gastrocnemius muscle. (**a**) shows the inversion transfer experiment with selective inversion of the PCr signal. At $t = 0$ an inversion of about 70% was achieved. The numbers next to the individual spectra indicate the mixing times in s, Ctl represents the control spectrum. The saturation transfer experiment with saturation of the γ-ATP resonance is shown in (**b**) and (**c**). The dynamic version with variation of the mixing time is displayed in (**b**), the steady state version for a long mixing time (6 s) in (**c**). In (**c**) the individual spectra are labelled with 'Sat' for γ-ATP saturated, 'Ctl' for irradiation symmetrical to the PCr signal, and 'Dif' for the difference spectrum Sat-Ctl. The arrows indicate the position for selective irradiation

Table 3. Advantages and disadvantages of different magnetization transfer experiments. The individual methods are I/IT: inversion transfer, I/ST: dynamic saturation transfer, I/PS: partial saturation, I/IR: inversion recovery, I/2DE: 2-dimensional exchange spectroscopy, and II/ST: saturation transfer

Method	Advantage	Disadvantage	Application
I/IT	—larger dynamic range than I/ST, I/PS —complete IT data set yields robust kinetic parameters	—spectra for several mixing times required —transfer depends on concentration of exchange partner	—exchange reactions in vivo
I/ST	—complete I/ST data set yields robust kinetic	—same as I/IT —smaller dynamic range than I/IT	—same as I/IT in vivo
I/PS	—same as I/ST —sensitivity superior to ST for low S/N	—same as I/ST	—same as I/IT
I/IR	—accurate measurement of relaxation rates	—same as I/ST —difficult to extract rate constants (correlations)	—not suited for study of exchange processes
I/2DE	—crosspeak proves exchange —analysis of complex reaction networks	—quantitative aspects: same as I/IT	—N-site exchange —analysis of unknown networks
II/ST	—transfer independent concentration of the exchange partner —measurement for the long mixing time sufficient for determination of rates, if T_1 is known	—spectra for several mixing times if T_1 is not known —unsuited for complex reaction networks	—exchange reactions in vivo (with good temporal resolution)

spectroscopy recently have been developed. In the "accordion" experiment [46] both the evolution and mixing times are incremented simultaneously ($t_m = \alpha t_1$) and the kinetic information is derived from the lineshape of the 2D peaks. Due to the inhomogeneously broadened resonance lines observed under in vivo conditions, this method is hardly suitable for such measurements. Recently, a 1D version of the 2D exchange experiment has been described [47, 48] which offers a considerable reduction of the measuring time. However, the method has not yet been applied to an in vivo example, and probably offers little advantage over the conventional 1D methods. In general, 1D methods such as saturation or inversion transfer, are the presently preferred techniques for in vivo kinetic investigations. This is certainly true for type-II experiments, in which the kinetic information (rate constants) may be derived from a single experiment with a long mixing time, provided the relaxation times are known from a different experiment. This can be seen from Eqs. (13, 15, 16, or 17), which all contain a constant term. Therefore, these methods have been extensively used for in vivo studies. A second important advantage of type-II experiments is, that in a two-site system A$\rightleftharpoons$B, upon saturation of B, the signal change of A due to the magnetization transfer depends only on $M_{zA,eq}$ and not on the equilibrium magnetization of its exchange partner. Therefore, chemical exchange with a substrate present at low concentration (e.g., ADP) may be studied (see below, adenylate kinase reaction).

5 Magnetization Transfer In Vivo: Selected Examples

It is not the purpose of this section to present a comprehensive review on in vivo magnetization transfer studies. Rather some problems which arise in the interpretation of the experimental data are discussed on the basis of some selected examples.

5.1 Phosphocreatine Kinase

Creatine kinase (CK) catalyzes the transfer of a phosphate group from phosphocreatine (PCr) to MgADP forming MgATP and creatine (Cr):

$$PCr + MgADP + H^{+} \rightleftharpoons Cr + MgATP.$$

High enzyme activity is found in brain, heart, and skeletal muscle [1]. Most evidence suggest that the reaction is near equilibrium. Magnetization transfer techniques have been extensively applied to measure forward and reverse flux through the CK reaction [1, 49–59].

Two roles for the CK reaction have been discussed. The first is that PCr acts as a "high-energy phosphate" store, designed to buffer changes in the ATP/ADP ratio. Alternatively, PCr may act as a shuttle for ATP between the sites of ATP production (mitochondria) and ATP utilization (cytoplasm) [60, 61]. In the latter case a strong coupling between the ATP utilization (workload) and the CK flux would be expected. Although there exist reports on such a correlation in the perfused rat heart [50] and rat brain [62, 63], this point is still controversial. Ugurbil et al. [58] have shown that pyruvate-perfused rat hearts can produce more work than glucose-perfused hearts, despite lower CK flux and ADP concentration. Measuring the metabolite diffusion constants and lifetimes in frog skeletal muscle, Yoshizaki et al. [64] came to the conclusion that the free mean diffusional pathway of all the substrate participating in the CK reaction is longer than the average length of the myocytes, indicating that the molecules can move freely from the sites of energy production and consumption. Therefore the shuttle hypothesis is not mandatory for energy transport between mitochondria and myofibrils, at least with respect to metabolite diffusion.

In skeletal muscle during steady-state isometric contraction the flux PCr → ATP decreases as the developed tension and the ADP concentration increase [65–67], which is inconsistent with the properties of the CK enzyme in solution. Shoubridge and Radda [65] have found that during a tetanus the flux PCr → ATP equals the rate of PCr hydrolysis, indicating that the CK is rate limiting for ATP synthesis. Since the ATP concentration does not change, there must be a coordinate control between ATP synthesis (via the CK) and ATP utilization.

A striking result of a number of magnetization transfer studies on the CK reaction is a discrepancy between the forward and reverse reaction flux. The most probable explanation for this phenomenon is that ATP is a substrate for a variety of processes. Therefore, the accurate determination of the CK reverse flux is not possible from conventional saturation transfer experiments using single frequency irradiation. Ugurbil et al. [58] have discussed the extension of the two-site model, commonly used for the CK reaction, to a three-site model:

$$PCr \rightleftharpoons ATP \rightleftharpoons P_i.$$

Using the multiple saturation transfer technique [18], they found that within error limits the forward and reverse CK fluxes were equal, in accordance with the near-equilibrium condition of the CK reaction.

Yet the three-site model is still an oversimplification, which does not account for subcellular compartmentation. Consider an extended reaction scheme with a cytoplasmatic and a mitochndrial compartment [58]:

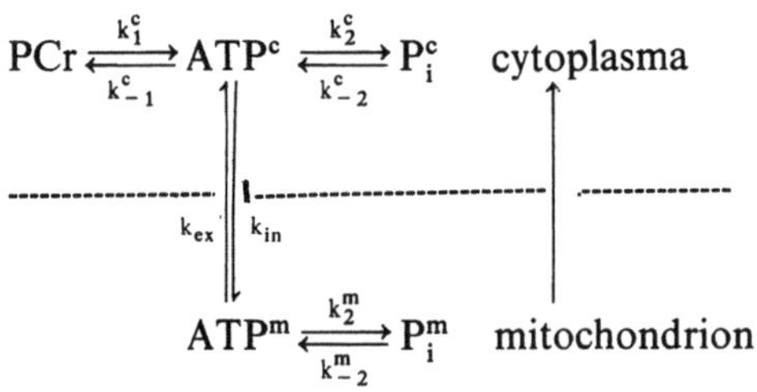

The signal of ATP is almost exclusively (>95%) due to cytoplasmatic ATP. However, both P_i pools are observable and cannot be distinguished. If the cytoplasmic and mitochondrial ATP are in slow exchange with respect to the spin lattice relaxation time, this constitutes an additional process to be considered in the original three-site model. Saturation of P_i may lead to saturation of the mitochondrial ATP, when the lifetime of ATP^m $(=1/(k_{ex}+k_2^m))$ is shorter than its spin–lattice relaxation time and the ATP translocase is working predominantly in the extrusion direction ($k_{ex}ATP^m \gg k_{in}ATP^c$). This has been shown to be true for energized mitochondria. Under these conditions the multisite system reduces to an effective two-site problem, allowing an accurate determination of the CK rate constants [58].

The multiple saturation transfer experiments assuming a three-site model yield identical values for the forward and reverse CK flux [58]. However, on the basis of model calculations, Brindle and Radda [59] have pointed out that the $P_i \rightarrow ATP$ exchange would have to be much higher than the values found experimentally, in order to account for a significant underestimation of the CK reverse flux.

The rate constants derived from the magnetization transfer experiments depend on the total CK activity of the tissue, despite the variations in the isoenzyme compostion and subcellular compartmentation [1]. This is expeccted based on previous considerations (Eq. (27)). Detailed analyses have shown that only about 10% of the total tissue capacity of the enzyme is used under conditions that approach basal metabolism [1].

5.2 F_1F_0 ATP Synthase

The mitochondrial F_1F_0 ATP synthase catalyzes the terminal step in the oxidative phosphorylation, $ADP + P_i \rightleftharpoons ATP$. The net flux $P_i \rightarrow ATP$ may be determined from magnetization transfer experiments in a straightforward way, e.g., by observing P_i with progressive saturation of the γ-ATP resonance [24]. The crucial question then is: Does the experimentally determined flux $P_i \rightarrow ATP$ reflect the activity of the mitochondrial ATPase? A number of problems have to be considered.

Besides the mitochondrial ATP synthesis via oxidative phosphorylation, ATP may be synthetized through the glycolytic pathway involving the enzymes glyceraldehyde-3-phosphate dehydrogenase (GAPDH) and 3-phosphoglycerate kinase (3-DPG) according to:

$$\mathrm{GAP + NAD + P_i \xrightleftharpoons{GAPDH} 1,3\text{-}DPG + NADH + H^+}$$

$$\mathrm{1,3\text{-}DPG + ADP \xrightleftharpoons{3\text{-}DPG} ATP + 3\text{-}PGA.}$$

The abbreviations denote glyceraldehyde-3-phosphate (GAP), nicotinamide-adenine dinucleotide (NAD), 1,3-diphosphoglycerate (1,3-DPG), and 3-phosphoglycerate (3-PGA). The importance of this parallel pathway to the mitochondrial ATP synthesis may be derived from experiments, in which GAPDH is completely inhibited with iodoacetate. Studying isolated perfused hearts, Kingsley-Hickman et al. [24] obtained a 50% reduction in the $P_i \rightarrow ATP$ flux in changing the perfusion medium from glucose as a carbon source to pyruvate/iodoacetate.

Changing the oxygen consumption (MVO_2) caused no effect on the flux $P_i \rightarrow ATP$ in glucose-perfused rat hearts, while with pyruvate as a carbon source an almost linear dependence rate $= \alpha\ MVO_2$ was obtained [24]. The slope α is the P:O ratio, i.e., the ratio of moles ATP synthetized per moles oxygen atoms consumed. Inhibition of glycolytic ATP synthesis with iodoacetate gave a P:O value of approximately 2.3 independent of the workload of the heart, while perfusion with glucose produced significantly higher P:O values, which decreased with increasing workload [24]. This reveals the importance of the glycolytic contribution to ATP synthesis in heart, especially at low and medium MVO_2 values. The P:O value of 2.3 determined in vivo is similar to that found in isolated mitochondria. It has been interpreted that the flux through the mitochondrial F_1F_0 ATP synthase is unidirectional despite the reversibility of the enzyme in isolated mitochondria [68].

A second complication arises because of compartmentation. In studying the formation of ATP from cytosolic P_i, transport processes across the mitochondrial membrane have to be considered in addition to the actual enzymatic process. Only if the transmembrane exchange of inorganic phosphate is rapid compared to the spin–lattice relaxation times of cytosolic and mitochondrial P_i, will meaningful values for the mitochondrial F_1F_0 ATP synthase flux be obtained [24]. The proportionality rate($P_i \rightarrow$ ATP) $= \alpha \cdot MVO_2$ shows that trans-membrane transport is of little importance in the pyruvate/iodoacetate perfused heart.

An additional control of the flux data is based on the following consideration. If the system has reached a steady state, then the forward and reverse flux have to be equal, irrespective of the workload and the actual values of the unidirectional reaction rates. Using the multiple saturation transfer technique, Ugurbil et al. [58] have measured the reverse flux ATP $\rightarrow P_i$ in perfused rat hearts. Within error limits, forward and reverse fluxes were equal. Even when the influence of the CK reaction has been eliminated using the multiple saturation transfer technique, complications may arise due to the interference of the adenylate kinase reaction (see below).

In contrast to the situation in perfused rat heart, the comparison of magnetization transfer derived ATP synthesis rates and oxygen consumption measurements in skeletal muscle indicates, that the flux $P_i \rightarrow$ ATP is predominantly due to mitochondrial F_1F_0 ATP synthase activity [67]. ATP synthesis depends linearly on the calculated concentration of free ADP up to ADP levels of about 90 μM. Mitochondrial ATP synthesis, therefore, seems to be controlled by the availability of ADP to the mitochondria [69, 70], with an apparent K_m of at least 30 μM [67]. In vitro enzymatic examinations in fast-twitch red and white muscle fibers during tetanic stimulation have displayed considerable heterogeneity of high-energy phosphate metabolism [71]. In vivo results derived from phase-modulated rotating frame experiments, however, revealed less heterogeneity than expected [31].

From the above statements it follows that ATP synthesis rates derived from magnetization transfer experiments have to be interpreted with care. Additional experimental data such as oxygen consumption of the tissue, effects of inhibition of the glycolytic ATP synthesis, measurement of ATP consumption (flux ATP $\rightarrow P_i$), all need to be generated in order to derive meaningful conclusions. Tissue specificity and heterogeneity should be considered in the analysis as well [31, 72].

5.3 *Adenylate Kinase*

Adenylate kinase is a ubiquitous enzyme catalyzing the reaction:

$$2\text{ADP} \rightleftharpoons \text{ATP} + \text{AMP}.$$

A single in vivo study [73] has been reported to date, besides several in vitro investigations [74, 75]. Gupta [73] measured the exchange rates by saturating the β-ATP resonance while observing the γ-resonance. This in fact is a two-step process:

$$\gamma\text{-ATP} \rightleftharpoons \beta\text{-ADP} \rightleftharpoons \beta\text{-ATP}.$$

A complication arises since the exchange between the β-phosphates of ADP and ATP is catalyzed by all enzymes which interconvert ADP and ATP.

5.4 *Membrane Transport*

The measurement of trans-membrane migration of a solute by MRS methods requires separate resonances for at least one nucleus of the solute, depending whether it is inside or outside of the compartment. For a review of membrane transport processes, the reader is referred to the literature [11, 76]. The present section is focussed on transport processes across the erythrocyte membrane.

Separation of the resonances may be an intrinsic property of the system. With many phosphoryl compounds, different ^{31}P-NMR frequencies are observed inside and outside the erythrocytes [77]. The origin of the phenomenon is attributed to the high protein content inside the cell relative to the outside, affecting the hydrogen bondings and thus the local environment of the ^{31}P nucleus, shifting its resonance to a lower frequency in the spectrum [77]. The measurement of the trans-membrane transport is therefore possible in a straightforward manner [78].

If the differentiation of resonances is not a property of the system, it may be induced by adding a chemical shift reagent, which is unable to cross the cell membrane. This technique is commonly applied for the study of ion transport processes (see J.S. Ingwall: Ion Transport Processes, in volume 3 of this series). Using 1D and 2D magnetization transfer techniques, Shungu and Briggs [79] studied the ionophore-mediated trans-membrane transport of sodium.

In other systems, the separation relies on different exents of magnetization transfer inside and outside the cell [80, 81]. For example, in the case of bicarbonate transport across the erythrocyte membrane, the intracellular exchange with carbon dioxide is catalyzed by the endogeneous carbonic anhydrase and therefore rapid relative to the extracellular spontaneous exchange.

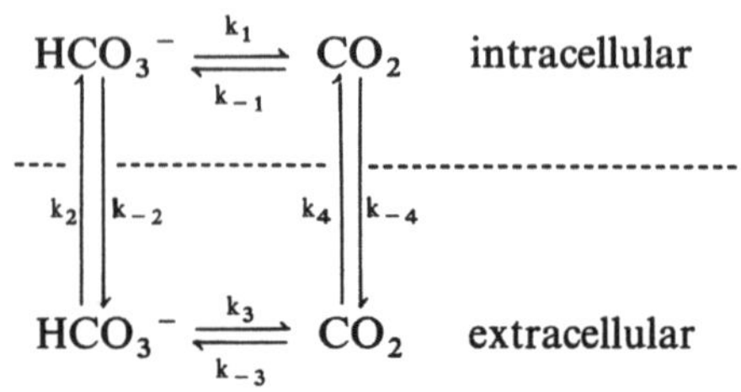

Thus the extent of saturation transfer differs in the two compartments. Separate experiments in packed cells, supernatant, and cell suspensions provided a sufficient data basis for the determination of bicarbonate trans-membrane exchange rate [80].

A similar method has been applied to measure glucose transport across the membrane of human erythrocytes. Differentiation of the intra- and extracellular compartment was achieved by addition to the extra-cellular space of mutarotase, which is not found in erythrocytes and which is unable to cross the cell membrane. Therefore, the exchange between the α- and β-anomer of glucose differed significantly between the intra- and extracellular space, and the trans-membrane migration of glucose could be measured [81].

5.5 Magnetization Transfer Between "Free" Tissue Water Protons and Protons Associated with Macromolecules

Contrast in proton-magnetic-resonance imaging (MRI) is essentially due to tissue specific relaxation times T_1 and T_2. A probable predominant mechanism of relaxation in biological tissue is cross-relaxation and/or chemical exchange between protons in "free", highly mobile water and those associated with macromolecules. Saturation transfer techniques may be applied in a straightforward manner to measure the exchange of magnetization between these two proton pools [36]. Type-II experiments are particularly suited for this purpose, since they do not depend on the spin concentration and hence magnetization of the saturated pool.

The principle of the method is illustrated in Fig. 12. The proton spectrum consists of two resonances one with a linewidth of a few Hz due to free water protons (H_f) and the other with a linewidth of typically > 10 kHz due to protons associated with macromolecules (H_r). Selective irradiation, typically 5 kHz off-resonance to the H_f signal, results in a specific decrease of the latter to a value $M_{f,ss}$. It is essential that the mixing time shall be enough to guarantee steady-state conditions. Values of $M_{f,ss}/M_{f,eq}$ of 0.30 ± 0.03 for kidney and 0.07 ± 0.01 for skeletal muscle have been measured in rabbits [36].

Recent studies on model systems suggest that the mechanism of the magnetization transfer between H_r and H_f is dipolar coupled cross-relaxation, while chemical exchange seems to be of less importance, at least in vitro [82].

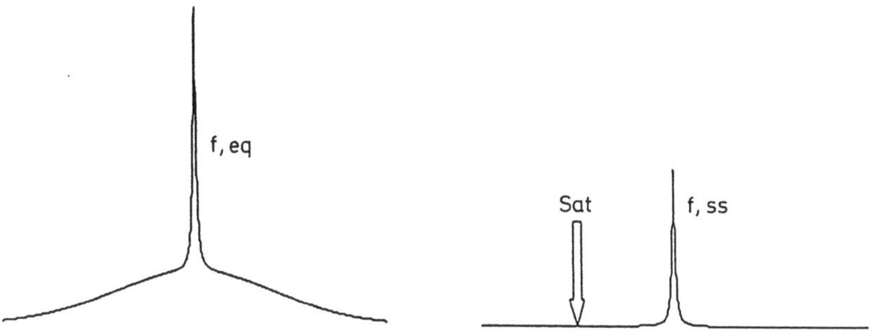

Fig. 12. Schematic representaton of the magnetization transfer between protons of 'free' tissue water (H_f) and protons associated with macromolecules (H_r). The broad resonance due to the H_r pool (linewidth >10 kHz) is saturated by selective irradiation typically 5 kHz off-resonance to the free water signal. As a consequence of magnetization exchange, the intensity of the latter is decreased to a value $M_{f,ss}$. The rate constant for the exchange H_f/H_r may then be derived according to Eq. (29)

In addition it has been found that the presence of hydroxyl groups on the macromolecule increases the efficiency of the exchange [83].

As the H_r signal is symmetrical with respect to the H_f resonance, the extent to which H_f is directly affected by the selective irradiation on the H_r cannot be experimentally determined in the usual way, i.e., by symmetrical irradiation with respect to the H_f resonance. Wolff and Balaban [36] have estimated that, for experimental parameters typically used, the decrease of the H_f signal due to direct saturation effects should be less than 8%.

Saturation transfer may be combined in a straightforward manner with conventional MRI sequences, e.g., spin-echo or gradient recalled echo, to provide images with magnetization transfer contrast (MTC) [36]. If in addition T_1 (in the presence of saturation) is known for each pixel, then maps representing the true pseudo-first order rate constants of the xchange H_f/H_r may be obtained according to:

$$\mathrm{k}[i,j] = (1 - M_{f,ss}[\mathrm{i},j]/M_{f,eq}[i,j])/T_{1,sat}[i,j], \tag{29}$$

where $[i,j]$ indicates the respective pixel. Eng et al. [84] have recorded rate constant maps of perfused rabbit kidneys and obtained significant differences between renal cortex ($\mathrm{k} = 0.91 \pm 0.08\,\mathrm{s}^{-1}$) and medulla ($\mathrm{k} = 0.24 \pm 0.04\,\mathrm{s}^{-1}$).

Figure 13 shows MTC images of a rat brain in comparison with conventional T_2 weighted images. It is obvious that the rate constant maps qualitatively resemble the T_2 images. This might be expected based on the equations:

$$1/T_{i,obs} = 1/T_{i,f} + x_r/(T_{i,r} + 1/k_r) \tag{30}$$

where the index i stands for 1 or 2, and f and r indicate the H_f and H_r pool [36]. The mole fraction of the latter is designated x_r. Since $T_{2,r} \ll T_{1,r}$ (rigid lattice approximation), the effect of the exchange is much more pronounced in T_2. Studies in multiple sclerosis patients indeed showed a better correlation of the $M_{f,ss}/M_{f,eq}$ values with tissue T_2 than with T_1 [85].

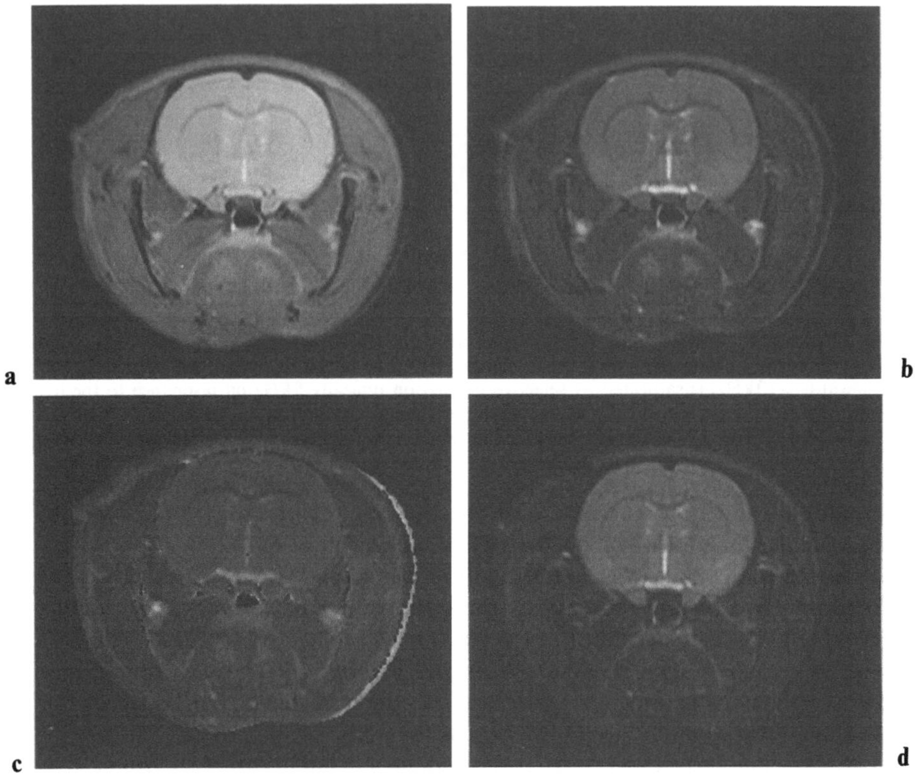

Fig. 13 a–d. MTC images of a rat brain. (**a**) shows a conventional spin-warp image with TR = 4.1 and TE = 20 ms. In (**b**) the image was collected with identical parameters as (**a**), but with 4.0 s of 1 W irradiation 5 kHz below the H_f resonance frequency. (**c**) shows the ratio of image (**b**) divided by (**a**), corresponding to the ratio $M_{f,ss}/M_{f,eq}$. (**d**) represents a conventional T_2 weighted image obtained with TR = 4.1 s and TE = 60 ms. The images have been recorded on a Bruker Biospec BMT 47/15 spectrometer using and Alderman-Grant [87] type resonator of 40 mm inner diameter. The field of view was 40 mm, the slice thickness 1 mm. Data acquisition time was 15 min per image

MTC may provide, besides T_1 and T_2, additional information valuable for the tissue characterization. As an example, Ordidge et al. [86] have observed in rats 24 h after unilateral occulsion of the middle cerebral artery that the infarcted brain tissue, due to an increased fraction of mobile protons, was less affected by the saturation pulse than the normal.

6 Conclusion

Kinetic parameters such as reaction rates and fluxes provide important additional information on the metabolic state of an organ. In many circumstances they are even more sensitive indicators of tissue function than the steady-state concentrations of the metabolites. For instance, the tissue concentration of ATP remains essentially constant despite large variations in the energy demand of a tissue (e.g., heart [50]) due to an efficient coupling of ATP synthesis and consumption. The ATP flux, however, is tightly coupled to the energy demand.

NMR magnetization transfer techniques are unique tools for kinetic measurements both in vitro and in vivo, since they are noninvasive, do not interfere with the chemical equilibrium, and allow the determination of unidirectional fluxes. Typical in vivo applications are the study of enzymatic reactions and membrane transport.

The observation of magnetization transfer between two substrates is good evidence that the respective chemical exchange occurs. The derivation of quantitative information, however, is not always trivial. This holds especially true for in vivo systems, where the exchange under study may be part of a complex reaction network. Therefore, in general, complementary experiments such as oxygen consumption measurement, inhibition of parallel reactions, modulation of the reaction under study, etc., are required in order to avoid misinterpretations.

Acknowledgements: The authors are grateful to Drs NS Cook and H Widmer for the critical reading of the manuscript.

7 References

1. Bittl JA, DeLayre J, Ingwall JS (1987) Biochemistry 26: 6083
2. Reo NV, Siegfried BA, Ackerman JJH (1984) J Biol Chem 259: 13664
3. Laughlin MR, Petit WA, Dizon JM, Shulman RG, Barrett EJ (1988) J Biol Chem 263: 2285
4. Forsen S, Hoffman RA (1963) Acta Chem Scand 17: 1787
5. Forsen S, Hoffman RA (1963) J Chem Phys 39: 2892
6. Forsen S, Hoffman RA (1964) J Chem Phys 40: 1189
7. Brown TR, Ugurbil K, Shulman RG (1977) Proc Natl Acad Sci USA 74: 5551
8. Alger JR, Shulman RG (1984) Quart Rev Biophys 17: 83
9. Brindle KM, Campbell ID (1987) Quart Rev Biophys 19: 159
10. Koretsky AP, Weiner MW (1984) in: James TL, Margulis AR (eds) Biomed magn reson. Radiology Research and Educational Foundation, San Francisco, p 209
11. Kuchel PW (1990) NMR in Biomed 3: 102
12. Ernst RR, Bodenhausen G, Wokaun A (1987) Principles of nuclear magnetic resonance in one and two dimensions. Clarendon Press, Oxford, p 285
13. Meier BH, Ernst RR (1979) J Am Chem Soc 101 (1979)
14. Jeener J, Meier BH, Bachmann P, Ernst RR (1979) J Chem Phys 71: 4546
15. Schotland J, Leigh J (1983) J Magn Reson 51: 48
16. McConnell HM (1958) J Chem Phys 28: 430
17. McConnell, Thompson DD (1957) J Chem Phys 26: 958
18. Ugurbil K (1985) J Magn Reson 64: 207
19. Morrison JF, Cleland WW (1966) J Biol Chem 241: 673
20. Kupriyanov VV, Steinschneider AY, Ruuge EK, Kapelko VI, Zueva MY, Lakomkin VL, Smirnov VN, Saks VA (1984) Biochim Biophys Acta 805: 319

21. Groen AK, Wanders RJA, Westerhoff HV, van der Meer R, Tager JM (1982) J Biol Chem 257: 2754
22. Walsh K, Koshland DE (1985) Proc Natl Acad Sci USA 82: 3577
23. Campbell-Burk SL, Jones KA, Shulman RG (1982) Biochemistry 21: 207
24. Kingsley-Hickman PB, Sako EY, Mohanakrishnan P, Robitaille P, From AHL, Foker JE, Ugurbil K (1987) Biochemistry 26: 7501
25. Morris GA, Freeman R (1978) J Magn Reson 29: 433
26. Ackerman JJH, Grove TH, Wong GG, Gadian DG, Radda GK (1980) Nature 283: 167
27. Haase A, Frahm J, Matthaei D, Haenicke W, Merboldt KD (1986) J Magn Reson 67: 258
28. Ugurbil K, Garwood M, Rath AR (1988) J Magn Reson 80: 448
29. Gonzales-Mendez R, Moseley ME, Murphy-Boesch J, Chew WM, Litt L, James TL (1985) J Magn Reson 65: 516
30. Styles P (1988) NMR Biomed 1: 61
31. Challiss RAJ, Blackledge MJ, Radda GK (1988) Am J Physiol 254: C417
32. Blackledge MJ, Cadoux-Hudson T, Radda GK (1989) Book of Abstracts. Eight Ann Meeting Soc Magn Reson Med, Amsterdam, p 460
33. Bottomley PA, Hardy CJ (1989) Bookk of Abstracts. Eigth Ann Meeting Soc Magn Reson Med, Amsterdam, p 253
34. Hseih PS, Balaban RS (1987) J Magn Reson 74: 574
35. Mora BN, Narasimhan PT, Barker PB, Ross BD, Allman JM (1990) Book of Abstracts. Ninth Ann Meeting Soc Magn Reson Med, New York, p 1020
36. Wolff SD, Balaban RS (1989) Magn Reson Med 10: 135
37. Rydzy M, Deslauriers R, Smith ICP, Saunders JK (1990) Magn Reson Med 15: 260
38. Rudin M (1989) Magn Reson Med Biol 2: 167
39. Led JJ, Gesmar H (1982) J Magn Reson 49: 444
40. Malloy CR, Sherry AD, Nunnally RL (1985) J Magn Reson 64: 243
41. Gadian DG, Radda GK, Brown TR, Chance EM, Dawson MJ, Wilkie DR (1981) Biochem J 194: 215
42. Balaban RS, Kantor HL, Feretti JA (1983) J Biol Chem 258: 12787
43. Zahler R, Bittl JA, Ingwall JS (1987) Biophys J 51: 883
44. Brindle KM, Radda GK (1985) Biochim Biophys Acta 829: 188
45. Marion D, Wuethrich K (1983) Biochem Biophys Res Comm 113: 967
46. Bodenhausen G, Ernst RR (1982) J Am Chem Soc 104: 1304
47. Engeler RE, Johnston ER, Wade CG (1988) J Magn Reson 77: 377
48. Bulliman BT, Kuchel PW, Chapman BE (1989) J Magn Reson 82: 131
49. Shoubridge EA, Briggs RW, Radda GK (1982) FEBS Lett 140: 288
50. Bittl JA, Ingwall JS (1985) J Biol Chem 260: 3512
51. Nunally RL, Hollis DP (1979) Biochemistry 18: 3242
52. Matthews PM, Bland JL, Gadian DG, Radda GK (1982) Biochim Biophys Acta 721: 312
53. Seymour AL, Keough JM, Radda GK (1983) Biochem Soc Trans 11: 376
54. Degani H, Laughlin M, Campbell S, Shulman RG (1985) Biochemistry 24: 5510
55. Meyer RA, Kushmerick MJ, Brown TR (1982) Am J Physiol 242: C1
56. Morris PG, Feeney J, Cox DWG, Bachelard HS (1985) Biochem J 227: 777
57. Degani H, Alger JR, Shulman RG, Petroff OAC, Prichard JW (1987) Magn Reson Med 5: 1
58. Ugurbil K, Petein M, Maidan R, Michurski S, From AHL (1986) Biochemistry 25: 100
59. Brindle KM, Radda GK (1987) Biochim Biophys Acta 928: 45
60. Meyer RA, Sweeney HL, Kushmerick MJ (1984) Am J Physiol 246: C365
61. Koretsky AP, Basus VJ, James TL, Klein MO, Weiner MW (1985) Magn Reson Med 2: 586
62. Rudin M, Sauter A (1989) J Pharm Exp Ther 251: 700
63. Rudin M, Sauter A (1989) Experientia 45: A55
64. Yoshizaki K, Watari H, Radda GK (1990) Book of Abstracts. Ninth Ann Meeting Soc Magn Reson Med, New York, p 173
65. Shoubridge EA, Radda GK (1984) Biochim Biophys Acta 805: 79
66. Shoubridge EA, Radda GK (1987) Am J Physiol 252: C1
67. Brindle KM, Blackledge MJ, Challiss RAJ, Radda GK (1989) Biochemistry 28: 4887
68. LaNoue KF, Jeffries FMH, Radda GK (1986) Biochemistry 25: 7667
69. Jacobus WE, Moreadith RW, Vandegear KM (1982) J Biol Chem 257: 2397
70. Radda GK (1986) Science 233: 640
71. Hintz CS, Chi MMY, Fell RD, Ivy JL, Kaise KK, Lowry CV, Lowry OH (1982) Am J Physiol 242: C218

72. Robitaille PM, Lew B, Merkle H, Sublett E, Lindstrom P, From AHL, Garwood M, Bache RJ, Ugurbil K (1989) Mag Reson Med 10: 108
73. Gupta RK (1979) Biochim Biophys Acta 586: 189
74. Brown TR, Ogawa S (1977) Proc Natl Acad Sci USA 74: 5551
75. Kantor HL, Feretti JA, Balaban RS (1984) Biochim Biophys Acta 789: 128
76. Kirk K (1990) NMR Biomed 3: 1
77. Kirk K, Kuchel PW (1988) Biochemistry 27: 8803
78. Price WS, Kuchel PW (1990) NMR Biomed 3: 59
79. Shungu DC, Briggs RW (1988) J Magn Reson 77: 491
80. Kuchel PW, Bulliman BT, Chapman BE, Kirk K (1987) J Magn Reson 74: 1
81. Kuchel PW, Chapman BE, Potts JR (1987) FEBS Lett 219: 5
82. Ceckler TL, Balaban RS (1990) Book of Abstracts. Ninth Ann Meeting Soc Magn Reson Med, New York, p 354
83. Wolff SD, Fralix TA, Simon SA, Balaban RS (1990) Book of Abstracts. Ninth Ann Meeting Soc Magn Reson Med, New York, p 350
84. Eng J, Ceckler TL, Balaban RS (1990) Book of Abstracts. Ninth Ann Meeting Soc Magn Reson Med New York, p 351
85. Tanttu JI, Sepponen RE, Foust RJ, Kinnunen E (1990) Book of Abstracts. Ninth Ann Meeting Soc Magn Reson Med, New York, p 356
86. Ordidge RJ, Helpern JA, Knight R, Welch KMA (1990) Book of Abstracts. Ninth Ann Meeting Soc Magn Reson Med, New York, p 357
87. Alderman DW, Grant DM (1979) J Magn Reson 36: 447

Author Index Volumes 21–27

MIX
Papier aus verantwortungsvollen Quellen
Paper from responsible sources
FSC® C105338
www.fsc.org

If you have any concerns about our products,
you can contact us on
ProductSafety@springernature.com

In case Publisher is established outside the EU,
the EU authorized representative is:
Springer Nature Customer Service Center GmbH
Europaplatz 3, 69115 Heidelberg, Germany

Printed by Libri Plureos GmbH
in Hamburg, Germany